GAME THEORY
AND POLITICS

Steven J. Brams

GAME THEORY AND POLITICS

THE FREE PRESS
A Division of Macmillan Publishing Co., Inc.
New York

Collier Macmillan Publishers
London

The Free Press
A Division of Macmillan Publishing Co., Inc.
866 Third Avenue, New York, N.Y. 10022

Collier-Macmillan Canada Ltd.

Library of Congress Catalog Card Number: 74-15370

Printed in the United States of America

printing number
1 2 3 4 5 6 7 8 9 10

Library of Congress Cataloging in Publication Data

Brams, Steven J
  Game theory and politics.

  Bibliography: p.
  Includes index.
  1. Political science—Mathematical models. 2. Game
theory. 3. Coalition (Social sciences) I. Title.
JA73.B73   320'.01'84   74-15370
ISBN 0-02-904550-9

To my parents

CONTENTS

5. VOTING POWER   157

6. COALITION GAMES   199

7. ELECTION GAMES   243

PREFACE

Game theory is a branch of mathematics that was created in practically one fell swoop with the publication in 1944 of *Theory of Games and Economic Behavior* by John von Neumann and Oskar Morgenstern.[1] This was a monumental intellectual achievement—certainly one of the outstanding mathematical contributions of the twentieth century—which seemed initially to hold great promise for the mathematization of social situations that involve conflict, as well as possible cooperation, among two or more actors (or players). Indeed, the theory was hailed by some as a creation as important to the theoretical development of the social sciences as the calculus had been to the development of classical mechanics and physics at the close of the seventeenth century.

Unfortunately, this promise has gone mostly unfulfilled over the past thirty years. Although the contributions of game theory to economics have been considerable,[2] game theory has had little impact on the other social sciences. This is true despite the splendid review and synthesis of applied game theory by R. Duncan Luce and Howard Raiffa in 1957,[3] as well as later expository works on game theory that highlight its social science applications.[4] In political science, game theory has been frequently misunderstood and misapplied at the same time that its concepts (e.g., "player," "payoff," "side payments," and "zero-sum") have become part of the working vocabulary of modern political scientists.

My purpose in this book is not so much to try to correct the loose usage of game-theoretic terminology or to rehearse its heuristic value. Rather, I shall try to demonstrate, mostly by example, the relevance of the mathematical theory to the explication of strategic features of actual political situations. No purely formal theory can possibly gain wide acceptance as a scientific theory unless its abstract concepts can be related

---

1. John von Neumann and Oskar Morgenstern, *Theory of Games and Economic Behavior* (Princeton, N.J.: Princeton University Press, 1944). The most recent revised edition of this classic work appeared in 1953.

2. L. S. Shapley and M. Shubik, *Game Theory in Economics* (forthcoming). See also Martin Shubik, *Strategy and Market Structure: Competition, Oligopoly, and the Theory of Games* (New York: John Wiley & Sons, 1959); and Lester G. Telser, *Competition, Collusion and Game Theory* (Chicago: Aldine-Atherton, 1972).

3. R. Duncan Luce and Howard Raiffa, *Games and Decisions: A Critical Survey* (New York: John Wiley & Sons, 1957).

4. Anatol Rapoport, *Two-Person Game Theory: The Essential Ideas* (Ann Arbor, Mich.: University of Michigan Press, 1966); Anatol Rapoport, *N-Person Game Theory: Concepts and Applications* (Ann Arbor, Mich.: University of Michigan Press, 1970); and Morton D. Davis, *Game Theory: A Nontechnical Introduction* (New York: Basic Books, 1970).

to real-world phenomena. For this purpose, one usually tries to construct a *theoretical model* of these phenomena, or a simplified representation that abstracts the essential elements one wants to study and links them through a set of axioms or assumptions. When the concepts in this representation are operationally defined, consequences (i.e., theorems) derived from the assumptions of the model can be tested.

In most chapters I have tried to develop in some depth, using only rudimentary mathematics, a relatively few game-theoretic models that have obvious political significance. This approach seemed preferable to presenting a smorgasbord of imaginary games that may cover many different kinds of hypothetical decision-making situations but which have no real analogues in politics. For the reader interested in broader coverage, rather extensive, though by no means comprehensive, citations to the mathematical and social science literature are given in the footnotes.

The intended audience for this book is students and teachers of politics, though I hope it also proves attractive to social scientists generally, as well as laymen interested in the serious study of politics. Although this is emphatically not a work of mathematics, I would also hope that mathematicians and scientists might find the material interesting, especially insofar as unsolved mathematical problems are briefly indicated.

Because only a background in high school mathematics is assumed, very little symbolic notation is used. Nevertheless, the book is not exactly light reading—a thorough understanding of abstract, deductive models requires careful study and thoughtful reflection, especially to grasp subtle points. Proofs of a few theorems that seem especially significant in politics are given in some detail, although no assertion is made that the mathematical development is completely rigorous.

If the somewhat informal treatment of game theory in this book is not totally reconcilable with the strict canons of mathematical rigor, neither is the formalism of game theory fully reconcilable with the investigation of important substantive questions and problems in politics. Generally speaking, the better developed the mathematics, the more trivial are the situations in politics to which it is applicable. This is probably the main reason why the early and bright hopes for game theory as a tool for political analysis have become tarnished.

A recent upsurge of interest in serious applications of game theory to politics may help change this image and dispel some of the disillusionment about game theory's unfulfilled promise.[5] Although most of the recent

---

5. There has been a parallel increase in interest among Russian scholars, especially as game theory may be useful in analyzing international affairs, which in part seems based on the fear that unless studied, a powerful new mode of analysis may be exploited exclusively by Western social scientists and mathematicians and used as a weapon against Marxism. See Thomas W. Robinson, "Game Theory and Politics: Recent Soviet Views," *Studies in Soviet Thought*, 10 (1970), pp. 291–315. Perhaps the coming-of-age of game theory is best epitomized by the publication, beginning in 1971, of the *International Journal of Game Theory* (Vienna: Institute for Advanced Studies), which is devoted exclusively to research in this discipline.

models that have been developed relate to relatively well-structured situations that one finds in legislative bodies and electoral politics, problems in international politics and other less rule-bound political arenas have also been analyzed.

Most of the analysis in this book is *descriptive* or *positive*, by which is meant that the models describe or explain events and processes as they actually occur in the world. Toward this end, considerable space is devoted to trying to lay bare the empirical correlates of the models and to test the truth of their predictions in real settings (evidence from experimental games has been mostly ignored).

The models may also be given another interpretation: to prescribe how actors should behave in particular situations, given the rational pursuit of certain goals (e.g., winning). This *prescriptive* or *normative* interpretation of game-theoretic models is quite consistent with their positive interpretation; it differs only in assuming that the goals or ends sought are desirable and ought to be pursued. At certain points in the book I offer an avowedly normative interpretation of the goals assumed in some of the models, where, it seems to me, the analysis uncovers particularly desirable or undesirable effects engendered by players pursuing their self-interests under the rules of a game. It is the interests of individual members of society—not the fiction of class or other collective interests, which almost invariably exclude the interests of certain benighted individuals—that I believe one should try to satisfy in constructing an ethical ideal for society.

ACKNOWLEDGMENTS

I would like to thank Roman Frydman, Richard D. McKelvey, Kenneth A. Shepsle, and Frank C. Zagare for carefully reading a first draft of this book, making many helpful suggestions, and detecting several errors; all should be absolved of deficiencies that remain in the book. I also want to thank Donna S. Welensky for quickly and skillfully typing the manuscript. Arthur L. Iamele of The Free Press offered valuable support from the beginning of this project. Ellen Simon, also of The Free Press, was very helpful in the editing of the manuscript. I am indebted to several publishers and journals, specifically acknowledged in the footnotes, for permission to use material from previously published sources. Finally, I am grateful to my wife, Eva, who gracefully endured a six-month period of relative isolation, in the process of raising a family, while this book was being written.

INTRODUCTORY NOTE

Games have three different formal representations that highlight different strategic aspects of decision-making situations; all are developed here and applied to the analysis of various strategic questions in political science. Common to all these representations is the assumption that players in a game are *rational*, a mischievous concept that has generated so much controversy and confusion in political science (and elsewhere) that a few introductory remarks seem in order to make more perspicuous the assumptions on which the subsequent analysis is based.

Very generally, behaving rationally in game theory means acting to maximize the achievement of some postulated goal, where the outcome depends not only on chance events and "nature" but also on the actions of other players with sometimes cooperative and sometimes conflicting interests. (Since exactly how a player can affect outcomes depends on the specific institutional context, we postpone a more precise specification of the concept of rationality and notions of a solution to a game.) What distinguishes game-theoretic models from other models of rational choice is that the outcome is assumed to be contingent on the choices of more than one player. Thus, the preferences of other players, and choices consistent with these preferences, must be explicitly taken into account when one chooses an optimal course of action (e.g., how to vote, what coalition to join, and so on).

Where a conflict exists, it is assumed to arise from the genuinely different interests of players—a craving for material possessions versus a desire for the simple life, a proclivity for combat versus a preference for compromise, an urge to dominate versus a willingness to share power—and not because these interests are irrational or pathological. Game theory takes players' interests as given and focuses on the logical implications that their satisfaction has on players' choices. It does not inquire into their origins, which have been characteristic concerns in both sociology and psychology with their emphasis on the study of socialization processes, roles, values, beliefs, personality, and so forth.

An epistemological assumption underlying all rational-choice models is that actors are purposeful decision makers, whether they be individuals or collectivities acting with a unitary purpose. Even if an individual or collective actor does not actually make, or is not able to understand, the rationalistic calculations the theorist imputes to him, the rationality assumption may still satisfactorily account for his behavior by providing an explanation of his actions *as if* he pursued some postulated goal. A model need not be completely realistic to open up novel ways of thinking about, and suggest new insights into, old and previously obscure questions and problems.

In many situations, the rationality assumption is *not* an unrealistic assumption to make about the behavior of political decision makers, at least as a first approximation. Yet in spirit it runs against the grain of much so-called behavioral political science, which implicitly assumes that forces outside the decision maker (social, economic, cultural, historical)—or sometimes inside (instinctual, psychoanalytic)—impinge on his being and control and direct his behavior. Because propositions from this school are usually arrived at inductively from observations, rather than derived deductively from the assumptions of a model, they often together constitute no more than a set of disconnected "findings." Uncoordinated by any logical structure, they are not theoretically compelling; based only on observed regularities, their empirical appeal is undermined when this regularity is broken, as for example has occurred when a group in the electorate upset a long-standing generalization about how it would vote in the next election. Questions about why members of this group deviated (irrationally?) from their normal voting behavior are largely foreclosed when one assumes, as an epistemological precept, that man marches unthinking and robotlike to the tune of external or internal forces that control his actions, rendering him incapable of making free choices.[1]

No satisfactory theory of politics, in my opinion, can ignore the fact that political actors consistently make choices to further certain ends. To be sure, they are constrained not only by the choices of other actors but also by rules, norms, incomplete information, problems of communication, collusive arrangements, and so forth. These constraints, however, can be incorporated into game-theoretic models, just as different goals can be postulated for actors that may rationalize behavior previously thought irrational. In a sense, the beauty of game theory, at least to nonmathematicians, is its incompleteness, its different viewpoints, its multiplicity of solution concepts. Although this richness in the theory permits a good deal of flexibility in the construction of specific models, the overall mathematical structure still provides one with a reasonably parsimonious, rigorous, and consistent way of formulating, analyzing, and solving strategic problems in politics, as the subsequent analysis seeks to show.

---

1. Curiously, game theory is sometimes accused of making this same kind of assumption, though the usual image evoked is not that of a robot but an automaton of superhuman intelligence.

LIST OF TABLES

LIST OF FIGURES

The next stage [in the evolution of a theory] develops when the theory is applied to somewhat more complicated situations in which it may already lead to a certain extent beyond the obvious and familiar.

John von Neumann and Oskar Morgenstern,
Theory of Games and Economic Behavior

GAME THEORY
AND POLITICS

1

INTERNATIONAL
RELATIONS GAMES

1.1. INTRODUCTION

Most discussions of game theory in the international relations literature
hardly go beyond a presentation of a few hypothetical examples used in
general expository treatments of the subject. Although some vague
analogies are usually drawn to situations of conflict and cooperation
in international relations, there seems to be little in the *substance* of
international relations that has inspired rigorous applications of the
mathematics of game theory. An exception is the considerable work (some
classified) that has been done on specific problems related to the use and
control of weapons systems (e.g., targeting of nuclear weapons, violation
of arms inspection agreements, search strategies in submarine warfare),
but these are not problems of general strategic interest.

Given this dichotomy between general but vague and rigorous but
specialized studies in international relations, we shall try to tread a middle
ground in our discussion of applications. At a mostly informal level, we
shall indicate with several real-life examples the relevance of game-
theoretic reasoning to the analysis of different conflict situations in the
international arena. At a more formal level, we shall develop some of the
mathematics and prove one theorem in two-person game theory that seem
particularly pertinent to the identification of optimal strategies and
equilibrium outcomes in international relations games.

Several of our examples relate to military and defense strategy, the
area in international relations to which game theory has been most
frequently (and successfully) applied. This is not to say, however, that
only militarists and warmongers find in game theory a convenient rationale
for their hard-nosed and occasionally apocalyptic views of the world.
On the contrary, some of the most interesting and fruitful applications
have been made by scholars with a profound interest in, and concern for,
the preservation of international peace, especially as this end may be
fostered by more informed analytic studies of arms control and disarmament policies, international negotiation and bargaining processes, and so
forth.

If there is anything that serious game-theoretic analysts of international
relations share, whatever their personal values, it is the assumption that

actors in the international arena are rational with respect to the goals they seek to advance. Although there may be fundamental disagreements about what these goals are, the game-theoretic analyst does not assume events transpire in willy-nilly uncontrolled and uncontrollable ways. Rather, he is predisposed to assume that most foreign policy decisions are made by decision makers who carefully weigh the advantages and disadvantages likely to follow from alternative policies.[1] Especially when the stakes are high, as they tend to be in international politics, this assumption does not seem an unreasonable one.

In our review of applications of game theory to international relations, we shall adopt a standard classification of games. Much of the terminology used in the book will be introduced in this chapter, mostly in the context of our discussion of specific games. This may make the discussion seem a little disjointed as we pause to define terms, but pedagogically it seems better than isolating all the new concepts in a technical appendix that offers little specific motivation for their usage. For convenient reference, however, technical concepts are assembled in a glossary at the end of the book.

Unlike later chapters, where we shall develop one or only a few different game-theoretic models in depth, we shall analyze several different games and discuss their applications in this chapter. From the point of view of the substance of international relations, our hopscotch run through its subject matter with examples will probably seem quite cursory and unsystematic. In part, however, this approach is dictated by the heterogeneity of the international relations literature and the lack of an accepted paradigm, or framework, within which questions of international conflict and cooperation can be subsumed. It is hoped that an explicit categorization of games in international relations will provide one useful touchstone to a field that presently lacks theoretical coherence.[2]

1.2. TWO-PERSON ZERO-SUM GAMES WITH SADDLEPOINTS

Although the concept of a "game" usually connotes lighthearted entertainment and fun, it carries no such connotation in game theory. Whether

1. This characteristic of foreign policy decisions, according to John C. Harsanyi, makes them "eminently susceptible to game-theoretical analysis." Harsanyi, "Game Theory and the Analysis of International Conflict," *Australian Journal of Politics and History*, 11 (Dec. 1965), p. 293. More modestly, another proponent of game-theoretic analyses of international conflict suggests that game theory's "present contributions must be regarded primarily in terms of the conceptualization and structuring it provides to the possible treatment of . . . problems." Thomas L. Saaty, *Mathematical Models of Arms Control and Disarmament: Application of Mathematical Structures in Politics* (New York: John Wiley & Sons, 1968), p. 61.

2. If this judgment seems exaggerated or unfair, the reader might check the following sources for the views of students of the subject: *Contending Approaches to International Politics*, ed. Klaus Knorr and James N. Rosenau (Princeton, N.J.: Princeton University Press, 1969); and James E. Dougherty and Robert L. Pfaltzgraff, Jr., *Contending Theories of International Politics* (Philadelphia: J. B. Lippincott Co., 1971).

a game is considered to be frivolous or serious, its various formal representations in game theory all connect its participants, or *players*, to *outcomes*—the social states realized from the play of a game—through the rules. Players need not be single individuals but may represent any autonomous decision-making units that can make conscious choices according to the *rules*, or instructions for playing the game. As we shall show, there are three major ways of representing a game, two of which are described in this chapter.

Although it is customary to begin with a discussion of one-person games, or games against nature, these are really of little interest in international relations. A passive or indifferent nature is not usually a significant force in international politics.[3] If "nature" (as a fictitious player) bequeaths to a country great natural resources (e.g., oil), it is the beliefs held by leaders about how to use these resources, not the resources themselves, which usually represent the significant political factor in its international relations (as the policies of oil-producing nations made clear to oil-consuming nations in 1973–74).

When we introduce a second player in games, many essentially bilateral situations in international relations can be modeled. Most of these, however, are not situations of pure conflict in which the gains of one side match the losses of the other. Although voting games of the kind we shall discuss in Chapter 3 can usefully be viewed in this way, there is no currency like votes in international relations that is clearly won by one side and lost by the other side in equal amounts. In the absence of such a currency, it is more difficult to construct a measure of value for adversaries in international conflict.

The analysis of the World War II battle described below illustrates one approach to this problem.[4] In February 1943 the struggle for New Guinea reached a critical stage, with the Allies controlling the southern half of New Guinea and the Japanese the northern half. At this point intelligence

---

3. For an interesting application of game theory in anthropology, where nature (in the incarnation of a chooser of water currents that affect fishing catches) is assumed to be pitted against Jamaican fishermen who select fishing locations whose yield depends in part on these currents, see William Davenport, "Jamaican Fishing: A Game Theory Analysis," in *Papers in Caribbean Anthropology*, Yale University Publications in Anthropology, Nos. 57–64, ed. Sidney W. Mintz (New Haven, Conn.: Department of Anthropology, Yale University, 1960), pp. 3–11. An alternative decision-theoretic approach to explaining the choices of the Jamaican fishermen, where nature is assumed to be a passive or indifferent player, is proposed in Robert Kozelka, "A Bayesian Approach to Jamaican Fishing," in *Game Theory in the Behavioral Sciences*, ed. Ira R. Buchler and Hugo G. Nutini (Pittsburgh: University of Pittsburgh Press, 1969), pp. 117–25. As another example of a one-person game, see Robert L. Birmingham, "A Model of Criminal Process: Game Theory and Law," *Cornell Law Review*, 56 (Nov. 1970), pp. 57–73, where the sanctions that a prospective criminal faces are conceptualized as a game against nature; and for examples from geography, see Peter R. Gould, "Man against His Environment: A Game Theoretic Framework," *Annals of the Association of American Geographers*, 35 (Sept. 1963), pp. 290–97.

4. This example is taken from O. G. Haywood, Jr., "Military Decision and Game Theory," *Operations Research*, 2 (Nov. 1954), pp. 365–85.

FIGURE 1.1 BATTLE OF THE BISMARCK SEA: STRATEGIES AND PAYOFF MATRIX[a]

		JAPANESE STRATEGIES		
		Sail north	*Sail south*	*Minima of rows*
	Search north	2 days	2 days	②
KENNEY'S STRATEGIES	*Search south*	1 day	3 days	1
	Maxima of columns	②	3	

[a] From Haywood, "Military Decision and Game Theory."

reports indicated that the Japanese were assembling a troop and supply convoy that would try to reinforce their army in New Guinea. It could sail either north of New Britain, where rain and poor visibility were predicted, or south, where the weather was expected to be good. In either case, the trip was expected to take three days.

General Kenney, commander of the Allied Air Forces in the Southwest Pacific Area, was ordered by General MacArthur as supreme commander to inflict maximum destruction on the convoy. As Kenney reported in his memoirs, he had the choice of concentrating the bulk of his reconnaissance aircraft on one route or the other; once the Japanese convoy was sighted, his bombing force would be able to strike.

Since the Japanese wanted to avoid detection and Kenney wanted maximum possible exposure for his bombers, it is reasonable to view this as a *strictly competitive game*. That is, cooperation between the two players is precluded by the simple fact that it leads to no joint gains; what one player wins has to come from the other player. The fact that nothing of value is added to, or subtracted from, a strictly competitive game means that the *payoffs*, or numbers associated with the outcomes for each pair of strategies of the two players, necessarily sum to some constant. (These numbers, called *utilities*, indicate the degree of preference that players attach to outcomes.)

We refer to such games as *constant-sum*; if the constant is equal to zero, the game is called *zero-sum*. In the above-described game, which came to be known as the Battle of the Bismarck Sea, the *payoff matrix*, whose entries indicate the expected number of days of bombing by Kenney following detection of the Japanese convoy, is shown in Figure 1.1 for the two strategies of each player. Because the payoffs to the Japanese are equal to the negative of the payoffs to Kenney (i.e., the payoffs multiplied by -1), the game is zero-sum. Since the row player's (Kenney's) gains are the column player's (Japanese's) losses and vice versa, we need list only one entry in each cell of the matrix, which conventionally represents the payoff to the row player.[5]

The commanders of each side had complete freedom to select one of their two alternative strategies, but the choice of neither commander alone could determine the outcome of the battle that would be fought.[6] If we view Kenney as the maximizing player, we see that he could assure himself of an outcome not less than the minimum in each row. Consider the minimum values of each row given in Figure 1.1: Kenney's best choice was to select the strategy associated with the maximum of the row minima (i.e., the value 2, which is circled), or the *maximin*—search north—guaranteeing him at least two days of bombing. Similarly, the Japanese commander, whose interest was diametrically opposed to Kenney's, would note that the worst that could happen to him is the maximum in any column. To minimize his exposure to bombing, his best choice was to select the strategy associated with the minimum of the row maxima (i.e., the value 2, which is circled), or *minimax*—sail north—guaranteeing him no more than two days of bombing.

The least amount that a player can receive from the choice of a strategy is the *security level* of that strategy. For the maximizing player, his security levels are the minima of his rows, and for the minimizing player they are the maxima of his columns. For the players to maximize their respective security levels, the maximizing player (in this example, Kenney) should choose a strategy that assures him of at least two days of bombing, and the minimizing player (the Japanese) a strategy that insures him against more than two days of bombing.

Note that there is no arbitrariness introduced in assuming that Kenney is the maximizing player and the Japanese the minimizing player. If we defined the entries in the payoff matrix to be the number of days required for detection of the convey *before* commencement of bombing, then the roles of the two players would simply be reversed but the strategic problem would remain the same.

The above strategy choices are predicated on the conservative assumption that a player will seek to foreclose the selection of his least desirable outcomes rather than try for his best outcome (three days of bombing in the case of Kenney), should there be a conflict between the two goals.

---

5. If we represented the payoff to Kenney as the expected number of days of bombing, and the payoff to the Japanese as the expected number of days without bombing, then the payoffs in each cell of the matrix would sum to three days (i.e., the length of the voyage). Strategically, this constant-sum representation of the game is equivalent to the zero-sum representation given in the text since the payoffs in the two representations differ only by a constant amount. All constant-sum games can be converted to zero-sum games by subtracting the appropriate constant from the payoffs to the players.

6. Here a "strategy" simply means one of the two alternative courses of action that each commander could choose. In more complicated situations, strategies (and metastrategies, to be discussed later in the chapter) may involve courses of action contingent on the courses of action (or expectations about these) of other players. As such, they provide a complete set of instructions of what choice to make for every contingency that may arise.

A player's choice of such a strategy will be reinforced if he anticipates that his opponent will apply the same assumption to his own choice of a strategy. For example, the pessimistic expectation of Kenney that the Japanese would sail north, where visibility was expected to be bad, gives him added impetus to search north (ensuring two days of bombing). On the other hand, the pessimistic expectation of the Japanese that Kenney would search north, where their chances of evasion were greatest, offers no incentive for them to sail north (whether they sail north or south, they can expect two days of bombing).[7] If their expectation should turn out to be incorrect, however, they would suffer a penalty for sailing south (namely, three days of bombing), which is sufficient to dictate their choice of the northern route. Thus, the outcome of the game is, in a sense, determined, and we refer to it as a *strictly determined game*.

More precisely, if a matrix game contains an entry that is simultaneously the minimum of the row in which it occurs and the maximum of the column in which it occurs (i.e., if the maximin is equal to the minimax), the game is strictly determined and the entry is a *saddlepoint*. The word "saddlepoint" derives its name from the fact that the surface of a saddle curves upward in one direction from the center (i.e., the line of motion of the horse), corresponding to a row minimum, and downward in the other direction, corresponding to a column maximum. Thus, an entry in a matrix game which is simultaneously the minimum of a row and the maximum of a column is analogous to the center of a saddle-shaped surface (though the matrix entries do not possess the property of continuity that characterizes a smooth surface). A two-person zero-sum game may have several saddlepoints, but in such a case all of them will have the same value.

The *value* of a game has the property that it is the best outcome that either player can assure himself of, and in a strictly determined game it is always equal to the saddlepoint.[8] A strategy assuring a player that he will obtain a payoff at least equal to this amount is an *optimal strategy*, and a player who selects his optimal strategy is said to be *rational*. In the Battle of the Bismarck Sea, the maximin and minimax strategies that intersect the saddlepoint are optimal strategies and are *in equilibrium*: It is not to the advantage of either player to change his optimal strategy if the other player does not change his. This can be seen from the fact that the row player (Kenney) cannot gain by unilaterally changing his maximin strategy since the saddlepoint is the largest entry in the Japanese "sail north" column; and the column player (the Japanese) cannot gain by unilaterally changing his minimax strategy since there is no larger entry in the Kenney

---

7. If the payoff to Kenney when the Japanese sail north and he searches north were only $1\frac{1}{2}$, instead of 2, days of bombing, then such a reinforcing incentive to sail north would also exist for the Japanese.

8. It is not necessarily true, however, that an entry in a payoff matrix that is equal to the value of a game is a saddlepoint. For an example, see Daniel P. Maki and Maynard Thompson, *Mathematical Models and Applications* (Englewood Cliffs, N.J.: Prentice-Hall, 1973), p. 216.

"search north" row. Therefore, a player's knowledge of his opponent's optimal strategy provides no inducement for him to switch his own choice of an optimal strategy; on the contrary, as we have shown, such knowledge may reinforce a player's choice of his optimal strategy.

Thus, it is not true, as Anatol Rapoport asserts, that "players *must* take the reasoning of the opponent" to determine their optimal strategies in a strictly determined game.[9] Indeed, in choosing his optimal strategy, a player can ignore the strategy choice of his opponent. By maximizing *his own* security level, he automatically assures himself of a payoff at least equal to the value of the game; if both players maximize their security levels, the saddlepoint will always be the outcome selected. To be sure, the logic of maximizing security levels is certainly prudent when playing against a rational opponent applying the same logic, but the strategy choice implied by this logic can be determined independently of an opponent's strategy choice.

Neither is it true, as Karl W. Deutsch asserts, that a minimax strategy "cannot take advantage of any mistakes he [an adversary] may make."[10] In games with a saddlepoint, a minimax strategy may indeed enable one to exploit an adversary's mistakes; a glance at Figure 1.1 will show that if the Japanese had selected their minimax strategy of sailing north, Kenney would have suffered his worst outcome (1 day of bombing) had he selected his nonoptimal strategy of searching south.

It is true that optimal strategies hold up best against the most damaging choices of an adversary. Furthermore, as the best "defensive" strategies, they are consonant with conventional military doctrine. As prescribed in United States military service manuals, a commanding officer is enjoined to take account of an enemy's capabilities (what he is able to do) and not his intentions (what he is going to do), assuming there may be a conflict.[11] Not surprisingly, the commanders in the Battle of the Bismarck Sea selected their optimal maximin/minimax strategies, which are consistent with this doctrine. The Japanese convoy sailed the northern route and was spotted by Kenney's reconnaissance aircraft, concentrating on the north, one day after its departure, which allowed Kenney two days of bombing. As it turned out, the Battle of the Bismarck Sea ended in a disastrous defeat for the Japanese, who did not know that Kenney, by modifying some of his aircraft for low-level bombing, had developed a technique that proved deadly in the battle.

---

9. Anatol Rapoport, *Fights, Games, and Debates* (Ann Arbor, Mich.: University of Michigan Press, 1960), p. 135; italics added.

10. Karl W. Deutsch, *The Analysis of International Relations* (Englewood Cliffs, N.J.: Prentice-Hall, 1968), p. 117.

11. The implication, of course, is that one should never assume that an opponent will act stupidly, which the game theorist translates to mean that an opponent's powers of reasoning should be considered the same as one's own. If this is the case, however, there can be no conflict between intentions and capabilities because both players, given the same information on each other's capabilities—as measured by the payoffs they can obtain for given pairs of strategy choices—would behave in an identical manner.

Although the Japanese's lack of intelligence might be viewed as a failure on their part, they did not err in choosing a strategy that would minimize their maximum losses. Because the game itself was unfair, they could not avoid some losses, whatever strategy they adopted. Formally, a game is *fair* if the choice of optimal strategies by both players results in the same zero payoff to each (neither player wins anything from the other—that is, the value of the game is zero).

1.3. INFORMATION IN GAMES

In section 1.2 we indicated that we would be concerned not with games against nature but with games that include one or more other players with freedom to make independent choices. In the Battle of the Bismarck Sea example, we showed that both players abided by the security-level principle—that is, they chose strategies that maximized their security levels. We refer to such games that involve rational players who invoke their optimal strategies as *games of strategy*, as contrasted with *games of chance*, whose outcomes do not depend at all on the strategy choices of the players but instead on some random or stochastic process determined by a probability distribution.

Many ordinary parlor games like chess and poker are games of strategy. Such games may usefully be divided into two classes. In games of *perfect information*, like chess, the players move alternately, and at each move a player is fully informed about the previous moves in the game. This is not the case in a game like poker, which is not a game of perfect information since a player does not know all the cards dealt to the other player(s). In the Battle of the Bismarck Sea, because both players had to make their strategy choices simultaneously and, therefore, in ignorance of each other's choices, that also is not a game of perfect information.

We earlier represented the Battle of the Bismarck Sea in matrix, or *normal, form*. In Figure 1.2 we represent this game by a *game tree*, in which the vertices of the tree represent choice points, or moves, and the branches, the alternatives that can be chosen. A *move* does not denote a particular course of action taken but a point at which one from a given set of alternatives is selected, either by a player or by chance (e.g., when a coin is flipped). A *play* denotes one selection of alternatives in a game, the last alternative being chosen at a move prior to a "termination point" (see the following paragraph).

Let us arbitrarily assign the first move to the Japanese when they set sail; they have a choice between two alternatives—to sail north or sail south. After the Japanese have made their choice, then Kenney, too, has a choice between two alternatives—to search north or search south. The payoffs to Kenney (the payoffs to the Japanese are the negative of these) are shown at the endpoints of the four lower branches. These are the *termination points* of the game; associated with each is a sequence of

FIGURE 1.2   GAME TREE OF BATTLE OF THE BISMARCK SEA

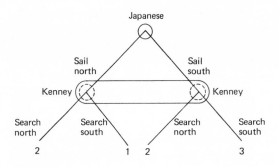

choices that traces a path through the game tree and uniquely characterizes the play of the game that led to that point.

The portrayal of the Battle of the Bismarck Sea as a series of alternate moves by the players, where the Japanese have two strategies (sail north, sail south) and Kenney, apparently, has four strategies (search north if the Japanese sail north, search north if the Japanese sail south, search south if the Japanese sail north, search south if the Japanese sail south), provides a representation of this game in *extensive form*. Every "finite" game (see footnote 15) in extensive form can be converted into an equivalent matrix game in normal form.[12]

For parlor games of sustaining interest, this extensive-form representation is not generally possible. Even in the simple game of tic-tac-toe, there are almost a trillion strategies (i.e., different sets of instructions specifying what square to play for every possible set of choices of one's opponent), but in fact these can be condensed into a few general strategic principles that make its solution trivial (a draw, if both players choose their optimal strategies).[13] The determination of optimal strategies in a game like chess, on the other hand, is far from trivial, though it can be shown that such strategies exist (see below). Whether the choice of these strategies leads to a draw, or a win by either white or black, is not known. To determine the answer to this question, we would have to investigate a fantastically large (though finite) number of strategies, which is well beyond the capabilities of present-day computers and hence makes the exhaustive examination of strategies impossible.

With the extensive-form representation of the Battle of the Bismarck Sea, we can now state precisely why this is not a game of perfect information. First note in Figure 1.2 that we have circled in solid lines (1)

---

12. For an illustration of how this is done, see Edna E. Kramer, *The Nature and Growth of Modern Mathematics*, vol. 1 (Greenwich, Conn.: Fawcett Publications, 1970), pp. 384–87.
13. Rapoport, *Fights, Games, and Debates*, pp. 146–47.

the upper vertex, and (2) the two lower vertices. The enclosed vertices are called *information sets* and indicate that the moves of players within each set are indistinguishable. Specifically, General Kenney, whose moves are represented by the two lower vertices, cannot distinguish which vertex he is at—that is, whether the Japanese sailed north or south—because these vertices are enclosed in a single information set.[14] Therefore, he has in fact only two strategies (search north, search south), rather than the four strategies previously indicated.

This emendation in the game-tree representation more faithfully mirrors the situation as it actually occurred, for in reality Kenney determined his course of action (to search north or south) simultaneously with the Japanese (i.e., in ignorance of their strategy choice). Although we had arbitrarily assumed that the Japanese had the first move—as forced upon us initially by the game-tree representation—we now correct this misrepresentation by enclosing the two vertices that represent Kenney's moves in a single information set. In effect, this means that Kenney does not know whether he is at one vertex or the other within his information set on the game tree. If he did—that is, if he were informed of the course of action taken by the Japanese prior to deciding on his own course of action—we would enclose his two moves (i.e., vertices) in separate information sets, as indicated by the dashed circles in Figure 1.2. We do enclose the first (and only) move of the Japanese in an information set by itself, because the Japanese, as the initially specified first-moving player, necessarily act in ignorance of the strategy choice of General Kenney.

In this manner, information sets tell us when players learn about events, whose occurrences in parlor games are prescribed by the rules. While a game in extensive form, depicted by a game tree, shows details of the sequencing of moves, in the Battle of the Bismarck Sea both players selected alternatives without knowledge of each other's choices. Inasmuch as both players had to make decisions in the dark, the temporal sequence in which moves are depicted is inconsequential. What is consequential is captured entirely in the normal form, depicted in the matrix game, where each player is assumed to make his choice simultaneously and independently of the other player's choice.

To be sure, in many international relations situations intelligence is available, leaks of information occur, signals are transmitted. We shall consider game-theoretic models of such situations later, but at this point it is useful to give a precise definition of *perfect information* so that we can unambiguously distinguish deviations from it. Information in a game in extensive form is perfect if every information set on the game tree contains only a single move (i.e., vertex). If this is the case, a player is fully informed of exactly where he is on the game tree at every move in the game. Tic-tac-toe and chess are both examples of games of perfect information.

---

14. The moves in an information set must always be those of only one player, and they must include the same number of alternatives.

FIGURE 1.3 A NONSTRICTLY DETERMINED GAME

		B'S STRATEGIES	
		b_1	b_2
A'S STRATEGIES	a_1	Win	Lose
	a_2	Lose	Win

It can be shown that every two-person zero-sum game of perfect information is strictly determined.[15] We shall not prove this result here, but we shall demonstrate that it holds for all such games that end in a win for one player and a loss for the other player. This excludes games like tic-tac-toe and chess, which can end in a draw, but the argument can be extended to cover these games, as well as games of perfect information with numerical payoffs (e.g., where the entries in a payoff matrix are +1, 0, and −1, representing win, draw, and lose).[16]

Now a game that terminates in a win for one player and a loss for the other player is strictly determined if one player always has a winning strategy (i.e., a sequence of choices at each vertex that always terminates in a win). The game in normal form in Figure 1.3 between players A and B is not strictly determined because neither player has a strategy that will guarantee him a win whatever his opponent does. Obviously, if either A or B had advance information on his opponent's strategy choice, he could always select a winning strategy; the fact that he does not means that this is not a game of perfect information.

We shall offer an indirect proof, based on the extensive form of a game, that a two-person finite game of perfect information can never be represented by a matrix like that in Figure 1.3. To prove this result we begin by assuming that neither player has a winning strategy (i.e., the game is not strictly determined) and show that this assumption leads to a contradiction.

Perfect Information Theorem

Every two-person finite game of perfect information, which results in a win for one player and a loss for the other, is strictly determined (i.e., one of the players always has a winning strategy).

---

15. This statement applies only to *finite games*, where each move includes a finite number of alternatives and the game itself ends in a finite number of moves. In Chapter 7 we shall analyze a two-person zero-sum game in which each player can choose from an infinite number of strategies.

16. For a rigorous proof, see Ewald Burger, *Introduction to the Theory of Games*, translated by John E. Freund (Englewood Cliffs, N.J.: Prentice-Hall, 1963), Theorem 1, pp. 25–28.

Proof. Suppose that the game is *not* strictly determined: Whatever strategy player A selects, there is *at least one* strategy that player B can choose that will lead to a win for B; and whatever strategy B selects, there is *at least one* strategy that player A can choose that will lead to a win for A. In other words, assume neither player *initially* has a winning strategy—each can be blocked by the other player.

Now if one of the players is *eventually* to win (say, A), as assumed, he must be able to go from his initial position (i.e., move) to some subsequent position where he has an abbreviated winning strategy from that position. But this subsequent position cannot be a position he can choose next, because if he has an abbreviated winning strategy from that position, he would have a winning strategy at the start (which would be to select the alternative that leads to this subsequent position and then use the abbreviated winning strategy at that position), which we assumed was not the case. Continuing in this manner, one can show that there is *no* subsequent position that he can reach in a finite sequence of choices where he has an abbreviated winning strategy. Furthermore, A can prevent B from winning in a subsequent position, for if he could not, B would have a winning position from the start (by the reasoning given above for A), which we assumed was not the case.

Thus, if the two players cannot win from their initial positions, they cannot win from their next subsequent positions, or for similar reasons from positions subsequent to these, and so on. Hence, the game can never end—by definition, this occurs only when one of the players reaches a winning position—which contradicts the assumption that the game ends in a finite number of moves. Therefore, a two-person finite game of perfect information, which results in a win for one player and a loss for the other player, must be strictly determined (i.e., one player must always have a winning strategy). This completes the proof.

Remark. Note that the proof rests on the fact that if there is a winning position, a player in a game of perfect information would be able to get to it eventually; he would always know exactly what position he occupied in the game tree and what sequence of choices would get him to the winning position, whatever his opponent does. (As we indicated earlier, if his opponent could block him, this would imply that the opponent himself had a winning strategy.) Therefore, if a game is one of perfect information, where the outcome can be a win only for one player and a loss for the other player, there is no way to stop the player with the winning strategy from adopting it from the beginning, which means that the game is strictly determined.

It should be noted that while perfect information is a sufficient condition for a game to be strictly determined, it is not necessary. The Battle of the Bismarck Sea was not a game of perfect information, yet it had a saddlepoint and so was strictly determined.

It is customary in game theory to distinguish between "perfect" and "complete" information. As we have seen, a game is one of perfect information if each player's location on the game tree can be ascertained from the previous choices made by players in the game. To be fully informed of the state of affairs at every point in the *course* of the game, however, does not imply that players have full knowledge of the state of affairs at the *beginning* of the game.

We say that a player has *complete information* when he knows from the start the rules of the game, which include full knowledge of the possible moves that can be made and the payoffs associated with every outcome that can occur. Clearly, a player may be fully informed of the rules of a game, but if the game (say, poker) includes chance moves, his information will not be perfect. Indeed, an optimal strategy in a two-person zero-sum game may prescribe chance moves precisely to ensure that a player's choice of strategies is not entirely predictable by his opponent, as we shall show in the section 1.4.

1.4. TWO-PERSON ZERO-SUM GAMES WITHOUT SADDLEPOINTS

We showed in section 1.2 that if a two-person zero-sum game has a saddle-point, it is a simple matter to find optimal strategies in the game. One locates in the payoff matrix the entry which is simultaneously the minimum of its row and the maximum of its column. The row and column strategies associated with this entry, which is the saddlepoint, are the optimal strategies of the two players. For games without a saddlepoint, the determination of optimal strategies for the players is not so simple.

As an example of such a game, we consider another World War II battle.[17] In August 1944, just after the Allied invasion of Normandy, the Allies broke out of their beachhead near Avranches and threatened the German Ninth Army. The German commander, General von Kluge, faced a choice between the following two alternatives: (1) to attack or (2) to withdraw and take up a more tenable defensive position. The Allied commander, General Bradley, reported in his memoirs that he considered the following three alternatives, all of which involved utilization of his four reserve divisions: (1) to use his reserves to reinforce the gap at Avranches, through which part of his forces had already slipped; (2) to send his reserves eastward to harass or possibly cut off withdrawal of the German Ninth Army; or (3) to leave his reserves uncommitted for one day, and then decide whether to use them to reinforce the gap or push eastward.

Bradley's estimate of the outcomes that would occur for each pair of strategy choices of the two commanders are shown in Figure 1.4. Unlike the Battle of the Bismarck Sea, there was no reasonable way to assign numerical payoffs to each of the outcomes. Bradley was able, however, to rank the six outcomes from best to worst (indicated by the numbers 1 to

17. Haywood, "Military Decision and Game Theory."

FIGURE 1.4   BATTLE OF AVRANCHES: STRATEGIES AND PAYOFF MATRIX[a]

	VON KLUGE'S STRATEGIES	
BRADLEY'S STRATEGIES	*Attack*	*Withdraw*
Reinforce gap	Gap holds (2)	Weak pressure on German withdrawal (3)
Move eastward	Gap cut (1)	Strong pressure on German withdrawal (5)
Hold back	Gap cut, possibly Germans encircled (6)	Moderate pressure on German withdrawal (4)

[a] From Haywood. "Military Decision and Game Theory."

6 in parentheses following the descriptions in Figure 1.4). Assuming that the higher the number, the greater the value to Bradley, we have indicated his most-preferred outcome with the number "6," his next-most-preferred with the number "5," and so on; presumably, von Kluge's preference ranking is the reverse of Bradley's.

The first thing to observe about the game matrix in Figure 1.4 is that Bradley's third strategy choice (hold back) is superior to his first strategy choice (reinforce gap), whichever strategy von Kluge chooses. In game-theoretic language, Bradley's third strategy *dominates* his first strategy— that is, it is no worse, and in at least one contingency better, than the strategy it dominates. Since in this situation Bradley's third strategy is better for *both* contingencies he faces (von Kluge attacks or he withdraws), we say that Bradley's third strategy *strictly dominates* his first strategy.

If we eliminate Bradley's first strategy from further consideration on the ground that it would obviously be irrational for him to adopt it, the game matrix reduces to the 2 × 2 submatrix shown in Figure 1.5. Provisionally assuming that the ranks can be treated *as if* they were numerical payoffs (we shall illustrate the consequences of this assumption in sections 1.5 and 1.6), it can be seen from the circled figures in Figure 1.5 that the maximin (4) and the minimax (5) are not equal. This game, therefore, has no saddlepoint.

Because the payoff matrix contains no entry that is simultaneously the minimum of a row and the maximum of a column, it is not strictly determined. Hence, this is not a game of perfect information (by the Perfect Information Theorem). Unlike the Battle of the Bismarck Sea, where neither player could benefit from knowing that the strategy choice of the other player would be that associated with the saddlepoint, in the absence of a saddlepoint in the Battle of Avranches each general *could*

FIGURE 1.5 REDUCED PAYOFF MATRIX FOR BATTLE OF AVRANCHES

VON KLUGE'S STRATEGIES

BRADLEY'S STRATEGIES	Attack	Withdraw	Minima of rows
Move eastward	1	5	1
Hold back	6	4	④
Maxima of columns	6	⑤	

benefit from advance information on the other's plans. If Bradley knew what von Kluge's strategy choice would be, he could guarantee himself a payoff of at least 5, whereas the best that he can ensure for himself without this information is 4 (the maximin). Similarly, if von Kluge knew what Bradley's strategy choice would be, he could guarantee himself a payoff of not less than −4, whereas the best that he could insure himself against without this information is −5 (that Bradley receives the minimax of 5).

It is easy to show that the strategy choices of the two players that maximize their security levels (i.e., that guarantee to Bradley at least the maximin and von Kluge not less than the negative of the minimax) are not in equilibrium. Specifically, if von Kluge adopts his minimax strategy (withdraw), there is an incentive for Bradley to depart from his maximin strategy (hold back) and instead choose his other strategy (move eastward), realizing a payoff of 5 rather than 4. Although there is no such incentive of von Kluge to depart from his minimax strategy (withdraw) when Bradley adopts his maximin strategy (hold back)—he would only worsen his situation by going from a payoff of −4 to −6—it is not difficult to demonstrate that it would be foolish for von Kluge to choose with certainty his minimax strategy, or at least lead Bradley to think that he would.

The reasoning is as follows. If we assume that Bradley is an intelligent player, viewing the payoff matrix he, too, would know that von Kluge could only worsen his situation by departing from his minimax strategy, given that he (Bradley) selects his maximum strategy. If Bradley concludes that von Kluge will, therefore, stick to his minimax strategy for want of a better choice, then he (Bradley) could safely depart from his maximin strategy and select his other strategy (move eastward), realizing the payoff of 5 rather than 4. Yet, there is no reason to assume that von Kluge, as an equally intelligent player, could not reach the same conclusion. Hence, it would follow that von Kluge should depart from his minimax strategy and, with Bradley committed to move eastward, himself attack and realize his best possible payoff of −1. Now if Bradley in turn

**FIGURE 1.6   SUCCESSIVE CHOICES OF PLAYERS
IN BATTLE OF AVRANCHES**

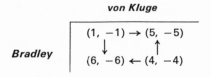

von Kluge

Bradley

$(1, -1) \rightarrow (5, -5)$
$\downarrow \qquad \uparrow$
$(6, -6) \leftarrow (4, -4)$

anticipated this result, it would obviously be in his interest to change his plans again. And so it goes if no outcome is in equilibrium for both players.

In Figure 1.6 we have depicted the way in which the payoffs associated with each pair of strategy choices would successively be chosen if the two players alternately anticipated the choices of each other. For example, if Bradley anticipated that von Kluge would choose his minimax strategy (second column), he would select his strategy that yields the higher payoff [i.e., first row, since he prefers $(5, -5)$ to $(4, -4)$, where the first number in the ordered pair represents the payoff to the row player—Bradley—and the second number represents the payoff to the column player—von Kluge]. Thus, if Bradley had previously decided upon his second-row strategy, yielding the outcome $(4, -4)$, he would be disposed to shift it to $(5, -5)$ by choosing his first-row strategy. This means that $(5, -5)$ dominates $(4, -4)$ for Bradley, which we indicate by the arrow from $(4, -4)$ to $(5, -5)$. It is also evident that $(6, -6)$ dominates $(1, -1)$ for Bradley.

For von Kluge, $(4, -4)$ dominates $(6, -6)$, since he prefers a payoff of -4 to -6; similarly, $(1, -1)$ dominates $(5, -5)$. It will be observed that the vertical arrows indicate Bradley's strategy preferences, and the horizontal arrows von Kluge's strategy preferences.[18]

Since the outcomes associated with all pairs of strategy choices by both players form a cycle, there is no position in equilibrium. As suggested by the preceding discussion, each player's anticipation of the other player's choices will render all outcomes unstable because there is always one player who can improve his lot once the other player has committed himself—or it is anticipated that he will. If and only if the matrix game has a saddlepoint will the maximin and minimax strategic choices of the row and column players be in equilibrium. In games where there is no such stable outcome, the dynamic of the system suggests that different pairs of strategy choices will be up for grabs, depending on the players' anticipatory reflexes about each other's probable choices.

---

18. This representation of dominant choices by a directed graph is used in Kenneth E. Boulding, *Conflict and Defense: A General Theory* (New York: Harper & Brothers, 1962), chap. 3.

In fact, it appears, the prescience of Bradley and von Kluge did not extend beyond their maximin and minimax strategies that maximized their respective security levels. Consistent with the play-it-safe notion of estimating capabilities rather than intentions, Bradley chose to hold his reserves uncommitted for one day, and von Kluge chose to withdraw, yielding the outcome (4, −4). Unfortunately for von Kluge, after he made his prudent decision to withdraw to a better defensive position, Hitler overruled him and ordered him to attack, thus assuring Bradley of his best possible outcome, (6, −6). The gap at Avranches held for one day without reinforcement, after which Bradley committed his four reserve divisions. The reserves succeeded in cutting off withdrawal of the German Ninth Army, thereby enabling Bradley's forces almost completely to surround it. Only remnants of this army escaped the encirclement; as a historical footnote, General von Kluge committed suicide after the battle.

1.5. PURE AND MIXED STRATEGIES

In the Battle of the Bismarck Sea, the optimal strategies of General Kenney and the Japanese had the property that if one player announced in advance the strategy he intended to use, the other player could not exploit this information to reduce his opponent's payoff to less than that of the saddlepoint. This was not the case in the Battle of Avranches, which did not have a saddlepoint that guaranteed both players the same (except for sign) payoff value; for this reason, as we have shown, that game did not have any stable (i.e., equilibrium) outcome.

What, then, are the optimal strategies in the Battle of Avranches? If knowledge of one player's strategy choice will put him at a disadvantage vis-à-vis his opponent, then his only alternative is not to choose one strategy with certainty. The one sure way he has to keep secret his choice is to select a strategy at random. For if there were any system or pattern in the way he makes a selection, it is reasonable to assume that it would be discoverable by a rational opponent. One insures against this possibility by not knowing oneself what the choice will be.

Recall that it was not necessary to make any assumptions about an opponent's rationality or intelligence in formulating a solution to strictly determined games; for each player, the solution consisted of choosing the strategy associated with the saddlepoint in the game, whatever the choice of the other player. But in games without a saddlepoint, we call upon the rationality assumption *to eliminate* obviously undesirable choices of strategy if played against a rational opponent (though, as we shall show, it is not necessary to assume that an opponent is rational to guarantee oneself the value of such a game). These undesirable choices include all those that involve the certain selection of any single, or *pure*, strategy.

In games without a saddlepoint, a player's optimal strategy is to select a probability distribution over his set of pure strategies, which is called a *mixed strategy*. A mixed strategy specifies the probabilities with which each pure strategy is to be chosen, but the actual selection of a pure strategy on any play of a game is made randomly according to these probabilities. For example, if a row player's optimal mixed strategy is to choose his first-row strategy with probability $\frac{2}{3}$, and his second-row strategy with probability $\frac{1}{3}$, he might determine his selection of a strategy on any play of the game by choosing one card at random from a set of three cards, where two of the cards are marked "choose row 1" and one of the cards is marked "choose row 2." Using this random device, he would choose on the average his first-row strategy two-thirds of the time and his second-row strategy one-third of the time.

To illustrate how these probabilities are calculated in a simple 2×2 matrix game, consider a game with the following payoff matrix (whose entries, as usual, designate the payoffs to the row player). Although the numerical payoffs are different, this game preserves the same ranking of outcomes as the reduced payoff matrix for the Battle of Avranches in Figure 1.5:

$$\begin{bmatrix} 0 & 2 \\ 3 & 1 \end{bmatrix}.$$

Since there is no entry that is both the minimum of a row and the maximum of a column, this game has no saddlepoint, and each player must therefore mix his strategies to ensure that this selection of a particular strategy cannot be predicted in advance by his opponent.

How should he do this? We start off by assuming that a rational player desires to maximize his *expected payoff*, which is defined to be the payoff he will receive for every possible outcome of the game times the probability that outcome will occur. Assuming that the row player chooses his first row with probability p_1 and his second row with probability p_2, and the column player chooses his first column with probability q_1 and his second column with probability q_2, and that the row and column players make their choices independently (i.e., simultaneously), then the probability that each entry in the payoff matrix will be chosen is given by the following "probability matrix":

$$\begin{bmatrix} p_1 q_1 & p_1 q_2 \\ p_2 q_1 & p_2 q_2 \end{bmatrix}.$$

Multiplying the corresponding entries in the payoff and probability matrices, and summing the products, we obtain the expected payoff to the row player R:

$$E(R) = 0p_1 q_1 + 2p_1 q_2 + 3p_2 q_1 + 1p_2 q_2.$$

Since $p_1 + p_2 = q_1 + q_2 = 1$ (i.e., each player must play one or the other of his strategies with certainty), where $p_1, p_2, q_1, q_2 \geq 0$ (i.e., the probabilities are nonnegative numbers),

$$p_2 = 1 - p_1 \quad \text{and} \quad q_2 = 1 - q_1.$$

Thus,

$$E(R) = 0p_1 q_1 + 2p_1(1 - q_1) + 3(1 - p_1)q_1$$
$$+ 1(1 - p_1)(1 - q_1).$$

By algebraic manipulation one can obtain the equivalent expression,

$$E(R) = -4(p_1 - \tfrac{1}{2})(q_1 - \tfrac{1}{4}) + \tfrac{3}{2}.$$

If $p_1 = \tfrac{1}{2}$ is substituted in the above expression,

$$E(R) = \tfrac{3}{2},$$

which means that the row player can assure himself of an expected payoff of $\tfrac{3}{2}$ no matter what strategy (q_1, q_2) the column player adopts. Likewise, if $q_1 = \tfrac{1}{4}$, the column player can ensure that the row player's expected payoff will be $\tfrac{3}{2}$ no matter what strategy (p_1, p_2) the row player adopts. In other words, if the row player chooses the strategy $(\tfrac{1}{2}, \tfrac{1}{2})$, he eliminates the possibility of getting a long-term gain of less than $\tfrac{3}{2}$; if the column player chooses the strategy $(\tfrac{1}{4}, \tfrac{3}{4})$, he eliminates the possibility of suffering a long-term loss of less than $-\tfrac{3}{2}$. Since these strategies maximize the minimum gain for the row player, and minimize the maximum loss for the column player, they are the *optimal* (*mixed*) *strategies* in this game, whose (unique) value is equal to $\tfrac{3}{2}$.[19] Moreover, these two mixed strategies are in equilibrium since one player cannot gain from changing his strategy if the other holds his strategy fixed.

This is not to say that either player could not do better if he departed from his optimal (mixed) strategy. But it would require that his opponent also depart from his optimal (mixed) strategy and that he be able, through prescience or luck, to exploit this fact.[20] Without foreknowledge of his opponent's strategy choice, however, there is nothing that a player can do to alter the expected payoff of the game in his favor if his opponent plays his optimal mixed strategy. For even though it is in a player's power to figure out his opponent's optimal mixed strategy as well as his own, this still does not tell him what strategy choice his opponent will make on any particular play of the game.

By the Fundamental Theorem of Game Theory, or Minimax Theorem, every two-person zero-sum game has a solution in pure or mixed

---

19. For a simple proof that the value of a two-person zero-sum game is unique, see John G. Kemeny, J. Laurie Snell, and Gerald L. Thompson, *Introduction to Finite Mathematics*, 2nd ed. (Englewood Cliffs, N.J.: Prentice-Hall, 1966), Theorem 1, p. 355.

20. A player cannot exploit his opponent's choice of a (nonoptimal) strategy if he himself plays his optimal mixed strategy, for as is evident in the "equivalent expression" for $E(R)$ in the text, $E(R) = \tfrac{3}{2}$ if *either* player chooses his optimal mixed strategy. Thus, no matter how poorly an opponent plays, since a player can realize no additional gains when he chooses his optimal mixed strategy, this strategy is most sensible to use against a capable and informed opponent. (A player who chooses his optimal pure strategy in a strictly determined game, however, may realize additional gains from an inferior—i.e., nonoptimal—strategy choice by his opponent, as we showed earlier.)

strategies.[21] Formally, an optimal pure strategy is simply a special case of an optimal mixed strategy, where the probability distribution over the set of pure strategies for a player assigns a probability of one to a single pure strategy and zero to all the rest. Qualitatively, however, the interpretation of a mixed strategy in the single play of a game is somewhat murky since the concept of "expected payoff" is most closely approximated only over several plays of a game.

There is a second problem in applying the concept of a mixed strategy to a situation like the Battle of Avranches. The ranking of the six alternative outcomes by Bradley from best to worst offers no clue as to *how much* he, and presumably von Kluge applying an opposite preference scale, preferred one outcome to others. Without knowing the "distance" between ranks, as measured on an interval scale, we cannot calculate the probabilities with which each pure strategy should be chosen.

To illustrate this problem, one need only compare the reduced payoff matrix for the Battle of Avranches, as given in Figure 1.5, with the payoff matrix for which we calculated optimal mixed strategies. The payoffs in both matrices are consistent with Bradley's ranking of the outcomes, but each assigns different numerical values, or utilities, to the outcomes (as shown in Table 1.1). We showed that the optimal mixed strategies for the second assignment are $(\frac{1}{2}, \frac{1}{2})$ and $(\frac{1}{4}, \frac{3}{4})$ for the row and column players, respectively; the optimal mixed strategies for the first assignment are $(\frac{1}{3}, \frac{2}{3})$ and $(\frac{1}{6}, \frac{5}{6})$ for the row and column players. By the first assignment, then, both the row (Bradley) and column (von Kluge) players should play their first strategies (move eastward for Bradley, attack for von Kluge) somewhat more frequently than by the second assignment, though in neither case with a probability exceeding 0.50.

1.6. INTERPRETATION OF MIXED STRATEGIES

To illustrate how these mixed strategies can be interpreted geometrically, we have graphed in Figure 1.7 Bradley's expected payoff as a function of the probability that he chooses his first strategy (move eastward) or second strategy (hold back) in the reduced payoff matrix in Figure 1.5. Associated with each of Bradley's pure strategies is a payoff that depends

---

21. For a relatively elementary proof, see Maki and Thompson, *Mathematical Models and Applications*, pp. 222–26. The intuitive reasonableness of optimal minimax strategies, as manifested in going from the extensive to the normal form of a game, has recently been attacked—and defended—but the arguments are not central to our consideration of applications and will therefore not be discussed here. See R. J. Aumann and M. Maschler, "Some Thoughts on the Minimax Principle," *Management Science*, 18 (Jan., Part 2, 1972), pp. P-54–P-63; and Morton D. Davis, "Some Further Thoughts on the Minimax Principle," and "Communications to the Editor" by Michael Maschler, Robert J. Aumann, and Guillermo Owen, *Management Science*, 20 (May 1974), pp. 1305–10, 1316–17.

TABLE 1.1 DIFFERENT UTILITY ASSIGNMENTS TO OUTCOMES IN BATTLE OF AVRANCHES

BRADLEY'S PREFERENCE RANKING OF OUTCOMES FROM BEST TO WORST	UTILITY ASSIGNMENTS TO OUTCOMES	
	Assignment 1	*Assignment 2*
1	6	3
2	5	2
3	4	1
4	1	0

on von Kluge's strategy choice. For example, if von Kluge attacks and Bradley always chooses his first (pure) strategy $(p_1, p_2) = (1, 0)$, Bradley's payoff will be 1 by the first assignment of utilities and 0 by the second assignment. If, however, Bradley always chooses his second strategy $(p_1, p_2) = (0, 1)$, his payoff will be 3 by the first assignment of utilities and 6 by the second assignment.

Since Bradley's expected payoff is a linear combination of the probability that he chooses his first (p_1) and second $(p_2 = 1 - p_1)$ strategies, his *expected* payoff for any strategy *mix* can be determined from the straight lines connecting his payoffs at (1, 0) and (0, 1). Each of von Kluge's strategies (attack or withdraw) is represented in Figure 1.7 by solid lines for the first assignment of utilities and dashed lines for the second assignment of utilities.

The points along the vertical axis where the two solid lines and the two dashed lines intersect give the maximum expected payoffs, for each assignment of utilities, that Bradley can assure himself of winning (and von Kluge can insure himself against losing), which are also equal to the values of the two games ($\frac{3}{2}$ for the first assignment of utilities, $1\frac{3}{3}$ for the second assignment). The points along the horizontal axis where the two solid lines and the two dashed lines intersect give the optimal mixed strategies of Bradley for each assignment of utilities [$(\frac{1}{2}, \frac{1}{2})$ for the first assignment of utilities, $(\frac{1}{3}, \frac{2}{3})$ for the second assignment]. These strategies for the two assignments of utilities are optimal because there are no higher expected payoffs than the intersections of the solid lines and the dashed lines for *both* pure strategies of von Kluge (i.e., attack, or withdraw, with a probability equal to one). Moreover, it can be shown that this is the maximum that Bradley can ensure for himself even if von Kluge mixes his strategies.[22]

---

22. For a general proof, see Maki and Thompson, *Mathematical Models and Applications*, pp. 220–21.

FIGURE 1.7   EXPECTED PAYOFFS AND MIXED STRATEGIES OF BRADLEY IN BATTLE OF AVRANCHES FOR DIFFERENT UTILITY ASSIGNMENTS

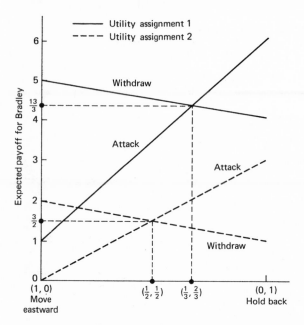

Mixed strategy $(p_1\ p_2)$ for Bradley

The search for optimal mixed strategies is more difficult in larger games—the Minimax Theorem only guarantees their existence—though it can be demonstrated that in $2 \times n$ and $n \times 2$ games, where $n > 2$ (i.e., where either the row or the column player has only two pure strategies but his opponent has more than two), the player with more than two pure strategies has an optimal mixed strategy that uses only two of his pure strategies.[23] However, we shall not consider such complications further, but instead indicate how the solution to a mixed-strategy game like the Battle of Avranches can be interpreted. After all, how the two commanders employ mixed strategies is not at all clear.

First, let us consider the question of evaluating the military worth of each of the outcomes. In discussing this example, Otomar J. Bartos observed:

---

23. These two strategies, however, need not be unique. For an example, see Kemeny, Snell, and Thompson, *Introduction to Finite Mathematics*, example 5, p. 372.

One could urge the commander to attempt to go beyond a mere ranking and to assign specific payoffs to each outcome, but it is doubtful whether this procedure should really be recommended. If it amounts to assigning specific numbers just for the sake of having such numbers, then one is engaging in an elaborate self-deception which may be very dangerous, precisely because it is elaborate. One may be dead wrong in the choice of strategy, but wrong with a precision which may be impressive to the uninitiated.[24]

Bartos also considered that the commanders might simply forego the use of mixed strategies and select the strategies that maximized their security levels, which is of course what each chose (though von Kluge's choice was countermanded by Hitler). The outcome that results, to be sure, is unstable, but in a one-shot game where each player must make his choice ignorant of his opponent's choice, he will never have a chance to capitalize on this instability, and change his response, since there are no subsequent plays of the game.

There is still another interpretation of the Battle of Avranches that does not deny the concept of a mixed-strategy solution but yet is largely consistent with the choices made by the commanders in this battle. A clue to this interpretation can be found in the graph of Figure 1.7, where it will be observed that Bradley's probability of holding back increases from $\frac{1}{2}$ for the second assignment of utilities to $\frac{2}{3}$ for the first assignment of utilities. Generally speaking, the greater utility that Bradley ascribes to his maximin strategy of holding back—regardless of the strategy choice of von Kluge—and the less difference the strategy choices of von Kluge (attack or withdraw) have on Bradley's maximin strategy, the more Bradley's optimal mixed strategy will tend toward the choice of his maximin strategy as an optimal pure strategy.

Analogously, similar considerations will push von Kluge toward the choice of his minimax strategy (withdraw) with a high probability. Indeed, the probability that von Kluge should withdraw, as prescribed by his optimal mixed strategy, increases from $\frac{3}{4}$ for the second assignment of utilities to $\frac{5}{6}$ for the first assignment.

We see, then, that the utilities associated with each outcome influence the predisposition of a player for one pure strategy, even if the game contains no saddlepoint and the optimal strategies of the two players are, therefore, mixed. It seems reasonable to assume that, consistent with military doctrine stressing that one's actions should be based on the capabilities, rather than the intentions, of the enemy, both commanders in a situation like the Battle of Avranches would be inclined to assign comparatively great weight to achieving the outcome that maximizes their security level; further gains would be frosting on the cake but not

---

24. Otomar J. Bartos, *Simple Models of Group Behavior* (New York: Columbia University Press, 1967), pp. 208–209.

something for which to risk a great deal. The choice of maximin and minimax strategies—with a high probability—is consonant with this conservative assignment of utilities in games, like the Battle of Avranches, that are not strictly determined.

There probably are situations, particularly those that extend in space or time, wherein genuinely mixed strategies are both applicable and desirable. Haywood suggests that mixed strategies may be optimal in the planning of numerous small-scale military actions, where what counts is the overall effect of all actions and not the outcome of a single encounter.[25] Also, optimal pursuit and evasion strategies in reconnaissance games, in which the players want over time to maximize the amount of information they gain about the behavior of an enemy and minimize his information about their own behavior, frequently involve the use of randomized strategy choices.[26]

Certainly, mixed strategies offer the element of surprise, which *does* seem to characterize the actions of some nations and their leaders in international politics. In fact, it seems, some of the most effective leaders are those who time their actions to create the greatest surprise effect, which may be a calculation akin to that based on the use of optimal mixed strategies. A bizarre illustration of such a calculation, replete with qualitative probabilistic estimates, is provided in the following analysis of the World War II Battle of the Bulge by the London *Daily Express* in 1945:

> The Allies were not surprised, because they knew the possibility of a surprise attack. What surprised them was that the Germans thought it worthwhile to make a surprise attack in spite of the fact that such an attack, though deemed possible, was not deemed probable, in view of the fact that we knew they would try to surprise us.[27]

Although this analysis smacks mostly of an attempt to avoid blame for a major Allied intelligence failure, it nevertheless reveals a keen intuitive appreciation of the possible effect of mixed strategies in fouling up attempts to second-guess an opponent's actions.

Naturally, this is by no means hard evidence that Hitler, consciously or unconsciously, resorted to mixed strategies in ordering an offensive strike by his forces, though he certainly was a master of surprise—if sometimes foolish and desperate—attacks. By the very nature of the random element in such strategy choices, the use of mixed strategies by actors

---

25. Haywood, "Military Decision and Game Theory," pp. 381–82. See also W. W. Fain and J. B. Phillips, "Applications of Game Theory to Real Military Decisions," in *Theory of Games: Techniques and Applications*, ed. A. Mensch (New York: American Elsevier Publishing Co., 1967), pp. 363–72.

26. For examples, see Melvin Dresher, *Games of Strategy: Theory and Applications* (Englewood Cliffs, N.J.: Prentice-Hall, 1961).

27. Quoted in Harold Wilensky, *Organizational Intelligence: Knowledge and Policy in Government and Industry* (New York: Basic Books, 1967), n. 64, p. 64.

in international politics is difficult to verify empirically. It would seem, though, that much that passes for inscrutable, or even irrational, behavior may be interpretable as the randomized strategic choices of players in a nonstrictly determined game.

We hasten to point out that no claim is made that such players think of their behavior as "random"; quite the contrary, they are likely to view their strategic choices as "calculated risks" precisely because they do not wholly rely on the security-level principle. To the extent that players hedge their bets by taking one course of action on one occasion and another course on another (apparently similar) occasion, their choices may be interpreted *as if* they selected strategies at random. Of course, a player may rationalize his particular choices as a response to the anticipated actions of another player, but if these actions were truly an unknown in his decision, then he can properly be said to be using a mixed strategy to gain the advantage of surprise—and perhaps a future reputation for being stupidly erratic or cunningly brilliant, depending on the outcome. For like bluffing in poker, not all attempts to be unpredictable are successful.[28] Yet, the willingness to bluff is exactly what ensures the advantage of surprise when one wants it to count.

As we have demonstrated, the calculation of mixed strategies depends on the numerical payoffs that players associate with outcomes. Although von Neumann and Morgenstern developed an axiomatically based procedure for deriving utility functions for players from their preferences for pairs of alternatives (actually, gambles or lotteries over the alternatives),[29] it seems inapplicable to most real-world situations, where access to decision makers is limited or nonexistent for the purpose of translating their subjective preferences into "utiles," or the units in which utility functions are expressed. We can usually make only an informed guess of the subjective value that actors attach to particular outcomes, which clearly circumscribes the validity of conclusions based on these estimates. (In Chapter 2, the analysis is based directly on the ordinal preferences of actors and we thereby avoid this problem.) Admittedly, this is a serious problem in many potential applications of game theory, and decision theory generally, to empirical situations. Nevertheless, in some situations sufficiently good approximate estimates can be made so as to render one's analysis useful, as we have tried to show.[30]

---

28. For a game-theoretic analysis of a simplified version of poker, see Kemeny, Snell, and Thompson, *Introduction to Finite Mathematics*, pp. 378–83.

29. For an excellent nontechnical discussion of utility theory that emphasizes the pitfalls to be avoided in interpreting the concept of utility, see Morton D. Davis, *Game Theory: A Nontechnical Introduction* (New York: Basic Books, 1970), chap. 4. For a more technical discussion, see R. Duncan Luce and Howard Raiffa, *Games and Decisions: Introduction and Critical Survey* (New York: John Wiley & Sons, 1957), chap. 2.

30. For some persuasive arguments to this effect, see Howard Raiffa, *Decision Analysis: Introductory Lectures on Choices under Uncertainty* (Reading, Mass.: Addison-Wesley, 1968), chaps. 4 and 5. Although this book is not a work on game theory, the arguments Raiffa makes are relevant to the subjective estimation of utility.

1.7. TWO-PERSON NONZERO-SUM GAMES

Continuing with our classification of games, we now turn to two-person nonzero-sum (or variable-sum) games. Considering only the ways in which two players can rank the four outcomes in 2×2 matrix games, Anatol Rapoport and Melvin Guyer showed that there are no less than 78 such distinct games, after eliminating those that can be derived from one of the 78 through the transposition of rows and columns or the relabeling of players.[31]

Obviously, we cannot possibly consider the potential relevance of all of these different games to international relations. Although it would probably be possible to conjure up hypothetical situations that mirror the different rankings of payoffs by the players in many of these very simple games, they would tell us little about their occurrence in reality, how numerical payoffs might be assigned to the different ranks, relationships among the games, and so forth. Fortunately, a relatively small number of games has been studied in depth, some of which seem to reflect general and persistent characteristics of many conflict-of-interest situations. Two of these games seem to have particularly important implications for the study of international conflict, and we shall accordingly limit our later discussion of applications to them.

Less fortunately, solutions in two-person nonzero-sum games tend to be less compelling than the minimax solution in zero-sum games. When we drop the zero-sum assumption that whenever the payoff to one player is an amount a, the payoff to the other player is the amount $-a$, we encounter situations in which one player can gain and the other player can lose different amounts, or both players can gain or lose at the same time. Whereas in two-person zero-sum games the players have no common interests, in two-person nonzero-sum games the players typically have both competitive and complementary interests. For this reason, such games are sometimes referred to as *mixed-motive games*.

These games generally provide a more realistic representation of complex political situations, which, except in war and a few other situations of pure conflict, rarely involve absolutely no cooperation between the players. Even in war the combatants may tacitly agree not to use certain very destructive weapons (e.g., poison gas in World War II, nuclear weapons more recently), pledge to treat civilians and prisoners-of-war humanely, abide by truces, and honor cease-fires.

To point up some difficulties that arise in the analysis of two-person nonzero-sum games—more will be spelled out in the examples discussed in subsequent sections—consider the game pictured in Figure 1.8. Notice that we list payoffs for *both* players in this game, as we generally did not do in zero-sum games (except in Figure 1.6), where the first number in the

---

31. Anatol Rapoport and Melvin Guyer, "A Taxonomy of 2 × 2 Games," *General Systems: Yearbook of the Society for General Systems Research,* 11 (1966), pp. 203–14.

FIGURE 1.8   A TWO-PERSON NONZERO-SUM GAME

	b_1	b_2
a_1	(2, 1)	(0, 0)
a_2	(0, 0)	(1, 2)

ordered pair represents the payoff to the row player and the second number the payoff to the column player. There are four properties of this game that distinguish it from zero-sum games, the first two of which render concepts attractive in zero-sum game theory unattractive in nonzero-sum game theory:

1. *Unsatisfactoriness of equilibrium pairs of strategies.* The game in Figure 1.8 has two equlibrium pairs in pure strategies, (a_1, b_1) and (a_2, b_2): There is no advantage for one player to deviate from these strategies if the other player does not. The problem is that the row player prefers pair (a_1, b_1) and the column player prefers (a_2, b_2). This problem cannot crop up in zero-sum games, because if there is more than one equilibrium pair, all pairs yield the same payoff. Thus, despite the fact that every two-person nonzero-sum game has an equilibrium pair in pure or mixed strategies,[32] there is no way to determine which player gets the larger, and which player the smaller, payoff in a game such as that shown in Figure 1.8.

2. *Unsatisfactoriness of optimal (mixed) strategies.* Applying the reasoning developed earlier for zero-sum games, a player might begin by asking what payoff he can ensure for himself. Considering the game from the row player's point of view, he can calculate his optimal strategy regardless of what the column player does (i.e., by ignoring what, if any, strategy choice the column player's own payoffs might prescribe for him). Following the method of calculation set forth in section 1.5, we find that the row player's optimal mixed strategy based on his payoffs alone is $(\frac{1}{3}, \frac{2}{3})$, which guarantees him an expected payoff of $\frac{2}{3}$. Similarly, from the symmetry of the game it is evident that the column player's optimal mixed strategy is $(\frac{2}{3}, \frac{1}{3})$, which guarantees him the same expected payoff of $\frac{2}{3}$.

Because this expected payoff is the value of the two (putative) zero-sum games, whose entries are the payoffs either to the row or to the column player, each player can prevent his opponent's getting more than this amount. Yet there is no compelling reason why a player should play *against* his opponent, rather than *for* himself, since the game is not zero-sum and a player does not benefit by putting an upper bound on his

---

32. For a proof, see T. Parthasarathy and T. E. S. Raghavan, *Some Topics in Two-Person Games* (New York: American Elsevier Publishing Co., 1971), Theorem 7.2.1, pp. 153–54.

opponent's payoff. Indeed, if the row player anticipates that the column player will choose his optimal mixed strategy $(\frac{2}{3}, \frac{1}{3})$, his best counter-strategy is to choose his pure strategy a_1. He thereby realizes an expected payoff of

$$2(1)(\tfrac{2}{3}) + 0(0)(\tfrac{1}{3}) = \tfrac{4}{3},$$

which is twice as great as the expected payoff he would obtain from playing his optimal mixed strategy $(\frac{1}{3}, \frac{2}{3})$. If the column player reaches the same conclusion that he could do better by deviating from his optimal mixed strategy and choosing his pure strategy b_2, the choice of the strategy pair (a_1, b_2) by both players would yield payoffs of 0 to each. Clearly, the optimal mixed strategies are not in equilibrium, leading to the kind of "he thinks that I think" regress discussed previously for zero-sum games.

3. *Advantages of irrevocable commitment.* In a two-person zero-sum game with a saddlepoint, a player neither hurts nor helps himself by revealing that he will choose his strategy associated with the saddlepoint; in a game without a saddlepoint, if a player discloses which pure strategy he will choose in a particular play of the game, his opponent can choose a pure strategy that may yield a higher payoff than does his optimal mixed strategy. Thus, there is never an advantage, and sometimes a penalty, in divulging one's strategy choice in a zero-sum game.

By contrast, in the nonzero-sum game in Figure 1.8, if the row player discloses that he will choose strategy a_1—and commits himself irrevocably to this choice whatever the column player does—the column player can do no better than choose strategy b_1, which yields the payoff (2, 1). (An analogous argument holds for the column player.) Of course, if the first player to make the "irrevocable commitment" were thought to be bluffing, his opponent could respond in kind, which would make both players worse off if both players remained unshakable in their commitments. Luce and Raiffa conclude from this situation:

> Thus, we see that it is advantageous in such a situation to disclose one's strategy first and to have a reputation for inflexibility. It is the familiar power strategy: "This is what I'm going to do; make up your mind and do what you want." If the second person acts in his own best interests, it works to the first person's advantage.[33]

4. *Advantages of cooperation.* As suggested by the third property, if the two players are to avoid the worst outcomes for both, certain kinds of "preplay" communication may be advantageous. Such communication, however, need not include only proclamations of defiance. For example, both players may, by communicating, agree to play strategy pairs (a_1, b_1) and (a_2, b_2) each one-half the time. By so agreeing, each would realize an expected payoff of $\frac{3}{2}$ over a series of plays. In zero-sum games, on the

---

33. Luce and Raiffa, *Games and Decisions,* p. 91.

FIGURE 1.9 BORDER DISPUTE

	Don't Attack	Attack
Attack	(2, 1)	(0, 0)
Don't Attack	(0, 0)	(1, 2)

other hand, cooperation is worthless since there is nothing to agree on that would redound to the benefit of both players.

In the game theory literature, the game we have just analyzed goes by the name Battle of the Sexes for reasons related to the defiance men and women show each other when they make irrevocable commitments. To illustrate this game's possible relevance to the study of international conflict—without trying to gauge its general applicability—consider a situation where two neighboring countries are involved in a border dispute (see Figure 1.9):

1. If one country attacks the other, the fighting will be brought to a quick cessation, perhaps with the aid of other parties, and the countries then will be able to agree on borders, favorable to the attacker, which brings about a permanent peace—(2, 1) and (1, 2) in the payoff matrix.

2. If both countries attack each other simultaneously, fighting will be prolonged and both countries will suffer—(0, 0) in the payoff matrix.

3. If neither country attacks the other, a simmering border dispute that does not erupt into war will continue, which is a prospect that both countries find no better than (2) above—(0, 0) in the payoff matrix.

Since both countries would prefer some agreement—even if precipitated by war and one-sided—to prolonged fighting or a continuing dispute, then it obviously pays for one country to seize the initiative and attack first, thereby obtaining more favorable treatment in the settlement. Such unstable situations are hardly unknown in international politics.

Having reviewed some of the complications that plague solution concepts that previously worked in the case of two-person zero-sum games, we now turn to the consideration of specific nonzero-sum games that seem to capture some important and enduring qualities of international confrontation and conflict. The two games we shall analyze will be viewed first as *noncooperative* games, in which preplay communication is not allowed and players cannot make binding and enforceable agreements; we shall then consider proposed "solutions" to them as *cooperative* games, wherein at least implicit communication occurs and the possibility of making binding and enforceable agreements exists.

1.8. PRISONER'S DILEMMA AND THE THEORY OF METAGAMES

There is a story, attributed to A. W. Tucker, that goes with the game we shall analyze in this section, whose payoff matrix is given in Figure 1.10. Two persons suspected of being partners in a crime are arrested and placed in separate cells so that they cannot communicate with each other. Without a confession from at least one suspect, the district attorney has insufficient evidence to convict them for the crime. To try to extract a confession, the district attorney tells each suspect the following consequences of his and his partner's actions:

1. If one suspect confesses and his partner does not, the one who confesses can go free (gets no sentence) for cooperation with the state but the other gets a stiff 10-year sentence—$(0, -10)$ and $(-10, 0)$ in the payoff matrix.

2. If both suspects confess, both get reduced sentences of 5 years—$(-5, -5)$ in the payoff matrix.

3. If both suspects remain silent, both go to prison for 1 year on a lesser charge of carrying a concealed weapon—$(-1, -1)$ in the payoff matrix.

What should each suspect do to save his own skin, assuming he has no compunction about squealing on his partner?[34] Observe first that, unlike the Battle of the Sexes, there is a single equilibrium pair of strategies that results in the outcome $(-5, -5)$: If either suspect confesses, it is rational for the other suspect to do likewise. The rub is that even though there is no conflict between different equilibrium pairs, as there was in the Battle of the Sexes, the idea of confessing and receiving a moderate sentence of 5 years is not at all appealing, even though neither suspect can assure himself of a better outcome.

A much more appealing strategy for both suspects would be not to confess, if one could be sure that one's partner would do the same. But without being able to communicate with him to coordinate a joint strategy, much less make a binding agreement, this could backfire and one could end up with 10 years in prison. The reason is that the better outcome than $(-5, -5)$ for *both* players, $(-1, -1)$, is unstable, for there is always the temptation for one player to doublecross his partner and turn state's evidence to achieve his very best outcome of being set free. Moreover, not only is there this temptation, but if one player thinks the other player might try to make a sucker out of him, he is left with no alternative but to confess. Thus, both suspects' strategies of confessing (strictly) dominate their strategies of not confessing, though the choice of the former strategies by both suspects results in a relatively undesirable 5 years in prison for each.

---

34. This is implicit in the utilities, which we assume incorporate moral and ethical considerations as well as more utilitarian factors.

FIGURE 1.10   PRISONER'S DILEMMA

		SUSPECT 2	
		Do not confess	*Confess*
SUSPECT 1	*Do not confess*	$(-1, -1)$	$(-10, 0)$
	Confess	$(0, -10)$	$(-5, -5)$

The dilemma lies in the fact that when both suspects play it safe by choosing their dominant strategies of confessing, they end up worse off— $(-5, -5)$—than had they trusted each other and both not confessed— $(-1, -1)$. In a sense, then, it may be rational to play this game irrationally—at least if one receives some cooperation from one's partner— by not confessing. But, as Rapoport points out, the choice between cooperation (C)—not confessing—and defection (D)—confessing—is not strictly a choice between an irrational altruism and a rational selfishness:

> The choice of C is not an act of self-sacrifice but rather an act of trust. But trust is not enough, because even convincing evidence that the other will choose C need not induce C. Not only must one be trusting; one must also be trustworthy; that is, one must resist the temptation to betray the other's trust.
>
> Is it "rational" to be trusting and trustworthy? Here the common usage of the term "rational" sometimes intrudes and beclouds the issue. The usual sense of this question is, "Is it safe to trust people?" But put in this way, the question is clearly an empirical one, to be answered by examining the behavior of a given sample of people in given circumstances.[35]

Before looking at empirical evidence on how this game is actually played, following Anatol Rapoport and Albert M. Chammah we symbolically represent the prospects of the two players in Figure 1.11, where R stands for "reward" (for cooperation), T for "temptation" (to doublecross the other player), S for "sucker's payoff" (to the player who is doublecrossed), and P for "punishment" (for playing it safe). A game is a Prisoner's Dilemma whenever $T > R > P > S$.[36]

In Figure 1.11 we also indicate the dynamics of this game, where, as in Figure 1.6, the vertical arrows signify the row player's preferences and

---

35. Anatol Rapoport, *Strategy and Conscience* (New York: Harper & Row, 1964), p. 50.

36. Anatol Rapoport and Albert M. Chammah, *Prisoner's Dilemma: A Study in Conflict and Cooperation* (Ann Arbor, Mich.: University of Michigan Press, 1965), pp. 33–34.

FIGURE 1.11   SYMBOLIC REPRESENTATION OF PRISONER'S DILEMMA

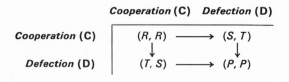

the horizontal arrows the column player's preferences. Unlike the Battle of Avranches, wherein there was no equilibrium pair of pure strategies, the arrows converge on the (P, P) outcome and show the strategies of defection (confessing) by both players to be in equilibrium.

One might think that if Prisoner's Dilemma were played repeatedly, perspicacious players could in effect communicate with each other by establishing a pattern of previous choices that would reward the choice of the cooperative strategy. But if the game ends after n plays, it clearly does not pay to play cooperatively on the final round since, with no plays to follow, the players are in effect in the same position that they would be if they played the game only once. If there is no point in trying to induce a cooperative response on the nth round, however, such behavior on the n-*1st* round would be to no avail since its effect could extend only to the nth round, where cooperative behavior has already been ruled out. Carrying this reasoning successively backward, it follows that one should not choose the cooperative strategy on *any* plays of the game.

Is there any escape from this dilemma? Many social-psychological experiments have been conducted to measure the effects of various parameters on the game's outcome for many different kinds of human subjects. We cannot review all the findings that relate to the conditions under which players cooperate, but some of the most significant include the absolute and relative sizes of payoffs, the number of trials the game is played, the personalities of the players, and the amount of communication between players that permits them to cooperate.[37]

---

37. The experimental literature on Prisoner's Dilemma is too extensive to cite here. Experimental research until 1965 is discussed in Rapoport and Chammah, *Prisoner's Dilemma*; for a more recent review, see Anatol Rapoport, "Prisoner's Dilemma—Reflections and Observations," in *Game Theory as a Theory of Conflict Resolution*, ed. Anatol Rapoport (Dordrecht-Holland: D. Reidel Publishing Co., 1974), pp. 17–34; and for a review of the experimental literature on two-person games generally, see Anatol Rapoport, "Experimental Games and their Uses in Psychology," 4044V00 (Morristown, N.J.: General Learning Corp., 1973).

One of the most interesting conclusions that emerges from the experimental research is that neither martyrs nor cynics do well in repeated trials of Prisoner's Dilemma.[38] That is, those willing to trust their opponents and always play their cooperative strategy, come what may, are mercilessly exploited; those unwilling to cooperate under any circumstances tend to provoke retaliation by the other player, which locks both players into the (P, P) outcome. On the other hand, many players who started out playing their noncooperative strategies were, after several trials, able to break out of the DD trap (both players defect) and achieve the rewards of CC (both players cooperate). In general, it seems, it takes some time for a mutual trust to develop between players that enables them to escape the compulsion to assume the worst and shift to the cooperative strategies that result in the mutually beneficial outcome, (R, R).

It is legitimate to ask at this point what relevance Prisoner's Dilemma has to international politics. If the seemingly paradoxical features of strategic choice for players in this game were limited only to hypothetical and artificial situations of the kind faced by the two criminal suspects in our example, we could easily pass this game off as an interesting curiosity but not of significance to real problems of social choice. In fact, however, instances of Prisoner's Dilemma have been identified in such diverse areas as agriculture, business, and law. Furthermore, it has been suggested that the problem it poses lies at the heart of a theory of the state; political philosophers at least since Hobbes have used the anarchy of a stateless society to justify the need for an enforceable social contract and the creation of government, by coercion if necessary.[39]

The standard example of Prisoner's Dilemma in international relations involves in an arms race two competing nations that may either continue the arms race or desist. If both nations desist (CC), they can devote the resources they would otherwise spend on armaments to socially useful projects, which presumably would make both better off and still preserve a balance of power between them.[40] If both arm (DD), both will be worse

---

38. The terminology *martyrs* and *cynics* is used in Deutsch, *Analysis of International Relations*, p. 122.

39. For different applications of this game, see Davis, *Game Theory*, pp. 93–103; Maki and Thompson, *Mathematical Models and Applications*, pp. 244–47. Prisoner's Dilemmas have even been uncovered in artistic works, as in the plot of Puccini's opera, *Tosca*. See Anatol Rapoport, "The Use and Misuse of Game Theory," *Scientific American*, Dec. 1962, pp. 108–18.

40. One exception to this statement comes to mind for a particular kind of arms expenditure. As argued by Oskar Morgenstern in 1959, once the United States had a strong nuclear *retaliatory* force, it was in her interest for the Soviet Union also to possess one. Otherwise, the Soviet Union would be tempted to initiate a preemptive first strike if she had the slightest fear that the United States might launch an offensive strike that would wipe out her capability to respond. Thus, it was in the interests of both sides to possess a second-strike retaliatory capability. In game-theoretic terms, this would lead to a stable equilibrium, whereas if only one side had the capability to strike back after a first strike (or worse, if neither side did), the situation would be obviously unstable. See Oskar Morgenstern, *The Question of National Defense* (New York: Random House, 1959).

off for their outlays on socially useless weapons systems and comparatively no stronger militarily. If one nation arms and the other does not (*CD* and *DC*), the nation that arms will develop the military superiority to defeat its adversary and thereby realize its best outcome; the nation that disarms will relinquish control over its fate and thereby suffer its worst outcome. Variants of this example relate to whether or not to adhere to the conditions of a treaty, cheat on a weapons inspection agreement, and so on, where it is understood that treaties or agreements are not completely enforceable. Central to all these examples is the question of trust: Under what conditions will the players be likely to trust each other sufficiently to risk adopting their cooperative, but unstable, strategies?

Although the experimental literature furnishes some clues about the kinds of conditions that might generate trust between countries, they are not always directly applicable to the international environment.[41] Take the condition of time, for example. Since the start of the Cold War, the United States and the Soviet Union have generally become more conciliatory toward each other and have even reached some agreements on limiting the testing and spread of nuclear weapons [e.g., the Treaty Banning Nuclear Weapon Tests in the Atmosphere, in Outer Space, and under Water (1963), and the Treaty on the Non-Proliferation of Nuclear Weapons (1970)]. At least in the recent period of détente, this would seem to indicate that time—as well as more accurate methods for detecting violations of these agreements—has tempered anxieties on both sides about the risks of abrogating agreements. By contrast, after four wars, time has, if anything, embittered relations between Israel and her Arab neighbors over roughly the same period. It would thus appear that if a recognition of the benefits of mutual cooperation depends on recurring experiences in Prisoner's Dilemma situations, it is also affected by other factors that are not well understood.

We can gain further insight into the conditions under which the cooperative outcome (*R, R*) is chosen by players in Prisoner's Dilemma from the *theory of metagames*, which has recently been developed by Nigel Howard.[42] This theory extends the concept of strategy to include one player's responses to possible strategy choices of his opponent, the opponent's responses in turn to the first player's conditional choices, and so forth. Because this concept involves choosing a rule to select a strategy conditional upon the strategy choice of one's opponent, it is called a *metastrategy*, which may be thought of as a strategy for selecting a strategy. For example, in Prisoner's Dilemma, a player has two strategy

41. Their inapplicability to one conflict is argued in Malvern Lumsden, "The Cyprus Conflict as a Prisoner's Dilemma Game," *Journal of Conflict Resolution*, 17 (March 1973), pp. 7–32.

42. Nigel Howard, *Paradoxes of Rationality: Theory of Metagames and Political Behavior* (Cambridge, Mass.: MIT Press, 1971). For an extension of the concept of a "metagame" to cover kinds of mutual prediction other than that indicated below, see Nigel Howard, "'General' Metagames: An Extension of the Metagame Concept," in *Game Theory as a Theory of Conflict Resolution*, ed. Anatol Rapoport, pp. 261–83.

FIGURE 1.12   SUSPECT 2'S METASTRATEGIES FOR STRATEGIES OF
SUSPECT 1

METASTRATEGIES OF SUSPECT 2

		C Regardless	D Regardless	Tit-for-tat	Tat-for-tit
STRATEGIES OF SUSPECT 1	C	(−1, −1)	(−10, 0)	(−1, −1)	(−10, 0)
	D	(0, −10)	(−5, −5)	(−5, −5)	(0, −10)

choices: Choose C or choose D. These two choices give rise to four meta-strategies:

1. Choose C, regardless of what the other player chooses;

2. Choose D, regardless of what the other player chooses;

3. Choose C if the other player chooses C, choose D if he chooses D (tit-for-tat);

4. Choose C if the other player chooses D, choose D if he chooses C (tat-for-tit).

Given that suspect 2 knew, or believed he could predict, suspect 1's strategy choices in Figure 1.10, we show in Figure 1.12 the outcomes associated with suspect 1's two strategies and suspect 2's four meta-strategies. For example, if suspect 1 chooses strategy C, and suspect 2 always chooses strategy C (i.e., regardless of the choice of suspect 1), the outcome will be (−1, −1) in this expanded game. Note that there is still only one equilibrium point, (−5, −5), from which it is not to the advantage of either player to shift his choice if the other player sticks to his. We have circled this point in Figure 1.12 and note that it is the same equilibrium point as we found in the original game.

Now suppose that suspect 1 has knowledge of, or can predict, the metastrategy choice of suspect 2. If he makes his strategy choice conditional on suspect 2's metastrategies, he can formulate his own metastrategies at a still higher level of analysis. This generates sixteen possible choices, depending on whether suspect 1 chooses C or D for each of suspect 2's four metastrategies. If, for example, suspect 1 chooses

(1) D whenever suspect 2 chooses C regardless,

(2) D whenever suspect 2 chooses D regardless,

(3) C whenever suspect 2 chooses tit-for-tat,

(4) D whenever suspect 2 chooses tat-for-tit,

FIGURE 1.13 METAGAME PAYOFF MATRIX FOR PRISONER'S DILEMMA

	SUSPECT 2			
	C *Regardless*	**D** *Regardless*	*Tit-for-tat*	*Tat-for-tit*
1. **C/C/C/C**	(−1, −1)	(−10, 0)	(−1, −1)	(−10, 0)
2. **C/C/C/D**	(−1, −1)	(−10, 0)	(−1, −1)	(0, −10)
3. **C/C/D/C**	(−1, −1)	(−10, 0)	(−5, −5)	(−10, 0)
4. **C/D/C/C**	(−1, −1)	(−5, −5)	(−1, −1)	(−10, 0)
5. **D/C/C/C**	(0, −10)	(−10, 0)	(−1, −1)	(−10, 0)
6. **C/C/D/D**	(−1, −1)	(−10, 0)	(−5, −5)	(0, −10)
7. **C/D/C/D**	(−1, −1)	(−5, −5)	⟨(−1, −1)⟩	(0, −10)
8. **D/C/C/D**	(0, −10)	(−10, 0)	(−1, −1)	(0, −10)
9. **C/D/D/C**	(−1, −1)	(−5, −5)	(−5, −5)	(−10, 0)
10. **D/C/D/C**	(0, −10)	(−10, 0)	(−5, −5)	(−10, 0)
11. **D/D/C/C**	(0, −10)	(−5, −5)	(−1, −1)	(−10, 0)
12. **C/D/D/D**	(−1, −1)	(−5, −5)	(−5, −5)	(0, −10)
13. **D/C/D/D**	(0, −10)	(−10, 0)	(−5, −5)	(0, −10)
14. **D/D/C/D**	(0, −10)	(−5, −5)	⟨(−1, −1)⟩	(0, −10)
15. **D/D/D/C**	(0, −10)	(−5, −5)	(−5, −5)	(−10, 0)
16. **D/D/D/D**	(0, −10)	⟨(−5, −5)⟩	(−5, −5)	(0, −10)

SUSPECT 1

we designate his higher-level metastrategy by $D/D/C/D$ in Figure 1.13.

For each of suspect 1's sixteen metastrategies, suspect 2 has four metastrategies, so there are sixty-four possible outcomes in the *metagame*, or the game derived from the original game that is defined by the payoff matrix in Figure 1.13. Each of these outcomes corresponds to one of the four outcomes in the original game.

To illustrate how these outcomes are found, consider the outcome at the intersection of suspect 1's metastrategy $D/D/C/D$ (no. 14 in Figure 1.13) and suspect 2's metastrategy C regardless. Because these metastrategies correspond to the choice of D by suspect 1 and C by suspect 2, which gives outcome (0, −10) in the payoff matrix of the original game in Figure 1.10, the outcome associated with these metastrategies in the payoff matrix of the metagame in Figure 1.13 is also (0, −10).

Now comes the surprise! As indicated in Figure 1.13 by the circled outcomes, there are three equilibria in the metagame, and two are the outcomes $(-1, -1)$ associated with the strategy choices of C by both suspects in the original Prisoner's Dilemma game. In other words, the cooperative outcome $(R, R) = (-1, -1)$ in the original game, which was not in equilibrium, emerges as an equilibrium point in the metagame expansion. The noncooperative outcome $(P, P) = (-5, -5)$ is also an equilibrium point, as it was in the expansion of only suspect 2's strategies in Figure 1.12. All of these equilibria are stable in the sense that if both suspects choose metastrategies associated with one of them, neither suspect can do better, and may do worse, by unilaterally switching to a different metastrategy.

What do these new equilibria signify? For suspect 2, the metastrategy tit-for-tat is associated with both of the cooperative equilibria, and it would seem to imply: "I'll cooperate only if you will." For suspect 1, whose reply is conditional on suspect 2's choice, his metastrategy associated with one of the cooperative equilibria $(D/D/C/D)$ was previously discussed; it would seem to imply: "In that case [i.e., if you choose tit-for-tat], I'll cooperate, too." With each suspect making his cooperation conditional on the cooperation of the other suspect, together they are able to lock into the cooperative outcome.

One might wonder what would the consequence of further expanding the payoff matrix to include suspect 1's knowledge of suspect 2's sixteen metastrategies. Would there be still new equilibria? Fortunately, we need not continue the expansion process any longer, for Howard has proved that in an n-person game, an expansion beyond n levels will not reveal any new equilibria, nor will any be lost. Furthermore, it does not matter which player we start the expansion with; if we gave suspect 1 the four conditional strategies at the first level, and suspect 2 the sixteen conditional responses, their roles would be reversed but the same equilibria would be obtained.

We have observed that there are two cooperative equilibria, but we gave an interpretation only to the outcome at the intersection of suspect 2's tit-for-tat metastrategy and suspect 1's $D/D/C/D$ metastrategy. For the same tit-for-tat choice of suspect 2, suspect 1's $C/D/C/D$ metastrategy (no. 7 in Figure 1.13) also results in the cooperative equilibrium outcome $(-1, -1)$, but it is not as compelling as the former. The reason is that $D/D/C/D$ dominates $C/D/C/D$, that is, for one metastrategy of suspect 2 (C regardless), suspect 1 fares worse with metastrategy $C/D/C/D$ than with metastrategy $D/D/C/D$: He obtains a payoff of -1 rather than 0. Since $D/D/C/D$ is at least as good and in one contingency better than $C/D/C/D$, it is the dominant equilibrium metastrategy for suspect 1.[43]

From the vantage point of suspect 2, he maximizes his security level by choosing either the metastrategy D regardless, or the metastrategy

---

43. Metastrategy $D/D/C/D$ also dominates $D/D/D/D$, suspect 1's metastrategy containing his noncooperative equilibrium outcome, but this is not true in all games. In fact, in section 1.9 we shall give an example of a game wherein the metastrategy containing a noncooperative equilibrium outcome dominates the one containing the cooperative equilibrium outcome.

tit-for-tat; whatever the choice of suspect 1, his payoff is at least -5. Given that he would prefer an equilibrium with a larger payoff for his opponent, as well as himself (on the assumption that his opponent would choose his metastrategy containing it), he should choose his metastrategy containing the cooperative equilibria $(-1, -1)$, rather than the non-cooperative equilibrium $(-5, -5)$, which dictates the choice of tit-for-tat. Thus, the outcome $(-1, -1)$ at the intersection of metastrategies $D/D/C/D$ and tit-for-tat is (1) in equilibrium; (2) preferred by both players to the noncooperative equilibrium point $(-5, -5)$; and (3) the product of undominated metastrategies for both players.

This is a rather complicated route to take to get to the cooperative solution to Prisoner's Dilemma. Is the reasoning on which it is based justified? Insofar as players can accurately predict each other's strategies and formulate metastrategy choices, the reasoning is logically impeccable. Because such predictions are usually based on some form of communication between the players, however, it seems also fair to say that the dilemma is in essence resolved by permitting preplay communication in a cooperative game. For once the players are free to communicate, they can agree to play their cooperative strategies and—provided there is some means for enforcing such an agreement—the dilemma disappears.

In place of insisting that agreements be enforceable, the metagame approach requires foreknowledge of the metastrategy choices of one's opponent. Obviously, if one's forecasting ability or clairvoyance is infallible, one does not need a guarantee that an agreement will not be broken, because one knows this a priori. Hence, given that there is no risk associated with one's choice, the theory of metagames resolves Prisoner's Dilemma by substituting knowledge—or at least expectations (i.e., imperfect knowledge)—of the future for an enforceable agreement needed in its absence. It is reasonable to suppose that some form of communication or bargaining occurs between the players to obtain either knowledge or an agreement.

Although the theory of metagames would seem to provide less of an "escape" from a paradox[44] than a recasting of it in different terms, this is not a mean achievement.[45] We shall try to demonstrate this point in section 1.9 with a particular example. At this juncture, however, we note

---

44. Anatol Rapoport, "Escape from Paradox," *Scientific American*, July 1967, pp. 50–56. Two other "solutions" to Prisoner's Dilemma are offered by Martin Shubik, but neither seems as relevant to the study of international relations as does the theory of metagames. See Martin Shubik, "Game Theory, Behavior, and the Paradox of the Prisoner's Dilemma: Three Solutions," *Journal of Conflict Resolution*, 13 (June 1970), pp. 181–93. For a recent and related paradox for which no generally accepted solution has been found, see Martin Gardner, "Mathematical Games," *Scientific American*, July 1973, pp. 104–108; and Martin Gardner, "Mathematical Games," *Scientific American*, March 1974, pp. 102–108.

45. Neither is it a mean achievement for game theory to have drawn attention to the paradox in the first place, which is less a creation of game theory—it was surely anticipated before its creation—than a limitation of certain situations that the theory helps to clarify.

in passing that probably most actors in international politics do *not* operate from a metagame perspective. Whether they lack information about an opponent, the ability to forecast his behavior, or simply mistrust him, their focus is not on metastrategies but on strategies, which is after all what they choose in the end.[46] From a game-theoretic perspective, there is hardly anything perverse or irrational in choosing to play it safe, or, if you will, to select the noncooperative strategy of defection in Prisoner's Dilemma. Indeed, the players in *Realpolitik* games who fail to grasp this fact, and the nature of the equilibrium solution, are often the victims of preemptive attacks or losers in arms races; it is precisely these players who are irrational for trusting an adversary without good reasons. Although, perhaps, "there should be a law against such games," as Luce and Raiffa wistfully put it,[47] since there is none, it behooves us to try to understand under what circumstances, if any, the basic minimax logic can be transcended. Clearly, the power of a theory lies in its ability not only to point out paradoxes but to identify solutions as well.

1.9. CHICKEN AND THE CUBAN MISSILE CRISIS

Next to Prisoner's Dilemma, the game of Chicken probably better captures the difficulties that two actors in international politics face in reaching agreements than any other mixed-motive game. This game takes its name from the rather gruesome sport that apparently originated among California teenagers in the 1950s. As two teenage drivers approach each other at high speed on a narrow road, each has the choice of either swerving and avoiding a head-on collision, or continuing on the collision course. As shown in the payoff matrix of Figure 1.14, there are four possible outcomes:

1. The player who does not swerve when the other does gets the highest payoff of 4 for his courage (or recklessness).

2. The player who "chickens out" by swerving is disgraced and receives a payoff of 2.

---

46. This is the point made in Harris' critique of Rapoport's claim of an "escape from paradox." See Richard J. Harris, "Note on Howard's Theory of Meta-Games," *Psychological Reports*, 24 (1968), pp. 849–50; in rebuttal, see Anatol Rapoport, "Comments on Dr. Harris' 'Note on Howard's Theory of Meta-Games,'" *Psychological Reports*, 25 (1969), pp. 765–66; in counter-rebuttal, Richard J. Harris, "Comments on Dr. Rapoport's Comments," *Psychological Reports*, 25 (1969), p. 825. At this point Howard joined the debate. See Nigel Howard, "Comments on Harris' 'Comments on Rapoport's Comments,'" *Psychological Reports*, 25 (1969), p. 826; A. Rapoport, "Reply to Dr. Harris' Comments on My Comments," *Psychological Reports*, 25 (1969), pp. 857–58; Richard J. Harris, "Paradox Regained," *Psychological Reports*, 26 (1970), pp. 264–66; Anatol Rapoport, "Comments on 'Paradox Regained,'" *Psychological Reports*, 26 (1970), p. 272; and finally, Nigel Howard, "Note on the Harris–Rapoport Controversy," *Psychological Reports*, 26 (1970), p. 316.

47. Luce and Raiffa, *Games and Decisions*, p. 97.

FIGURE 1.14   CHICKEN

PLAYER B

Swerve (C)   *Not swerve* (D)

		Swerve (C)	*Not swerve* (D)
PLAYER A	*Swerve* (C)	(3, 3) ⟶ (2, 4)	
		Compromise B's victory	
	Not swerve (D)	(4, 2) (1, 1)	
		A's victory ⟶ Collision	

3. If both players lack the will to continue on the collision course to the bitter end, they both suffer some loss of prestige, obtaining payoffs of 3, but not as much as if only one player had played it safe by swerving.

4. If both players refuse to compromise, then they hurtle to their mutual destruction, which may be fine for martyrs but not for the players in this game, who (posthumously) receive the lowest payoffs of 1 each.

As with Prisoner's Dilemma, Chicken is defined by the ranking of outcomes, not the actual numerical payoffs that we have used for convenience in Figure 1.14.[48]

This game bears some resemblance to Prisoner's Dilemma, except that the worst outcome for both players in Chicken occurs when both players "defect" from cooperating. This was the next-to-worst outcome for both players in Prisoner's Dilemma, the worst going to the player who defected (*D*) when his opponent cooperated (*C*). Unlike the situation in Prisoner's Dilemma, two outcomes are in equilibrium, *CD* and *DC*. Like Prisoner's Dilemma, both players, by choosing *C*, can do better than *DD*, but the *CC* outcome is unstable since, given the choice of the *C* strategy by one player, the other player has an incentive to choose his *D* strategy, as indicated by the arrows in Figure 1.14.[49]

Chicken serves as a good analogue for some situations in international politics where threats to use military force figure prominently in bargaining strategies among nations.[50] Perhaps the most potentially devastating confrontation between major powers ever to occur was that between United States and the Soviet Union in October 1962. This confrontation, in what has come to be known as the Cuban missile crisis, was precipitated by Soviet installation in Cuba of medium-range and intermediate-range

---

48. The ranking of outcomes in Chicken simply reverses the payoffs we have used: "4" is the best outcome, "3" next, and so on.

49. Deutsch labels *CC* the "rational solution," but it is certainly not in a game-theoretic sense because of its instability. See Deutsch, *Analysis of International Relations*, p. 120.

50. Many examples of the use of military power as a bargaining tool can be found in Thomas C. Schelling, *Arms and Influence* (New Haven, Conn.: Yale University Press, 1966).

FIGURE 1.15   PAYOFF MATRIX OF CUBAN MISSILE CRISIS

		SOVIET UNION	
		Withdrawal **(W)**	*Maintenance* **(M)**
Blockade **(B)**		(3, 3) Compromise	(2, 4) Soviet victory
UNITED STATES			
Air strike **(A)**		(4, 2) U.S. victory	(1, 1) Nuclear war

nuclear-armed ballistic missiles that were capable of hitting a large portion of the United States.

The goal of the United States was to obtain immediate removal of the Russian missiles, and United States policy makers seriously considered two alternative courses of action to achieve this end:

1. A naval blockade (B), or "quarantine" as it was euphemistically called, to prevent shipment of further missiles, followed by possibly stronger action to induce the Soviet Union to withdraw those already installed;

2. A "surgical" air strike (A) to wipe out the missiles already installed, insofar as possible, followed possibly by an invasion of the island.

The less provocative blockade option was eventually chosen which left open to Soviet leaders essentially two alternatives:

1. Withdrawal (W) of their missiles;

2. Maintenance (M) of their missiles.

These alternative courses of action for both players are shown in Figure 1.15; the probable outcomes for each pair of strategy choices by both players are also indicated.

Needless to say, the strategy choices and probable outcomes as presented in Figure 1.15 provide only a skeletal picture of the crisis as it developed over a period of thirteen days. Both sides considered more than the two alternatives we have listed, and several variations on each. The Russians, for example, demanded withdrawal of American missiles from Turkey as a *quid pro quo* for withdrawal of their missiles from Cuba, a demand ignored by the United States. Furthermore, there is no way to verify that the outcomes given in Figure 1.15 were "probable," or valued in a manner consistent with the game of Chicken. For example, if the Soviet Union had viewed an air strike on their missiles as jeopardizing their vital national interests, the AW outcome may very well have ended in nuclear war between the two sides, giving it the same value as AM. Still another simplification relates to the assumption that the players chose their actions simultaneously, when in fact a continuous exchange in both words and deeds occurred over those fateful days in October.

Nevertheless, the basic conception most observers have of this crisis is that the two superpowers were on a "collision course," which is actually the title of one book recounting this nuclear confrontation.[51] Most observers also agree that neither side was eager to take any irreversible steps, such as the teenage driver in a game of Chicken might do by defiantly ripping off his steering wheel in full view of his adversary, thus foreclosing his alternative of swerving. Although in one sense the United States "won" by getting the Russians to withdraw their missiles, Khrushchev at the same time extracted from Kennedy a promise not to invade Cuba, which seems to indicate that the eventual outcome was a compromise solution of sorts. Moreover, even though the Russians responded to the blockade and did not make their choice of a strategy independently of the American strategy choice, the fact that the United States held out the possibility of escalating the conflict to at least an air strike would seem to indicate that the initial blockade decision was not to be considered final—that is, the United States considered its strategy choices still open after imposing the blockade.

What can the theory of metagames tell us about how a cooperative solution was achieved in this crisis, despite the instability of this outcome in the game of Chicken? One's first response might be that there is nothing paradoxical about the compromise reached because this crisis really was not a noncooperative game. There was extensive formal and informal communication between the two sides during the crisis, it might be argued, which explains how the two adversaries were able to arrive at the cooperative outcome even though a binding and enforceable agreement was never concluded.

The assertion that open lines of communication, and particularly the "hot line" between Washington, D.C., and Moscow, moderated the crisis seems certainly true, but this is not a very complete or satisfying explanation. To be so, such an explanation should indicate at least what the general nature of the communication must have been to assure each side that it could trust the other. Otherwise, simply "talking," letting the conflict cool down, would always put an end to such crises, which is patently not the case.

If we expand the game of Chicken in the same manner as we did Prisoner's Dilemma, we can derive a 16 × 4 payoff matrix that gives the metagame representation for Chicken. We have not illustrated in Figure

51. Henry M. Pachter, *Collision Course: The Cuban Missile Crisis and Coexistence* (New York: Frederick A. Praeger, 1963). Other books on this crisis include Elie Abel, *The Missile Crisis* (Philadelphia: J. B. Lippincott Co., 1966); Graham T. Allison, *The Essence of Decision: Explaining the Cuban Missile Crisis* (Boston: Little, Brown and Co., 1971); and Robert F. Kennedy, *Thirteen Days: A Memoir of the Cuban Missile Crisis* (New York: W. W. Norton & Co., 1969).

FIGURE 1.16   PARTIAL METAGAME PAYOFF MATRIX OF CUBAN
MISSILE CRISIS

		SOVIET UNION			
		W *Regardless*	M *Regardless*	*Tit-for-tat*	*Tat-for-tit*
UNITED STATES	B/B/B/B	(3, 3)	((2, 4))	(3, 3)	(2, 4)
	A/B/B/A	(4, 2)	((2, 4))	(3, 3)	(4, 2)
	A/A/B/A	(4, 2)	(1, 1)	((3, 3))	(4, 2)
	A/A/A/A	((4, 2))	(1, 1)	(1, 1)	((4, 2))

1.16 all sixteen metastrategies of the United States but only those four that contain equilibria, which would presumably be the basis for any solution viewed as stable by both sides. It will be observed that four of the five equilibria, which are circled, end in victories for the United States or the Soviet Union; only the outcome at the intersection of United States metastrategy $A/A/B/A$ and the Russian metastrategy tit-for-tat results in the compromise outcome, (3, 3).

In analyzing the Cuban missile crisis, Howard, whose representation of the players' strategies and metastrategies differs somewhat from ours, draws three conclusions:[52]

1. *For the compromise outcome* (3, 3) *to be stable, both sides must be willing to risk nuclear war.* This is surely the meaning of the Russian metastrategy of tit-for-tat. Since the only American response to this choice that results in a cooperative equilibrium outcome is $A/A/B/A$, which stipulates American cooperation (blockade) only if the Russians choose tit-for-tat, the American metastrategy $A/A/B/A$ can also be viewed as a tit-for-tat policy, though one step removed (i.e., conditional on the Russian selection of tit-for-tat).

2. *If one side but not the other is willing to risk nuclear war, that side wins.* For example, if the Russians choose metastrategy *M* regardless, the only American metastrategies that result in equilibrium outcomes, $B/B/B/B$ and $A/B/B/A$, are those that involve the American response of *B* that results in a Soviet victory.

3. *If neither side is willing to risk nuclear war, no stable outcome is possible.* For example, if the Russians choose metastrategy *W* regardless, and the American response is $B/B/B/B$, the outcome is compromise, (3, 3), but it is not in equilibrium: the Russians could do better by unilaterally switching to *M* regardless, the Americans by unilaterally switching to

---

52. Howard, *Paradoxes of Rationality*, p. 184.

$A/A/A/A$. Thus, it would appear that a policy of deterrence, by which each side promises retaliation for any untoward acts by the other, is not only desirable from the viewpoint of both players, but stable as well. This analysis suggests that if we live in an age of "balance of terror," the only way to ensure the maintenance of this balance is if both sides pursue retaliatory policies.

Superficially, this conclusion would seem to contradict the fact that the compromise outcome in the original game occurred when neither side was willing to risk nuclear war. It will be recalled, however, that this compromise outcome was unstable, as are all but one of these outcomes in the metagame. The only stable compromise outcome in the metagame occurs when each side refuses to give in unless the other side also does— it is better to incur nuclear war than be blackmailed. On the other hand, if one player thinks his opponent is really not bluffing but means business, then the analysis recommends that he respond by giving in—approach the brink but not overstep it, as implied by the strategy of "brinkmanship"— rather than provoking his opponent into a situation that terminates in a collective suicide pact.[53]

In sum, what the theory of metagames has revealed is the effect on the outcome of players' expectations—and sometimes knowledge, as sequential moves in the game occur—about an opponent's strategy choices. This effect may be considerable and salutory, changing a desirable but unstable outcome into a stable one from which the players will not be motivated to depart. Typically, these expectations are inferred from, and perceived through, a variety of communication channels, which means that the road to the cooperative solution in Prisoner's Dilemma and Chicken must be paved with more than the good intentions of one side. If anything, good intentions alone will probably ruin any possibility of stable peace.

Metagame theory specifies rather precisely, if indirectly, the *content* of the communications and the *nature* of the bargaining necessary to reach compromise.[54] At least in the two games we have analyzed, the players must convey tough (but not totally inflexible) images that indicate their

---

53. If a player is truly in the dark about his opponent's intentions—which was not the case in the Cuban missile crisis—then what Howard calls the "sure-thing" metastrategy is probably best, which for the United States is metastrategy $A/B/B/A$. This choice maximizes a player's security level, at least among his equilibrium metastrategies (i.e., those containing an equilibrium outcome); furthermore, it dominates the one other equilibrium metastrategy ($B/B/B/B$) that yields the same security level. If the Russians anticipate this choice, however, it has the unfortunate effect (for the United States) of inducing them (the Russians) to choose M regardless and thereby consummate a Soviet victory—i.e., outcome (2, 4). This kind of effect did not occur in Prisoner's Dilemma, where the dominant metastrategy of the row player containing the cooperative equilibrium outcomes (i.e., $D/D/C/D$) also dominated the metastrategy containing the noncooperative equilibrium outcome (i.e., $D/D/D/D$).

54. In this regard it offers a more "dynamic" picture of the *processes* that lead to various outcomes than does classical game theory, which offers no way for unifying cooperative and noncooperative solution concepts within a single framework. For other attempts to transcend the limitations of the original theory to include a theory of bargaining (and arbitration), see Luce and Raiffa, *Games and Decisions*, pp. 119–52; and Robert L. Bishop, "Game-Theoretic Analysis of Bargaining," *Quarterly Journal of Economics*, 77 (Nov. 1963), pp. 559–602.

readiness to cooperate but, if necessary, also to apply sanctions. A content analysis of the messages exchanged between the United States and the Soviet Union tends to support the proposition that both sides held similar perceptions of the threatening and conciliatory actions of each other during the crisis,[55] which is perhaps at least a partial explanation of why we are still around to tell about this momentous confrontation.

To conclude our analysis of the Cuban missile crisis, we describe below one calculation, suggested by Glenn Snyder, which in principle might be made by policy makers trying to decide what course of action to take in the game of Chicken (as well as other games).[56] We assume that they are not only capable of determining the utility of each outcome for themselves but also can estimate the probability that their adversary will choose to retreat or stand firm. Although such a probabilistic estimate of an adversary's resolve may sound absurd, in the Cuban missile crisis President Kennedy reportedly estimated that he thought that there was a one-third to one-half chance that the Russians would not withdraw their missiles from Cuba, the conflict would escalate, and the world would be plunged into a nuclear war.[57]

(If Kennedy had thought the Russians would make their choice of a strategy strictly according to some random device that would give this probability of escalation, obviously no amount of bargaining on his part would have been justified: The Russians would select a course of action independently of the course of action he chose. The fact that both sides went to great lengths to clarify their positions of what would be acceptable to resolve the crisis indicates that the communication of intentions—relating to both concessions and threats—played a prominent role in the final settlement. These exchanges enabled each side to estimate more accurately the probable reactions of the other side to the range of alternative actions they considered taking, which underscores the relevance of the previous metagame analysis elucidating the conditional choices through which the expectations of players are tied to outcomes. As Theodore Sorenson described American deliberations, "We discussed what the Soviet reaction would be to any possible move by the United States, what our reaction with them would have to be to that Soviet reaction, and so on, trying to follow each of those roads to their ultimate conclusion."[58])

---

55. Ole R. Holsti, Richard A. Brody, and Robert C. North, "Measuring Affect and Action in International Reaction Models: Empirical Materials from the 1962 Cuban Crisis," *Journal of Peace Research*, 1 (1964), pp. 170–89.

56. Glenn H. Snyder, "'Prisoner's Dilemma' and 'Chicken' Models in International Politics," *International Studies Quarterly*, 15 (March 1971), pp. 87–89; for a similar calculation, see Robert Jervis, "Bargaining and Bargaining Tactics," in *Coercion: Nomos XIV*, ed. J. Roland Pennock and John W. Chapman (Chicago: Aldine-Atherton, 1972), pp. 277–78.

57. Allison, *Essence of Decision*, p. 2.

58. Quoted in Holsti, Brody, and North, "Measuring Affect and Action in International Reaction Models," p. 188.

Using the payoffs to the United States given in Figure 1.14, and assuming that the probability is $\frac{2}{3}$ that the Russians will withdraw their missiles and $\frac{1}{3}$ that they will retain them (which was Kennedy's most optimistic estimate), we find the expected utility of blockade (B) for the United States to be

$$E(B) = \tfrac{2}{3}(3) + \tfrac{1}{3}(2) = \tfrac{8}{3},$$

and the expected utility of an air strike (A) to be

$$E(A) = \tfrac{2}{3}(4) + \tfrac{1}{3}(1) = \tfrac{9}{3}.$$

Since $E(A) > E(B)$, by this calculation an air strike (which, incidentally, was almost carried out) has a greater expected utility and is therefore to be preferred. (For Kennedy's pessimistic estimate of a fifty–fifty chance of war, the choice between the two options is a toss-up since $E(B) = E(A) = \frac{1}{2}$.) We hasten to add that the payoffs we have used were not the estimates of any policy makers, most of whom probably would have assigned a comparatively much lower value to the outcome of nuclear war than we did. This would have decreased the value of $E(A)$ relative to $E(B)$, which evidently was the case when the final choice of the blockade option was made.

We conclude this section by mentioning, but not developing, other approaches to the study of international conflict that smack of game-theoretic analysis but do not rigorously employ its panoply of concepts.[59] For example, problems of military deterrence, surprise attack, limited war, disarmament, and tacit coordination have been provocatively analyzed by Thomas C. Schelling,[60] but much of his analysis is based on a rather loose formulation of rational behavior, which incorporates cultural, psychological, and other factors extraneous to the formal structure of game theory; this leads to difficulties in analyzing the bargaining aspects of games, as John C. Harsanyi has shown.[61] Strategic analysis, perhaps best exemplified by the work of Herman Kahn,[62] is primarily devoted to the design of cost-effective weapons systems and the formulation of

---

59. We have ignored in this chapter games involving more than two players, which are treated in subsequent chapters, particularly Chapter 6.

60. Thomas C. Schelling, *The Strategy of Conflict* (Cambridge, Mass.: Harvard University Press, 1960).

61. John C. Harsanyi, "On the Rationality Postulates Underlying the Theory of Cooperative Games," *Journal of Conflict Resolution*, 5 (June 1961), pp. 179–96. For another critique, see Anatol Rapoport, *Strategy and Conscience*, pp. 112–24.

62. Herman Kahn, *On Thermonuclear War* (Princeton, N.J.: Princeton University Press, 1960). For a critique of this and other works of nuclear strategists, see Philip Green, *Deadly Logic: The Theory of Nuclear Deterrence* (Columbus, Ohio: Ohio State University Press, 1966); for a particularly biting critique of the ignorance shown by some critics of the nuclear strategists, see A. Wohlstetter, "Sin and Games in America," in *Game Theory and Related Approaches to Social Behavior*, ed. Martin Shubik (New York: John Wiley & Sons, 1964), pp. 209–25. More recently, the debate between strategists and their critics has been carried on in *Strategic Thinking and Its Moral Implications*, ed. Morton A. Kaplan (Chicago: University of Chicago Press, 1973).

strategies that optimize quantifiable ends. It employs a variety of techniques, including those that go under the rubric of "systems analysis," but most of these make no serious use of game theory. Whereas some of these techniques are based on methods of economic analysis,[63] others use simulation and gaming as means for discovering the consequences of strategic choices in situations more complex than the simple games discussed here.[64] In these more complex situations, rigor and generality are usually sacrificed for realism and specificity in order to analyze operational problems in concrete settings.

1.10. SUMMARY AND CONCLUSION

Every approach to the study of human conflict carries the baggage of certain built-in a priori assumptions, and it is useful to try to ferret these out. In international relations, which is a field bereft of any generally accepted conceptual focus, this task seems especially important in order to avoid throwing its study into a greater muddle than it is in already. The approach to the study of international conflict advanced here might be best appreciated by juxtaposing the views of two leading exponents of rigorous studies of international conflict. On the one hand, Lewis F. Richardson believed that his theory would describe

> what people would do if they did not stop to think. Why are so many nations reluctantly but steadily increasing their armaments as if they were mechanically compelled to do so? Because ... they follow their traditions ... and instincts ... and because they have not yet made a sufficiently strenuous intellectual and moral effort to control the situation.[65]

Labeling this an "oversimplified and dangerous view," John C. Harsanyi argued that

> it leads to serious underestimation of the political forces actually generating and maintaining international conflicts; and in particular it precludes understanding of the fact that ... non-cooperative behavior may often be the only rational and realistic response to a given international situation.[66]

---

63. See, for example, Charles J. Hitch and Roland N. McKean, *The Economics of Defense in the Nuclear Age* (Cambridge, Mass.: Harvard University Press, 1960).

64. See, for example, *Simulation in International Relations: Developments for Research and Teaching*, ed. Harold Guetzkow (Englewood Cliffs, N.J.: Prentice-Hall, 1963).

65. Lewis F. Richardson, *Arms and Insecurity: A Mathematical Study of the Causes and Origin of War* (Pittsburgh: Boxwood Press, 1960), p. 12.

66. Harsanyi, "Game Theory and the Analysis of International Conflict," p. 302.

Although our interpretation mirrors Harsanyi's view that it is fatuous to believe that wars, arms races, and other forms of international conflict are simply the product of policy makers' irrationality—and hence can be remedied with better information or a more enlightened view of one's opponent—this is not to deny that forbearance under certain conditions may ameliorate international conflicts and yield mutually beneficial outcomes for the parties involved.

Before summarizing these conditions, let us briefly review the contents of this chapter. We started out by defining two-person zero-sum games and distinguished between those that have saddlepoints and those that do not in the matrix, or normal-form, representation of a game. We proved that all two-person zero-sum games with perfect information that result in a win for one player and a loss for the other have saddlepoints and are therefore strictly determined, but games without perfect information may also have saddlepoints. The Battle of the Bismarck Sea was one such game, each of whose players made their strategy choices independently and without knowledge of the choice of the other player. We showed how this fact could be visualized by enclosing different vertices of the game tree in the same information set in the extensive-form representation of the game, where plays are considered to be sequential, not simultaneous and static.

Since the Battle of Avranches had no saddlepoint, there was no equilibrium solution in pure strategies. We were able, however, to eliminate one of Bradley's three strategies from consideration since it was dominated by another strategy. We illustrated both algebraic and geometric methods for finding optimal mixed strategies in the reduced game, but since the mixed-strategy solution depended on the numerical payoffs assigned to each outcome, and we had information only on Bradley's ranking of outcomes, it was impossible to prescribe an optimal strategy mixture. Even given such a prescription, its application to the selection of a strategy in a single encounter would present difficulties of interpretation, although we suggested alternative procedures that players might use to obviate these difficulties. We also discussed situations wherein the use of genuinely mixed strategies might prove quite sensible.

The concept of a solution in two-person nonzero-sum games raised a host of problems, most of which relate to the tension between equilibrium pairs of strategies and nonequilibrium pairs that yield greater payoffs to one or both players. The instability of the cooperative pair of strategies in both Prisoner's Dilemma and Chicken may lead both players to choose their noncooperative strategies, which is especially disastrous in the game of Chicken. In this game, Rapoport remarks, a player is motivated to instill *fear* in his opponent, perhaps by being the first to make some kind of irrevocable commitment to his noncooperative strategy; by contrast, in Prisoner's Dilemma a player has a strong incentive to establish credibility by creating *trust* that shows he intends to play cooperatively.[67]

---

67. Rapoport, *Strategy and Conscience*, p. 116.

The theory of metagames, however, suggests that these one-sided postures in Chicken and Prisoner's Dilemma may not be rational (or, more accurately, metarational), given that both players want to maximize their payoffs in these games. In the case of Prisoner's Dilemma, each player must also be ready to exact retribution if he expects the other player to act recalcitrantly; in the case of Chicken, each player must also be ready to cooperate if he expects the other player also to cooperate. In other words, the players in both these games must be both firm and forbearing—firm when their adversary is firm, forbearing when he is forbearing—and convey this image to him as well. This obviously requires that the players be able to communicate with each other, implicitly or explicitly, but not necessarily that they be parties to a binding and enforceable contract. What agreement or understanding they do reach, in effect, is rendered enforceable by the expectation that each player will be willing to abide by the terms of the unwritten contract and enforce it, if necessary, by "defecting" himself. This reduces the risks to the players of choosing their cooperative strategies.

In the games of Prisoner's Dilemma and Chicken, the metarational cooperative outcome for two players 1 and 2 was found by (1) expanding player 2's strategies into four metastrategies in anticipation of player 1's original strategies and (2) subsequently expanding player 1's original strategies into sixteen metastrategies in anticipation of player 2's four metastrategies. Harsanyi has questioned whether metastrategies really offer a justification for the choice of the cooperative strategies in the original game, arguing that in the final analysis players have only strategies, not metastrategies, open to them.[68]

This argument seems persuasive only if we assume that players never communicate with each other, which realistically is hardly ever the case. As soon as we admit the possibility of communication between players, then we allow for the sequential and dynamic buildup of expectations, which typically are formed from the words and actions of the players in bargaining and negotiation processes.

Metagame theory describes what these expectations must be to lead to the cooperative solution. In the case of the Cuban missile crisis, it would appear that each side's tit-for-tat expectations about what the other's behavior would be became sufficiently reassuring after a series of exploratory moves that a settlement satisfactory to both sides could in the end be reached. Testifying to the durability of the outcome, the solution worked out has proved stable for more than a decade, which—in the

---

68. Review of Howard, *Paradoxes of Rationality*, by John C. Harsanyi, in *American Political Science Review*, 67 (June 1973), pp. 599–600; in response to one assertion by Harsanyi, see N. Howard, "Comment on a Mathematical Error by Harsanyi," *International Journal of Game Theory*, 2 (1973), pp. 251–52; and Howard's comment and Harsanyi's rejoinder in "Communications," *American Political Science Review*, 68 (June 1974), pp. 729–31. Another game theorist, Guillermo Owen, has expressed skepticism about the testability of metagame theory in his review of Howard's book in *Behavioral Science*, 18 (March 1973), pp. 128–29.

absence of a binding and enforceable agreement—lends support to the proposition that it is a product of metastrategy, not strategy, choices.

At the same time that we make this argument, it is useful to echo the warning that William H. Riker and Peter C. Ordeshook offer about applications of game theory:

> Political scientists too often take examples (as well as the theorems of game theory) as fully operating models of real political processes. Thus, we read such sentences as "let China be the first decision maker with strategies a_1 and a_2, and let. . . ." But it is absurd to believe that any complex process can be so simply modeled, just as it is absurd to suppose that the design of an efficient gas turbine resides somewhere as a deduction from the basic laws of thermodynamics.[69]

This warning, however, should not be misunderstood as a bar to empirical studies; quite the contrary, Riker and Ordeshook assert that "the only way to generate equilibrium [in the mathematical theory] is to introduce more assumptions and institutional detail and to include other features of the decision makers or their environments."[70] By introducing players' expectations about the environments in which they interact, metagame theory provides one approach to specifying more precisely the bargaining processes that lead to different outcomes in mixed-motive games.

We have tried to demonstrate in this chapter that the concepts of game theory and metagame theory can illuminate the study of strategic choices of decision makers in real-life situations in international politics. As primitive as our analysis has been, it alerts one to both the opportunities and limitations for reasoned solutions to problems of international conflict based on the compatibility of expectations, the enforceability of outcomes, and so forth, which studies rooted in intuitive wisdom and armchair gamesmanship often ignore. If there is no pat or easy "solution" to games like Prisoner's Dilemma or Chicken, it seems more useful to expose their salient strategic features than to conceal the structure of such conflicts in a cloud of rhetoric.

---

69. Riker and Ordeshook, *An Introduction to Positive Political Theory* (Englewood Cliffs, N.J.: Prentice-Hall, 1973), p. 239.

70. Riker and Ordeshook, *Introduction to Positive Political Theory*, p. 236.

2

QUALITATIVE VOTING GAMES

2.1. INTRODUCTION

In this and the next chapter we shall be concerned with different kinds of voting systems and the strategies that voters use to achieve particular ends within the constraints of these systems. In this chapter we shall explore voting systems that permit voters only to choose among a set of alternatives—that is, make qualitative choices—whereas in Chapter 3 we shall examine a voting scheme that permits voters to express their intensities of preference for alternatives by making quantitative choices.[1]

To show how collective decisions, which are based ultimately on the preferences of individual voters, are shaped and circumscribed by constraints in the environment, it is useful to divide environmental constraints into three categories. The first category includes *procedural* constraints that specify conditions under which the voting body may take collective action, the order in which motions are voted upon, and so forth. The second category includes *information* constraints that determine the extent to which voters can take account of the possible actions of other voters. The third category includes *communication* constraints that specify the degree to which voters are able to coordinate their strategies and form coalitions that bind their members to a particular course of action.

To be sure, these categories may not be mutually exclusive, as when members of a voting body vote on the decision rule that they henceforth will adhere to (e.g., when they adopt a constitution).[2] In such a case, the decision rule is not a fixed procedural constraint but will depend at least in part on the preferences of voters on substantive matters, bargaining among coalitions, and so forth.

---

1. There is a rough correspondence between our use of the terms *qualitative* and *quantitative* and that used in Austin Blaquière, Françoise Gérard, and George Leitmann, *Quantitative and Qualitative Games* (New York: Academic Press, 1969), which is a mathematical treatment of two-person games.

2. For an analysis of such situations, see James M. Buchanan and Gordon Tullock, *The Calculus of Consent: Logical Foundations of Constitutional Democracy* (Ann Arbor, Mich.: University of Michigan Press, 1962).

When the decisions of a collectivity are made by a single individual (i.e., a dictator), then environmental constraints will have no bearing on its decisions since the collective choice reduces to a private choice of this individual. We shall not be concerned with such situations but rather with voting situations in which all voters have some, though not necessarily equal, weight. Rules for summing votes and arriving at a collective decision vary considerably, and in subsequent chapters we shall study in detail the effects of different institutional mechanisms for summing individual preferences. In this chapter, however, we shall devote most of our attention to simple plurality and majority procedures, though many of our conclusions will not be restricted to these.

In social situations where collective decisions are not the product of dictatorial choice or of pure whimsy (i.e., chance), the choices of voters will not in general be independent of each other but instead will depend on information and communication constraints. These are the constraints that link the choices of voters to each other and ultimately to the outcome of a social-choice process. They are akin to the traffic going in a perpendicular direction at an intersection that one must check after stopping at a stop sign at the intersection; unlike a stoplight at an intersection, which (usually) eliminates the necessity of checking traffic going in a perpendicular direction, the stop sign evokes a contingent response dependent on the actions of other drivers.

And so it is in any interdependent decision situation in which the outcome depends on the choices of other players in a game. There is no independently best choice one can make, for "best" will depend at least in part on what the other players choose. The web of relationships among players based on their preferences and environmental constraints must first be unraveled before better and worse choices for each player can be determined. Game theory provides tools for sorting out contingent choices of rational actors in such a way as to enable them to maximize the achievement of certain goals.[3]

2.2. VOTING PROCEDURES[4]

As a starting point for our analysis, we assume that there exists a set of *voters*, who may be individual or collective actors (e.g., members of a

---

3. We postpone a discussion of the concept of "rationality" in this chapter until the final section so that we can first develop some implications of the different environmental constraints. Following this development, we shall be in a better position to relate rational behavior to these constraints.

4. Several of the examples used to illustrate the voting procedures and types of voting in this and subsequent sections are adapted from Robin Farquharson, *Theory of Voting* (New Haven, Conn.: Yale University Press, 1969). Although Farquharson gives an interpretation to the outcomes different from that given below, our analysis follows quite closely the lines of his masterful theoretical treatment. Unfortunately, this treatment is marred by numerous errors; for a listing of these, see Richard D. McKelvey and Richard G. Niemi, "Strategic Voting: Some New Results and Some Old Questions" (University of Rochester, 1974).

committee or blocs in an assembly). We assume that each voter has *preferences* with respect to a set of *outcomes*, which are alternative social states that he can rank from best to worst.

Consider the following situation as an illustration of these concepts. A bill is introduced in a legislature, and three outcomes are possible: (1) the original bill passes (O); (2) an amended bill passes (A); and (3) no bill passes (N). We assume that these possibilities are exhaustive: Action on the bill may take only one of these three forms, so the set of possible outcomes is $\{O, A, N\}$.

In our hypothetical legislature we assume that there are three blocs of voters, whose members act in concert. The legislature may thus be thought of as effectively consisting of three distinguishable voters. We further assume that each of these collective voters ranks the set of outcomes as follows:

1. The *passers* prefer O to either of the other alternatives, but forced to make a choice between the other alternatives, would prefer some bill (A) to none (N). Thus, their preference scale is (O, A, N).

2. The *amenders* would most prefer A, but this failing would prefer the original bill (O) to none (N). Thus, their preference scale is (A, O, N).

3. The *defeaters*, if they cannot prevent passage of a bill (N), would prefer the amended version (A) to the original bill (O). Thus, their preference scale is (N, A, O).

Having tied the voters to the outcomes through *preference scales*, which are the set of outcomes ordered by the preference rankings of the voters, we shall now show how different voting procedures affect the choice of an outcome by the voters. We begin by defining a *voting procedure* to be a way of arranging the outcomes to be voted upon, examples of which are given below.

The first class of procedures we shall consider is *binary*, which divides the set of outcomes into two subsets, each subset in turn into two further subsets, and so on, until a single outcome is reached. This is a rather common method of voting in committees and legislatures, perhaps being a reflection of the fact that the voting options available to members often are just two (i.e., binary): approval ("yea") or disapproval ("nay").

We may distinguish two different types of binary procedures. To illustrate the first type, which we call the *successive* procedure, assume that the first vote (or division) is taken on the original bill. If the bill is defeated, then a second vote takes place on an amended bill; this vote may result either in the amended bill's passing (A) or being defeated, in which case no bill passes (N). The sequence of votes may be depicted as a tree, with each fork corresponding to a vote and the endpoints of each branch corresponding to an outcome, as shown in Figure 2.1.

Since the subsets of outcomes at the endpoints of the branches of each fork (O and $\{A, N\}$ at the endpoints of the branches of the first fork, A

FIGURE 2.1 SUCCESSIVE PROCEDURE

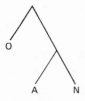

and N at the endpoints of the branches of the second fork)[5] have no members in common, we refer to this binary procedure as "disjoint." In this procedure, one outcome (O) is singled out for initial consideration while the remaining two outcomes $(A$ and $N)$ are given identical treatment. Clearly, either of the other two outcomes could occupy the place that outcome O does in the outcome tree, a fact we shall later exploit to contrast the effects produced by different arrangements of alternatives on the outcome tree.

The second type of binary procedure, which we call the *amendment* procedure, does not require that the subsets at a division have no members in common. In fact, the usual procedure for voting on amendments includes a division into nondisjoint subsets, and for this reason we refer to this binary procedure as "overlapping."

In our example, this procedure would pit the subset of outcomes $\{A, N\}$ against the subset $\{O, N\}$. If the amendment is adopted, the second vote is on the amended bill A versus no bill N. If the amendment is defeated, the second vote is on the original bill O versus no bill N. Defeat can thus occur in two different ways under this procedure, as shown diagramatically in Figure 2.2.[6]

Since each successive vote involves only two subsets of outcomes, this procedure, too, is binary; it is distinguished from the first binary procedure only by the fact that both subsets of outcomes at the first division have more than one member. As in the case of the disjoint binary procedure, it is possible to conceive of different orderings of outcomes: The outcome N, which appears twice among the endpoints, could be replaced by either O or A.

There are other voting procedures besides binary ones, including those which divide the set of outcomes into three, four, and more subsets. In

---

5. We enclose subsets of outcomes in braces only if they contain two or more members.

6. In this section, our outcomes $\{O, A, N\}$ correspond to Farquharson's outcomes $\{A, B, C\}$ in that order—except in this case where O corresponds to C and N corresponds to A; this makes the amendment procedure of Figure 2.2 Procedure IIc in Farquharson's treatment. Farquharson, *Theory of Voting*, pp. 11–12.

FIGURE 2.2   AMENDMENT PROCEDURE

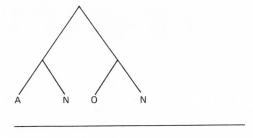

FIGURE 2.3   PLURALITY PROCEDURE

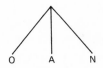

the interest of simplicity, we shall limit our subsequent discussion of voting procedures to two different kinds of *ternary* procedures, which divide the set of outcomes into three subsets. These will offer us some perspective on alternative voting schemes, whose effects we shall study in detail later in the chapter.

As with the class of binary procedures, ternary procedures may be either disjoint or overlapping. The disjoint case is depicted in Figure 2.3, where the outcomes are divided into three one-member subsets. Voting is not sequential but involves immediate selection of one of the three outcomes at a single division. We call this the *plurality* procedure, although we reserve for section 2.3 a discussion concerning *how* outcomes are actually selected.

A somewhat more complicated ternary procedure is that of *elimination*, in which the set of three outcomes is divided into three two-element subsets, with each element in each subset overlapping one element in another subset (see Figure 2.4). The first vote is taken over the three subsets, which eliminates two, and then a second vote is taken between the two elements of the chosen subset.

So far we have given examples of different voting procedures, each of which may be viewed as an alternative way of arranging the outcomes to be voted upon. We shall next consider different types of voting that may be used in deciding among subsets of outcomes at a voting division.

FIGURE 2.4   ELIMINATION PROCEDURE

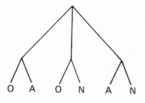

2.3. SINCERE VOTING

At the endpoints of every branch in the tree diagrams, the subsets of outcomes constitute each voter's available *choices*, from which he selects one subset. On what basis does a voter make a selection?

The simplest assumption we can make is that he votes directly in accordance with his preferences, which we call *sincere voting*. A voter votes sincerely if, when choosing among subsets at a fork of the outcome tree, he selects the subset containing his most-preferred outcome. If, for example, a voter's preference scale is (O, A, N), as are the passers in our example, under the successive procedure of Figure 2.1 he would vote for O over $\{A, N\}$ on the first vote, since outcome O is preferred to either A or N.

If no subset contains a voter's most-preferred outcome, a voter votes sincerely if he selects the subset containing his second-most-preferred outcome (or failing the second, the third-most-preferred, and so forth). Refer once again back to the successive procedure of Figure 2.1. A voter with preference scale (O, A, N) would vote for A over N on the second vote, since outcome A is preferred to outcome N.

Finally, if two or more subsets at a fork contain the same "top" outcome (i.e., the outcome most preferred is contained in more than one subset), a voter votes sincerely if he chooses the subset containing the highest-ranked second-to-top outcome (or failing the second, the third-to-top, and so on). Thus, in the elimination procedure of Figure 2.4, a voter with preference scale (O, A, N) would vote for $\{O, A\}$ over $\{O, N\}$ in the first vote, since both subsets contain O but A is preferred to N. (The subset $\{A, N\}$ would be eliminated from consideration at the start because it does not contain the top outcome, O.)

Having defined and illustrated the meaning of sincere voting for individual voters (or blocs), we shall now explicate the relationship between individual choices and the collective choice of a voting body. We start by assuming that none of the three voters or blocs in our previous example constitutes more than half the membership of the voting body. Assuming simple majority rule, this means that no individual voter is decisive. For an outcome, therefore, to be the collective choice of the voting body, it must be the choice of at least two voters at a terminal fork of the voting procedure.

For all of the voting procedures previously described a fork is *terminal* when a subset of outcomes chosen by a majority of voters contains only one member. This obviously follows from the fact that a one-member subset cannot be further subdivided.

To what extent do the results of sincere voting depend on the voting procedures adopted? To answer this question consider first the successive procedure of Figure 2.1. The first vote is taken on the original bill:

For original bill (O): Passers;

For amended bill (A) or no bill (N): Amenders and defeaters.

The original bill thus fails, and the second vote is between the amended bill and no bill:

For amended bill (A): Passers and amenders;

For no bill (N): Defeaters.

The amended bill thus passes.

The amendment procedure of Figure 2.2 leads to the same result—namely, passage of the amended bill:

First vote
For A or N: Amenders and defeaters;
For O and N: Passers.
Intermediate result: A or N.

Second vote
For A: Passers and amenders;
For N: Defeaters.
Final result: A.

The results of the plurality procedure of Figure 2.3 are indeterminate unless we assume that one voter (say, the passers) either possesses more votes than the other two voters—but still less than a majority—or, if all the voters are equally weighted, can break a tie. This assumption has no effect on the results given above for either the successive or amendment binary procedures.

Under the plurality procedure, each voter would have supported his own first preference. Assuming that the passers have the edge over the other voters, the result would be O, given that the alternative with the most votes wins.

Finally, under the elimination procedure of Figure 2.4, as under the two binary procedures, the amended bill would be victorious:

First vote
For O or A: Passers and amenders;
For O or N: No voters.
For A or N: Defeaters.
Intermediate result: O or A.

FIGURE 2.5 NEW ORDERING OF SUCCESSIVE PROCEDURE

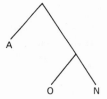

Second vote
  For *O*: Passers;
  For *A*: Amenders and defeaters.
  Final result: *A*.

In summary, three of the four voting procedures (successive, amendment, and elimination) that we have considered give the same result (*A*); the deviation of the one procedure (plurality) that leads to a different collective choice (*O*) depends on an additional assumption about the unequal size of voters.

This may not seem a very serious deviation, given the agreement of all but the plurality procedures on a single collective choice. This consistency is only accidental, however, because—as Farquharson shows—the three procedures that give the same outcome generate different collective choices for four of the eight possible preference scales for three voters choosing among three alternatives, where no two voters most prefer the same alternative.[7]

To illustrate the variable results of different voting procedures under sincere voting, consider again the successive procedure shown in Figure 2.1. By simply changing the *order* in which the three outcomes {*O, A, N*} are voted upon, the collective choice of the voters is changed. Specifically, if the first alternative considered were the amended bill (*A*)—that is, if what we have called the amended bill were moved first, and, it failing, what we have called the original bill were then moved—the voting procedure would be as depicted in Figure 2.5.[8]

---

7. Farquharson, *Theory of Voting,* Appendix I, pp. 64–67.

8. It is immaterial whether or not we consider this new ordering of outcomes to be a "new" voting procedure. The new ordering does change the collective choice of the voting body from that generated by the ordering in Figure 2.1, but structurally the procedures are identical (i.e., the outcome tree is the same—only the labeling of the endpoints of the branches is different).

This new ordering of outcomes results in passage of the original bill:

First vote
    For A: Amenders;
    For O or N: Passers and defeaters.
    Intermediate result: O or N.

Second vote
    For O: Passers and amenders;
    For N: Defeaters.
    Final result: O.

For the sake of completeness, we mention that if N were considered first—for example, if the first vote were on the motion that O or A be tabled or postponed—then the resultant outcome would be A, duplicating the result for the original ordering of the successive procedure given in Figure 2.1.

It should be evident from the foregoing examples and discussion that the voting procedure, and even the ordering of alternatives for a given outcome structure, may make a difference in the outcome selected by members of a voting body. In other words, as we indicated in section 2.1, the "constitutional" choice of a particular voting procedure or ordering of alternatives may not be independent of choices on substantive issues, given that voters can anticipate the results of voting under each procedure or ordering of alternatives.

When voters are aware not only of the voting procedure or ordering to be followed but also know the preference scales of the other voters, we say they have *complete information*. Given this information, it seems reasonable to assume that they will use it to predict the choices other voters will make, although these predictions need not be based only on sincere voting, as we shall show presently.

Some consequences of relaxing the assumption of complete information will be discussed later in the chapter, but for now we point out that if voters have no information on each other's preference scales, they would have no basis on which to predict each other's choices and formulate strategies—in short, to be players in a game. The empirical evidence presented at the end of this chapter suggests that through discussion, opinion polls, and other means voters *are* often able to acquire considerable knowledge about the preference scales of other voters and groups of voters—and act accordingly.

Assume that the outcomes posed in our three-voter example are the sole concern of members of the voting body—that is, voting on no other issues matters or can be anticipated—and the members possess complete information. Then, given a prior choice on how these outcomes are to be voted upon under the successive procedure, the amenders and defeaters would choose the ordering of Figure 2.1, the passers the ordering of Figure 2.5. This follows from the fact that the preference scale of the passers is (O, A, N), so they would prefer the ordering of Figure 2.5 to ensure adoption of O; the preference scale of the amenders is (A, O, N), so they would prefer the ordering of Figure 2.1 to ensure adoption of A; and the preference scale of the defeaters is (N, A, O), so in the absence of an ordering that ensures adoption of N, they would prefer the ordering of Figure 2.1 to ensure adoption of A. Thus, a majority of voters (the amenders and defeaters) would prefer the ordering of Figure 2.1 to that of Figure 2.5.[9] On a procedural vote, therefore, to determine whether to follow the Figure 2.1 ordering or the Figure 2.5 ordering, the Figure 2.1 ordering would win, with the consequence that A would be the collective choice of the voting body.[10]

All well and good for the amenders and defeaters—if they are given the opportunity to vote on, or otherwise determine, the ordering of alternatives. If they are not always successful in controlling the agenda, however, it may be advantageous for them *not* to vote sincerely, as we shall show in subsequent sections.

2.4. STRAIGHTFORWARD STRATEGIES AND VOTING PROCEDURES

Before considering alternatives to sincere voting, we must first define the strategies of a voter. By *strategy* we mean in this context a complete plan that specifies a voter's choice at each division of an outcome tree. In the case of the plurality procedure of Figure 2.3, there is only one division, at which each voter has three choices—to vote for O, A, or N. Thus, each voter has three strategies from which to choose in deciding upon a course of action.

A strategy is more complicated when there is more than one voting division in the outcome tree. In the successive procedure of Figure 2.1, for example, there are two divisions, at which each voter has two choices. Each voter therefore has four strategies, which specify all possible courses of action he might take at both divisions:

(a) Vote for O, that failing for A;

(b) Vote for O, that failing for N;

(c) Vote against O (i.e., for $\{A, N\}$), then for A;

(d) Vote against O (i.e., for $\{A, N\}$), then for N.

---

9. This is an example of how an overarching "procedural rationality"—the order of voting on alternatives—can be derived from, or revealed by, a "substantive rationality" that is based on the posited preferences of voters. For a discussion of these concepts, see William H. Riker and Peter C. Ordeshook *An Introduction to Positive Political Theory* (Englewood Cliffs, N.J.: Prentice-Hall, 1973), pp. 14–16.

10. Such procedural votes are in effect allowed in the Swedish parliament, or riksdag. See Dankwart Rustow, *The Politics of Compromise: A Study of Parties and Cabinet Government in Sweden* (Princeton, N.J.: Princeton University Press, 1955), p. 194.

We now consider the conditions under which a voter can choose a strategy that, no matter what the choices of the other voters are, leads to at least as desirable an outcome as any other strategy. Of course, this strategy will not necessarily guarantee a voter his most-preferred outcome but rather an outcome no worse than would result from his choice of another strategy. We call a strategy *straightforward* if, independently of the choices of other voters, it cannot be improved upon; we also refer to a voting procedure that affords a voter a straightforward strategy at all divisions as straightforward.

For the class of binary procedures discussed earlier, it is not difficult to establish that a procedure is straightforward for a voter if and only if every division of it separates his preference scale. A scale is *separated* by a division if the top outcome of one subset at the division is not ranked higher than the bottom outcome of the other. This is true for the passers—preference scale (O, A, N)—and defeaters—preference scale (N, A, O)—under the successive procedure of Figure 2.1. The first division divides the set of outcomes $\{O, A, N\}$ into the subsets O and $\{A, N\}$: For the passers, their top outcome A in the subset $\{A, N\}$ is not ranked higher than outcome O; similarly for the defeaters, outcome O is not ranked higher than their bottom outcome A in the subset $\{A, N\}$. This separation, however, is not possible for the preference scale of the amenders (A, O, N): Outcome O does rank higher than their bottom outcome N in the subset $\{A, N\}$.

Another way of stating the separability condition is that a preference scale is separated by a division if the subsets into which the set of outcomes is divided do not overlap on the scale. To illustrate, we can draw lines cutting the preference scales of the passers and defeaters, $(O/A, N)$ and $(N, A/O)$, such that the members of one subset at the first division in Figure 2.1 lie to the left, members of the other subset lie to the right (i.e., there is no overlap). On the other hand, no line can separate the preference scale of the amenders, (A, O, N), into O and $\{A, N\}$.

This separability notion is also applicable to binary voting procedures that divide the set of outcomes into overlapping subsets. In the amendment procedure of Figure 2.2, for example, the initial division of outcomes into the subsets $\{A, N\}$ and $\{O, N\}$ separates the preference scale (A, N, O), given that we allow one outcome (N in this case) to lie astride the cutting line: $(A, N/N, O)$. A division that "weakly" separates a scale in this manner is consistent with our definition of separability: for the preference scale (A, N, O), the top outcome (N) of one subset, $\{O, N\}$, is not ranked higher than the bottom outcome (also N) of the other subset, $\{A, N\}$. The fact that none of the three voters in our example has the preference scale (A, N, O)—or the scale (O, N, A), where N is also the middle-ranked outcome and hence is a scale that can be separated by the first division of Figure 2.2—means that no voter has a straightforward strategy under the amendment procedure of Figure 2.2. Given this notion of separability, we now are in a position to prove the following theorem.

Separability Theorem

A binary voting procedure separates a voter's preference scale at every division in the outcome tree if and only if he can make an unconditionally best choice at every division (i.e., has a straightforward strategy).

Proof. If a binary procedure separates a voter's preference scale at every division, he is able to choose the subset of outcomes at each division whose least desirable member is at least as preferred as any member of the other subset. Thus, he cannot regret his choice, for no other choice on his part, whatever choices are made by the other voters, could have secured for him a preferable outcome. This gives him an unconditionally best choice at every division and establishes that separability is a sufficient condition for their being a straightforward strategy under binary voting procedures.

To show that it is also necessary, we assume that a voter's scale is not separated at every division. Then there is a possibility that, whichever of the two subsets he selects at a division where his scale is not separable, its lowest-ranked member will eventually be selected. But it is possible that a higher-ranked member could have been chosen from the other subset. Now, if his choice were decisive in the selection of the first subset, this means that it is possible that he could have obtained a preferable outcome by voting differently (depending on how the other members voted). Since the same possibility exists no matter which subset he chooses, he has a straightforward strategy only if his scale is separable at every division in the outcome tree. This completes the proof.

If a voter does not have a straightforward strategy, he must, to make the best use of his vote, attempt to predict how the other voters are likely to vote (i.e., what strategies they are likely to adopt, given that he has complete information about their preference scales). This is easy for the amenders in our previous example, because, as we showed, both the passers and defeaters have straightforward strategies under the successive procedure of Figure 2.1:

The passers, with preference scale (O, A, N), will choose strategy (a)—Vote for O, that failing for A.

The defeaters, with preference scale (N, A, O) will choose strategy (d)—Vote against O (i.e., for $\{A, N\}$), then for N.

The best course of action for the amenders, with preference scale (A, O, N), taking the above strategies as fixed, is to choose $\{A, N\}$ over O on the first division, A over N on the second division, which is strategy (c)—Vote against O (i.e., for $\{A, N\}$), then for A. This follows from the fact that the amenders are the tie-breaking voter at each division [compare the opposite sides taken by the passers and defeaters choosing strategies (a)

and (*d*) above]. Whatever strategy the amenders adopt will, therefore, be decisive, so it is natural to assume that they will vote in such a way as to ensure adoption of their most-preferred outcome, *A*. This they accomplish by selecting strategy (*c*).

Coincidentally, if all three voters voted sincerely, they would have adopted precisely the "best" strategies specified here—and the outcome, of course, would be exactly the same (i.e., *A*) as the one we showed on page 57 for sincere voting under the successive procedure of Figure 2.1. But under certain conditions, the outcome of sincere voting *can* be upset by the kinds of strategic calculations sketched in this section, as we shall next show.

2.5. EQUILIBRIUM CHOICES AND VULNERABILITY

To extend the logic of the last section, we start by considering the conditions under which a voter would be motivated *not* to vote directly according to his preferences. It seems reasonable to suppose that such an incentive would exist if the voter could, by changing his vote, change the collective choice of the voting body to one that ranks higher on his preference scale.

To illustrate a sincere collective choice that is vulnerable to such strategic calculations, consider the successive procedure of Figure 2.5, where only the order of voting on outcomes has been changed from that of Figure 2.1. Here, it will be recalled (see page 59), the outcome of sincere voting was *O*, whereas the sincere outcome based on the ordering of Figure 2.1 was *A*.

We showed in section 2.4 that the sincere outcome *A* under the procedure of Figure 2.1 was not vulnerable to strategic calculations. But under the procedure of Figure 2.5, if the defeaters vote for *A* instead of {*O*, *N*} at the first division, they will change the collective choice of the voting body from *O* to *A*:

> *First vote*
> For *A*: Amenders and defeaters;
> For *O* or *N*: Passers.
> Intermediate result: *A*.

This intermediate result is in fact the final result, and no second vote is required, since *A* is a single outcome and thereby terminates the voting process.

The choice on the part of the defeaters, whose preference scale is (*N*, *A*, *O*), for *A* rather than {*O*, *N*} may not seem justified, given that {*O*, *N*} contains their most-preferred outcome. But in view of the facts that

> (1) their vote is decisive at the first division, since the passers and amenders split their votes if they vote sincerely, and

    (2) by voting "insincerely" they can effect a collective choice (*A*) that ranks higher on their preference scale than the sincere collective choice *O*,

the logic of their selection would appear to be sound. Yet, can we expect the passers, whose preference scale is (*O*, *A*, *N*), to sit idly by and continue to vote sincerely if the defeaters are able to change the sincere outcome to their own advantage and to the disadvantage of the passers?

    This kind of problem did not crop up in our earlier discussion of voting under the successive procedure of Figure 2.1, because two of the three voters (the passers and defeaters) had straightforward strategies that could not be improved upon whatever strategy was selected by the third member of the voting body (the amenders). With the strategies of the passers and defeaters fixed, the amenders' strategy involved simply voting in such a way as to effect their higher-ranking subsets of outcomes at each division; no account had to be taken of possible subsequent responses on the part of the two other voters to their choices.

    Under the successive procedure of Figure 2.5, however, the preference scale of only one voter, the amenders, is separated by the first division: (*A/O*, *N*). This means that the passers and defeaters must somehow "adjust" their voting not only to the straightforward strategy of the amenders but also to the strategies likely to be adopted by each other. How can we determine what strategies are "best," and therefore likely to be adopted, for each of these voters when neither strategy can be taken as fixed but instead must be thought of as a response to the other's choice of strategies?

    This is fundamentally a game-theoretic question and requires that we make additional assumptions that establish the bounds within which interdependent calculations can be made. Since the solution to this game is the set of strategies that it is reasonable to assume the voters will adopt, we begin by specifying all four strategies available to each of the voters under the successive procedure of Figure 2.5, which are analogous to the four strategies under the successive procedure of Figure 2.1 that were given in section 2.4 (to distinguish them from the previous strategies, we add primes):

    (*a′*) Vote for *A*, that failing for *O*;

    (*b′*) Vote for *A*, that failing for *N*;

    (*c′*) Vote against *A* (i.e., for {*O*, *N*}), then for *O*;

    (*d′*) Vote against *A* (i.e., for {*O*, *N*}), then for *N*.

    From the Separability Theorem, the straightforward strategy of the amenders can be read off directly from their preference scale, (*A*, *O*, *N*): vote for *A*, that failing for *O*, which is strategy (*a′*). Since the preference scales of the passers and defeaters are not separated by the successive procedure of Figure 2.5, however, they do not have straightforward

TABLE 2.1 SIXTEEN OUTCOMES ASSOCIATED WITH THE
DIFFERENT STRATEGIES OF PASSERS AND DEFEATERS UNDER
SUCCESSIVE PROCEDURE OF FIGURE 2.5, GIVEN AMENDERS
ADOPT THEIR STRAIGHTFORWARD STRATEGY (a')

PASSERS' STRATEGIES	DEFEATERS' STRATEGIES			
	a'	b'	c'	d'
a'	A (E)	A (E)	A	A
b'	A (E)	A (E)	A	A
c'	A (E)	A (E)	O	O (S)
d'	A	A	O	N

strategies. We can begin the search for their best strategies by examining the sixteen outcomes associated with the four strategies that independently can be chosen by each of these voters, given that the amenders choose their straightforward strategy (a') (see Table 2.1).[11]

The outcomes associated with the intersection of the passers' and defeaters' strategies are based on majority choices at each division. For example, if the passers select strategy (c') and the defeaters strategy (d'), the result will be as follows, given that the amenders adopt their straightforward strategy (a'):

First vote
  For A: Amenders;
  For O or N: Passers and defeaters.
  Intermediate result: O or N.

Second vote
  For O: Passers and amenders;
  For N: Defeaters.
  Final result: O.

This result is not unfamiliar: The strategies (a'), (c'), and (d') happen to be the sincere strategies of the amenders, passers, and defeaters, respectively. In Section 2.3 we showed that the sincere outcome under the successive procedure of Figure 2.5 is O, which we denote in Table 2.1 with the parenthetic expression (S).

This outcome is vulnerable to strategic calculations, however, as we showed at the beginning of this section. Specifically, if the defeaters vote

---

11. If the amenders did not possess a straightforward strategy, we would have to consider the sixty-four (4 × 4 × 4) combinations of strategies that could be chosen by all three voters.

for A instead of $\{O, N\}$ at the first division [i.e., choose either strategy (a') or (b')], they will change the collective choice from O to A, which ranks higher on their preference scale. The sincere outcome O is therefore not an *equilibrium choice*—there is at least one voter who can, by changing his strategy, obtain a better outcome for himself. Formally, a collective choice is *vulnerable* (i.e., not an equilibrium choice) if another choice

(1) can be obtained from the first by substituting a strategy of at least one voter;

(2) is preferred to it by that voter, or those voters.

We have indicated such equilibrium choices by the parenthetic expression (E) in Table 2.1.

Sincere outcomes may or may not be equilibrium choices.[12] The sincere outcome under the successive procedure of Figure 2.5 is not, as we have shown. Under the successive procedure of Figure 2.1, on the other hand, the sincere outcome is an equilibrium choice. For, as we showed in section 2.4, given that the passers and defeaters choose their straightforward strategies (a) and (d), which are also their sincere strategies, the amenders can do no better than choose strategy (c), which is their sincere strategy as well. The sincere outcome under this procedure, therefore, is invulnerable to the substitution of strategies that can lead to preferable outcomes for any voter.

If all voters have straightforward strategies under a particular voting procedure, their adoption will always produce an equilibrium choice. For, even if one or more voters could change the outcome by substituting a different strategy, none would have the incentive to do so since his straighforward strategy is unconditionally best. The converse obviously does not hold—an equilibrium choice is not necessarily the product of straightforward strategies—since outcomes generated by the adoption of other strategies (e.g., the unanimous selection of one strategy by all voters) will also be equilibrium choices.

Sincere strategies are not necessarily straightforward [e.g., strategy (c) for the amenders under the successive procedure of Figure 2.1], but straightforward strategies are always sincere [e.g., strategies (a) and (d) for the passers and defeaters under the Figure 2.1 procedure]. The reason that the latter statement is in general true is that if a voter has a straightforward strategy, one subset of outcomes at each division will always provide him with an unconditionally best choice. Since this subset necessarily contains his single most-preferred outcome at the division, its selection is equivalent to voting sincerely.

---

12. For necessary and sufficient conditions for sincere voting to produce an equilibrium choice, see Prasanta K. Pattanaik, "On the Stability of Sincere Voting Situations," *Journal of Economic Theory*, 6 (Dec. 1973), pp. 558–74.

FIGURE 2.6 RELATIONSHIPS AMONG STRATEGIES AND EQUILIBRIUM CHOICES

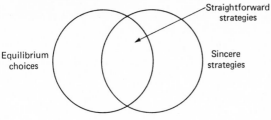

The results we have established above are summarized in the Venn diagram of Figure 2.6. Straightforward strategies are a subset of sincere strategies, but only the selection of the former by all voters (if available under a particular voting procedure) will invariably result in an equilibrium choice.

2.6. DESIRABLE STRATEGIES AND EQUILIBRIA

In previous sections we suggested two criteria for narrowing down the set of strategies voters would be likely to adopt. First, if a voter had a straightforward strategy, he would always adopt it since he could obtain no better outcome by choosing another strategy. For the class of binary voting procedures, we showed that voters whose preference scales are separated by such a procedure possess a straightforward strategy, which is necessarily sincere.

Second, if a collective choice is not an equilibrium choice, there will be at least one voter who will have an incentive not to choose a strategy associated with that choice and, moreover, can ensure that it will not be selected. Given that there is complete information about the preference scales of all voters, other voters could make the same strategic calculation. With every voter aware of the vulnerability of such an outcome, no set of strategies associated with a nonequilibrium choice would be adopted by all the voters.

But not all strategies associated with equilibrium choices will be appealing to voters. If we had not assumed that the amenders, under the successive procedure of Figure 2.5, adopt their straightforward strategy (a'), but instead had assumed that they might select one of their other three strategies, then the sixty-four possible outcomes for the four strategies of each voter would reveal still more equilibrium choices than those shown in the reduced array of Table 2.1. Because the amenders have a straightforward strategy, however, the strategies associated with the additional equilibria have nothing to recommend them except their "stability," which is a sufficient but hardly necessary condition for adoption. On the other hand, the fact that a straightforward strategy is the best that a voter can do no matter what circumstances arise seems a compelling reason for its adoption.

If we assume that the amenders adopt their straightforward strategy in the preceding example, we are able to pare down the number of equilibrium choices to the six shown in Table 2.1. Associated with these choices, which all result in the collective choice of outcome A, are three strategies of the passers and two strategies of the defeaters, which still leave multiple choices open to both voters. Are there any additional criteria which might enable the two voters to narrow down their strategy choices still further?

To begin with, even if a voter has no information about the preference scales of the other voters, he can rule out those strategies which result in outcomes inferior to, or at least not superior to, those he can obtain by voting sincerely. To illustrate, if the passers under the successive procedure of Figure 2.5 adopt strategy (d'), the outcomes they obtain for every strategy choice of the defeaters are identical to those they would obtain if they adopted their sincere strategy (c')—except when the defeaters themselves select strategy (d') (see Table 2.1). Then their sincere strategy (c') gives them a better outcome, since O ranks higher on their preference scale than N. Thus, it would obviously be foolish for the passers ever to choose strategy (d'), which could result in outcome N. Likewise, strategies (a') and $b')$ of the passers can also be eliminated, because they lead to an outcome (A) inferior to that yielded by strategy (c') (O) when the defeaters choose strategy (c') or (d'). Hence, the passers' strategy (c') dominates their other three strategies.

Similarly, we would expect the defeaters never to choose strategy (c'), because for every strategy of the passers it leads to identical outcomes as their sincere strategy (d')—except when the passers choose strategy (d'). Then the defeaters' sincere strategy (d') gives them a better outcome, since they prefer outcome N to outcome O. Because strategies (a'), (b'), and (d') for the passers, and strategy (c') for the defeaters, are thus dominated by the sincere strategies of each voter, we can delete them from the rows and columns of Table 2.1, leaving the three outcomes shown in Table 2.2.

Because the removed strategies in some cases lead to outcomes less desirable, and in no cases more desirable, than some other strategies, we call them *undesirable*. On the other hand, assuming voters with straightforward strategies always adopt them, a strategy is *desirable* (before deletion of the undesirable strategies) if there is no other strategy providing

(1) an outcome at least as good, whatever strategies are selected by the other voters without straightforward strategies;

(2) a better outcome for some selection of strategies of the other voters without straightforward strategies.

TABLE 2.2 (PRIMARILY) DESIRABLE STRATEGIES AND OUTCOMES UNDER SUCCESSIVE PROCEDURE OF FIGURE 2.5

PASSERS' STRATEGY	DEFEATERS' STRATEGIES		
	a′	b′	d′
c′	$A(E)$	$A(E)$	$O(S)$

Obviously, straightforward strategies are desirable (there are no other strategies that can lead to better outcomes), and for a voter who possesses a straightforward strategy, it will be his only desirable strategy (not more than one strategy can be unconditionally best). Voters who do not possess straightforward strategies will have one or more desirable strategies, which can be found through a process of elimination of undesirable strategies, as we have shown in the case of the successive procedure of Figure 2.5.

2.7. THE RELATIONSHIP BETWEEN DESIRABLE AND ADMISSIBLE STRATEGIES

The assumption that voters with straightforward strategies always adopt them initially facilitates elimination of undesirable strategies by removing from the consideration of other voters all but one of the strategies of each of the straightforward voters. Farquharson does not make this assumption in defining an *admissible* strategy, which is identical to our concept of a desirable strategy except that he does not limit it to voters "without straightforward strategies" in the definition of desirable strategies given above. Instead, he assumes that a voter will take account of *all* the strategies of all the other voters, including those with straightforward strategies, to determine whether there is any other strategy that satisfies the two conditions of a desirable strategy (minus the "without straight-forward strategies" restriction) specified in section 2.6.[13]

At a minimum, every voter will have one admissible strategy—sincere voting. To demonstrate this, consider Farquharson's first condition for an admissible strategy, analogous to our first condition for a desirable strategy (we distinguish it with a prime): A strategy is admissible if there is no other strategy providing

(1′) an outcome at least as good, whatever strategies are selected by the other voters.

---

13. Farquharson, *Theory of Voting*, pp. 28–29.

Assuming (for convenience) that there is an even number of "other voters," then "whatever strategies are selected by the other voters" will necessarily include some situations in which, at every division where a voter's most-preferred outcome appears in some subset of outcomes, this subset of outcomes is tied with another subset for the most number of votes of "other voters." (There may not be such tied situations if we immediately restrict voters with straightforward strategies only to the choice of these, as we do in the definition of a desirable strategy.) In these tied situations, a voter's choice at each division where his most preferred outcome appears will be decisive. If a voter votes sincerely at each of these divisions, he will choose the subset containing his most-preferred outcome, which eventually will result at a terminal division in the selection by the voting body of this outcome. Since in these tied situations a voter's sincere strategy uniquely ensures his very best outcome, there can be no other strategy providing an outcome at least as good (in these situations), which satisfies the first condition for admissibility.

The second condition for an admissible strategy, analogous to the second condition for a desirable strategy, is that there be no other strategy providing

(2′) a better outcome for some selection of strategies of the other voters.

Although there may be another strategy providing a better outcome for some selection of strategies of the other voters, the fact that sincere voting in the tied situations described previously assures one of one's best outcome means that such another strategy cannot also satisfy condition (1′), which is sufficient to establish that sincere voting is admissible.

We have shown that sincere strategies are always admissible, but admissible strategies are not always desirable. Rather, like sincere strategies, desirable strategies are a subset of admissible strategies. To illustrate the more inclusive nature of admissible strategies, consider the successive procedure of Figure 2.1, where we showed in section 2.4 that two of the voters in our example (the passers and defeaters) have straightforward strategies. Assuming that these strategies are adopted, we showed that the amenders could do no better than choose their sincere strategy (c), which is necessarily admissible.

But strategy (a) is also admissible for the amenders, because whatever strategies are selected by the passers and defeaters (see Table 2.3), no other strategy provides the amenders with

(1′) an outcome at least as good, whatever strategies are selected by the other voters: The outcomes of strategy (b) are sometimes worse (and never better), the outcomes of strategies (c) and (d) are sometimes worse (and sometimes better);

(2′) a better outcome for some selection of strategies of other voters: For the starred selections of strategies in Table 2.3, no other strategy provides a better outcome than strategy (a); for the nonstarred selections, another strategy does, but it does not also satisfy condition (1′).

**TABLE 2.3 SIXTY-FOUR OUTCOMES RESULTING FROM
AMENDERS' STRATEGIES FOR ALL SELECTIONS OF STRATEGIES
OF PASSERS AND DEFEATERS UNDER SUCCESSIVE PROCEDURE
OF FIGURE 2.1**

STRATEGIES	AMENDERS' STRATEGIES			
(Passers, defeaters)	a	b	c	d
(a, a)*	O (E)	O (E)	O	O
(a, b)*	O (E)	O (E)	O	O
(a, c)	O	O	A (E)	A
(a, d)	O	O	A (E)	N
(b, a)*	O (E)	O (E)	O	O
(b, b)*	O (E)	O (E)	O	O
(b, c)	O	O	A	N
(b, d)*	O (E)	O (E)	N	N
(c, a)	O	O	A	A
(c, b)	O	O	A	N
(c, c)*	A	·A	A (E)	A
(c, d)*	A	N	A (E)	N
(d, a)	O	O	A	N
(d, b)*	O	O	N	N
(d, c)*	A	N	A	N
(d, d)*	N	N	N	N (E)

* Selections of strategies of passers and defeaters for which no other strategy of amenders
is better than strategy (*a*). Note that the strategy choices of the passers and defeaters are
simply reversed in several rows and hence the outcomes in these rows are the same (though
not necessarily the equilibria). Preference scales: passers—(*O, A, N*); amenders—(*A, O, N*);
defeaters—(*N, A, O*).

Thus, since there is no other strategy that satisfies both conditions given
above, strategy (*a*) is an admissible strategy. [We could show in the same
manner that strategy (*c*) is also a strategy for which there is no other
strategy that satisfies these conditions, but since we know that it is the
sincere strategy of the amenders, this is sufficient to establish that it is an
admissible strategy, as we have demonstrated.] Yet strategy (*a*) is not a
desirable strategy for the amenders. For if we assume that the passers and
defeaters adopt their straightforward (and necessarily sincere) strategies,
represented by row (*a, d*) in Table 2.3, then of the four strategies open to
the amenders, only strategy (*c*) is desirable: none of the other three
strategies provides an outcome at least as good, or better than, *A*, given
that there are no other voters—without straightforward strategies—whose
selections of strategies they must take into account.

We can perhaps best summarize the relationships discussed in this and
previous sections by adding desirable and admissible strategies to the

FIGURE 2.7 FURTHER RELATIONSHIPS AMONG STRATEGIES AND EQUILIBRIUM CHOICES

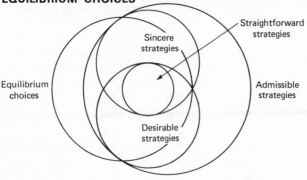

Venn diagram in Figure 2.6. Figure 2.7 contains these additions, with straightforward strategies shown as a subset of desirable strategies, which are in turn a subset of admissible strategies.

It is worth pointing out one special case of the general relationship between straightforward and admissible strategies: A voting procedure offers a voter only one admissible strategy if and only if this strategy is straightforward (and necessarily desirable). If there exists only one admissible strategy, all other strategies are necessarily inadmissible; with respect to these other strategies, therefore, the admissible strategy *does* provide

(1') an outcome at least as good, whatever strategies are selected by the other voters,

(2') a better outcome for some selection of strategies of the other voters,

which makes the admissible strategy unconditionally best and therefore straightforward. Conversely, a straightforward strategy is uniquely admissible (and desirable), because there are no other strategies which can lead to better outcomes whatever strategies are selected by the other voters.

The general relationships among the different strategies we have just delineated have now become rather complicated, but they help to clarify which set of strategies is most suitable for limiting the equilibrium choices to those that are likely to be adopted by members of a voting body. Clearly, we cannot limit the strategies of voters to those that are straightforward, because not all voters will always have such a strategy under a particular voting procedure. On the other hand, although all voters have at least one admissible strategy, the determination of such strategies is not based on the assumption that voters with straightforward strategies will adopt them. This unnecessarily broadens the concept of a strategy likely to be adopted by a voter without a straightforward strategy to one incompatible with the selection of straightforward strategies by those other voters who have them.

The sharpest concept of a strategy that is consistent with the assumption that voters with straightforward strategies will adopt them, and that voters without straightforward strategies will act on the assumption that they do, is a desirable strategy. Whereas admissible strategies can be determined without assuming that voters have any information on the preference scales of other voters, desirable strategies presume that voters without straightforward strategies have such information; otherwise, they could not determine which voters have straightforward strategies and thereby rule out the selection of nonstraightforward strategies on the part of these voters.

Although a desirable strategy is more demanding in its information requirements, it more quickly narrows down the strategies likely to be adopted by the voters without straightforward strategies. Indeed, since a voter has a single admissible strategy if and only if that strategy is straightforward, we know that voters without straightforward strategies will have at least two admissible strategies, but not necessarily more than one desirable strategy.

The fact that there are generally fewer desirable than admissible strategies means that the search for them can be carried out more efficiently. In the case of the successive procedure of Figure 2.1, for example, the search for admissible strategies required comparisons among the sixty-four outcomes shown in Table 2.3; by contrast, the search for desirable strategies required comparisons among only the four outcomes in row (a, d) of this table, which gives the straightforward strategies.

2.8. SOPHISTICATED VOTING

In section 2.7 we argued that the outcomes likely to be adopted by voters will be limited to those equilibrium choices associated with desirable strategies. In the case of the successive procedure of Figure 2.5, we showed in section 2.5 that there are two desirable equilibria (see Table 2.2)— the only ones of the original six equilibria listed in Table 2.1 to have survived the cut of the undesirable strategies (a'), (b'), and (d') of the passers and (c') of the defeaters.[14] Because this cut did not reduce the desirable outcomes to a single equilibrium, the defeaters are still left with more than one desirable strategy, associated with these equilibria, from which to choose. To reduce their desirable strategies still further,

---

14. Recall from section 2.5 that we exclude from the category of "original equilibria" those associated with the three undesirable (i.e., nonstraightforward) strategies of the amenders. If we had counted these equilibria as "original," of course, many more would have survived a cut of inadmissible strategies.

we now assume that each voter has not only knowledge of the preference scales of the other voters but also is able, on the basis of this information, to determine the desirable strategies of each of the other voters (as well as himself). In other words, we assume that he is able to convert the "raw information" he has on the preference scales of the other voters into new information about the strategies they are likely to adopt. In this manner, all voters can become aware of the reduced set of desirable strategies shown in Table 2.2.

But the process need not stop here, for now some of these desirable strategies in the reduced table of outcomes (Table 2.2) become less appealing. This is so because, with the elimination of the undesirable strategies, the basis of comparison changes. For example, comparing the passers' strategy (d') with strategies (a') and (b') in Table 2.2, we see that the choice of (d') on the part of the defeaters is inferior to (a') and (b'), since the sincere outcome O ranks lowest on the defeaters' preference scale.

If we treat the reduced set of desirable strategies in Table 2.2 as if they were the original set of strategies, we may now delete the dominated strategy (d') of the defeaters. This leaves them with strategies (a') and (b'). The intersection of strategy (c') of the passers, and strategies (a') and (b') of the defeaters, admits one equilibrium choice: outcome A.

It may seem a bit strange that at the end of this elimination process two strategies can both be "best" for the defeaters, even though they result in the same outcome. Informally, we can show that this result is plausible by observing that both the passers and amenders prefer outcome O (the defeaters' worst outcome) to outcome N (the defeaters' best). Outcome O would always win, therefore, if there should be a second vote between O and N under the successive procedure of Figure 2.5. With this outcome in effect foreordained if a second vote is required, it is of no import to the defeaters whether they choose strategy (a') or (b'), both of which involve an initial vote for outcome A that ensures its adoption—and that their worst outcome O will not be adopted—given that the amenders adopt their straightforward strategy, (a').

On the other hand, if a second vote is required, the passers, who also do not have a straightforward strategy, could make the difference between outcomes O and N, since the amenders and defeaters would split their votes at the second division. On this vote, the passers would obviously choose their most-preferred outcome O to ensure that their worst outcome N will not be adopted. Consistent with this choice would be a prior vote for $\{O, N\}$, which is strategy (c'), despite the fact that the passers can determine (for reasons given in the previous paragraph) that their next-to-worst outcome A will be adopted.

It is precisely this informal "backward" reasoning—designed to insure against (if possible) worst, then next-to-worst (and so forth) outcomes being adopted—which we systematized in the reduction process we have illustrated. To lay bare the structure of this process in somewhat more formal terms, it is helpful to define recursively a succession of desirable strategies:

A strategy is *primarily desirable* if it is either straightforward or if there is no other strategy which produces at least as good a result for every selection of strategies of the other voters with nonstraightforward strategies, and a better result for some selection.

A strategy is *secondarily desirable* if, on the assumption that all other voters use only primarily desirable strategies, it produces at least as good a result for every selection of strategies of the other voters with nonstraightforward strategies, and a better result for some selection.

In general, a strategy is *m-arily desirable* if it remains desirable on the assumption that all other voters use only $(m - 1)$-arily desirable strategies.

For the successive procedure of Figure 2.1, the primarily desirable strategies are the straightforward strategies (a) and (d) of the passers and defeaters, respectively, and the sincere strategy (c) of the amenders. For the successive procedure of Figure 2.5, the primarily desirable strategies are those shown in Table 2.2, the secondarily desirable strategies are (c') for the passers, (a') and (b') for the defeaters, with the straightforward strategy (a') of the amenders primarily and secondarily desirable.

At the highest level of desirability, all strategies are "ultimately" straightforward in the sense that they are unconditionally best, given the successive elimination of lower-level desirable strategies by all the voters. However, should one or more voters not adopt his "ultimately" straightforward strategy, only the straightforward strategy of the amenders remains *unconditionally* best and is therefore a dominant strategy. Whereas the admissible and the desirable strategies of the passers and defeaters are undominated, these voters have no straightforward strategies that are unconditionally best—except in the reduced game, but this is conditional on all voters' choosing successively more desirable strategies.

The succession of desirable strategies correspond to, but are not identical with, the succession of admissible strategies defined by Farquharson.[15] Like Farquharson, we assume that voters not only are aware of the preference scales of the other voters but also use this information to determine which voters have straightforward strategies. But while Farquharson assumes that voters with straightforward strategies will always adopt them, he does not assume that other voters will use this information in choosing their own admissible strategies, whose selection does not depend on it. By contrast, we assume that voters use this information to weed out admissible strategies that are undesirable at each stage; as we showed earlier, utilization of this information results in a

---

15. Farquharson, *Theory of Voting*, p. 39.

more efficient reduction process. The equilibrium choice(s) selected by both reduction processes, nonetheless, is (are) the same.[16]

In game-theoretic terms, the successive elimination of strategies by voters—and outcomes associated with the eliminated strategies—produces a different and smaller game in which some formerly desirable strategies in the larger game cease to be desirable. The removal of undesirable strategies at successively higher levels of desirability in effect enables the voters to determine what outcome eventually *will* be adopted by eliminating those outcomes that definitely *will not* be chosen by the voters.

Driving this process is the assumption that all voters will rule out undesirable strategies at each stage, which seems quite reasonable in view of the fact that such strategies by definition can never lead to a better outcome than that which results from the choice of a desirable strategy, and for at least one selection of strategies of the other voters must lead to an unequivocally worse outcome. Or, as we showed by our informal "backward" reasoning earlier, voters eliminate strategies that could result in their worst, then next-to-worst (and so on) outcomes' adoption, even if the strategy or strategies left remaining offer no guarantee that their best outcome will be chosen. The assumption that other voters will do likewise renders the choices of all voters interdependent.

We call voting *sophisticated* if it forecloses the possibility that a voter's worst outcomes will be chosen—insofar as this is possible—through the adoption of successively more desirable strategies, given that other voters do likewise. In the case of binary voting procedures, it has one important consequence, as given by the following theorem.

Binary Theorem

Under binary voting procedures, sophisticated voting produces a single outcome.

Proof. Assume that outcome A is pitted against outcome B at a terminal division. Consider the strategies of some voter, all of which include a choice at this division. Now for every strategy that includes a final choice of A, there is a matching strategy that is identical to it except that it includes a final choice of B at this terminal division.

Assume that the voter prefers outcome A to outcome B and chooses some strategy that involves the final choice of A. Call that strategy his A-strategy, and the matching strategy that is identical to it except that it involves a final choice of B, his B-strategy. Then the voter's A-strategy will always lead to the same result as his B-strategy (which will *not* necessarily be the adoption of either A or B, depending on the choices of

---

16. Whatever reduction process one uses, however, there may be complications in determining the ultimately admissible (and desirable) strategies. For details, see McKelvey and Niemi, "Strategic Voting: Some New Results and Some Old Questions," in which the authors suggest a different method for determining sophisticated strategies.

the other voters), except for those situations in which the selection of strategies by the other voters results in a tie between A and B at this division (assuming it is reached). In these situations, the voter's A-strategy will give a better outcome, which means that his B-strategy is not admissible. If the voting procedure is binary, this will leave him with just one admissible choice at the terminal division—that of A. Likewise, every other voter will have just one admissible choice, and that outcome which receives a majority of admissible choices (say, A) will be selected. This terminal division may thus be deleted and outcome A substituted in its place.

In similar fashion, all the other terminal divisions can be deleted and single outcomes substituted in their places. Deleting in turn the terminal divisions of the "pruned" outcome tree formed from the substitution of outcomes for divisions, the process can be continued until the first division is reached, where one admissible outcome will remain. This completes the proof.

We call any procedure *determinate* if it results in the selection of a single outcome if voting is sophisticated. Nonbinary procedures may or may not be determinate, depending on the preference scales of the voters.

The plurality procedure of Figure 2.3 is determinate for the preference scales of the three voters in our example: passers—(O, A, N); amenders—(A, O, N); defeaters—(N, A, O). To demonstrate this, recall that under this procedure, to avoid the possibility of a three-way tie, we assumed that the passers are most numerous, or at least have a tie-breaking vote. This fact is sufficient to ensure that their sincere strategy—vote for O—is straightforward: No other choice would be better if the other voters split their votes, worse if the other voters both chose a single outcome. Assuming that the passers adopt their straightforward strategy, the outcomes associated with all strategies of the amenders and defeaters are shown in Table 2.4.

The amenders have one primarily desirable strategy (vote for A), which is straightforward, whereas the defeaters have two primarily desirable strategies (vote for A, vote for N), both of which result in the solid boxed outcomes in Table 2.4. For the defeaters, however, only one strategy (vote for A) is secondarily desirable, which leaves outcome A (E) as the equilibrium choice of the voting body. Thus, the plurality procedure is determinate for the preference scales of the three voters in our example.

Suppose now that the preference scale of the amenders is (A, N, O) instead of (A, O, N), with the scales of the other two voters remaining the same as before [(O, A, N) for the passers, (N, A, O) for the defeaters]. The passers would still have a straightforward strategy, but the amenders would now have two primarily desirable strategies (vote for A, vote for N), instead of one (vote for A), as before. In this instance, the primarily desirable strategies of the amenders and defeaters cannot be further reduced, and the outcomes associated with these strategies include the three possible outcomes (O, A, and N) in the game (dashed boxed

TABLE 2.4 NINE OUTCOMES ASSOCIATED WITH THE DIFFERENT
STRATEGIES OF AMENDERS AND DEFEATERS UNDER PLURALITY
PROCEDURE OF FIGURE 2.3, GIVEN PASSERS ADOPT THEIR
STRAIGHTFORWARD STRATEGY (VOTE FOR O)

AMENDERS' STRATEGIES	DEFEATERS' STRATEGIES		
	Vote for O	Vote for A	Vote for N
Vote for O	O (E)	O	O
Vote for A	O	A (E)	O
Vote for N	O	O	N (E)[a]

[a] Equilibrium choice if preference scale of amenders is (A, N, O); see text for explanation.

outcomes in Table 2.4). Although two of these outcomes are now equilibrium choices, this confers no special status on them as far as the selection
of strategies by the voters is concerned. The amenders would most prefer
A, the defeaters N, but neither can avoid the possibility that their worst
outcome O will be selected if they are not able to communicate and must
choose strategies independently. Indeed, there is a fatal symmetry in this
set of choices identical to that faced by the players in the Battle of the
Sexes (see section 1.7): If both voters try for their most-preferred outcome,
both will end up with their worst outcome, O. Because this is only one
of the three possible outcomes at the intersection of the three voters'
primarily desirable strategies, the plurality procedure of Figure 2.3 is
obviously not determinate for the preference scales postulated above.

The "fatal symmetry" alluded to above is a paradox that can in principle
be resolved if the voters are able to communicate with each other and
coordinate their choices of strategies. Specifically, if the amenders and
defeaters are able to agree on the selection of either outcome A or outcome
N, then they can ensure its adoption by the choice of their appropriate
primarily desirable strategies (see Table 2.4 for the strategies that intersect
these outcomes). Whichever of these outcomes they settle on (given that
they can agree on one)—outcome A favors the amenders, outcome N the
defeaters—it is better for both voters than outcome O, which is the rueful
collective choice when both voters choose the strategy that offers them the
possibility of their best (as well as worst) outcomes.

2.9. COALITIONS AND INFORMATION

The example just discussed illustrates how a willingness to cooperate on
the part of some voters can lead to a better outcome for them—but for
the voters left out of the bargain, it can spell disaster. If the amenders and
defeaters settle on outcome N, for example, the passers will have to

acquiesce in the adoption of their worst outcome, despite the fact that they have a straightforward strategy. Moreover, because this outcome is an equilibrium choice, no voter—even the passers—can do anything to alter the situation, at least as an individual actor.

If the passers do not act alone, however, but instead propose to the amenders that they form a coalition and both vote for A, presumably the amenders would accept this offer because they (like the passers) would prefer outcome A to outcome N. This agreement would appear to resolve the paradox of fatal symmetry that the reduction of desirable strategies was not able to do: The equilibrium choice A will be preferred to N by a majority of voters (passers and amenders). Should members of the other two possible coalitions (passers and defeaters, amenders and defeaters) agree on an outcome other than A (O or N), at least one member of each coalition can be tempted to defect to the coalition of passers and amenders that proposes joint strategies that result in the adoption of A. Since there is no coalition that can block outcome A (i.e., whose members both would prefer some other outcome), it would appear to be the collective choice favored by a majority.[17]

Yet, this "resolution" of the paradox through the formation of coalitions assumes cooperation among voters. In contradistinction, sophisticated voting assumes no cooperation: Communication among voters that could lead to binding and enforceable agreements is not permitted. As we shall show later in this section, however, even if communication is allowed and voters can coordinate their choices of strategy, this may not be sufficient to ensure an enforceable agreement. In general, equilibrium choices may be vulnerable to subsets of members who form coalitions, which may in turn be vulnerable to other coalitions, and so forth.

It is possible to extend the concept of an equilibrium choice for individual members to an equilibrium choice for coalitions. Coalition equilibria would be invulnerable to subsets of voters of a specified size whose members can obtain a preferable outcome by agreeing to the joint selection of strategies. In the preceding example, outcome A is a coalition equilibrium of order two (i.e., not vulnerable to any subset of two voters)—as well as of order one—for whatever strategies are selected by the three voters that lead to outcome A, the members of no subsets of two voters can both obtain a preferable outcome by jointly altering their strategies.

In contrast, outcome N is an equilibrium of order one but not of order two. It is an equilibrium of order one since unilateral deviation by either the amenders or defeaters from the "vote for N" strategy always leads to outcome O, the worst outcome for each voter given the preference scales

---

17. For a formalization of the reasoning in this example in terms of the game-theoretic notion of "core," see Riker and Ordeshook, *Introduction to Positive Political Theory*, pp. 137–38.

postulated above. It is not an equilibrium of order two, since when both the passers and amenders select the strategy "vote for A," outcome A is the collective choice, which both voters prefer to outcome N.

Although communication may elevate some individual equilibrium choices to the status of higher-order equilibria—as is true for outcome A in the preceding example—it may also have the effect of destroying all equilibrium choices. Assume for purposes of illustration that our three voters have the following preference scales: passers—(O, A, N); amenders—(A, N, O); defeaters—(N, O, A). Under the successive procedure of Figure 2.1, outcome O will be the sophisticated equilibrium choice. Yet, if coalitions can form,

O is vulnerable to a coalition of amenders and defeaters that prefers N;

A is vulnerable to a coalition of passers and defeaters that prefers O;

N is vulnerable to a coalition of passers and amenders that prefers A.

Thus, there is no coalition equilibrium choice invulnerable to all challenges, which is an affliction that also affects vote trading, as we shall show in Chapter 4.

Sophisticated voting does not allow communication among voters for the purpose of coordinating strategies, but it does assume that voters have complete information on the preference rankings of all other voters. Obviously, this latter assumption is not satisfied in many real-life voting situations, which even in the case of binary voting procedures would seem to demolish any hope of predicting a single outcome. In fact, however, incomplete information may have no effect on the choice of outcomes and may even lead to a more precise specification of "best" strategies. To illustrate these strange effects of uncertainty, assume that our three voters have the preference scales that we originally postulated: passers—(O, A, N); amenders—(A, O, N); defeaters—(N, A, O). Assume now, however, that the defeaters have no information about the amenders' ranking of outcomes; we symbolize *their perception* of the amenders' preference scale—without commas—as (AON).

Under the successive procedure of Figure 2.5, we showed in section 2.8 that if there is complete information, the sophisticated outcome is A, and the defeaters have two secondarily desirable strategies associated with this outcome, (a') and (b'). But if the defeaters have no information about the preference scale of the amenders, then they cannot be sure that the amenders have a straightforward strategy, as was the case when information was assumed to be complete. Instead, all strategies are possible, and the defeaters must therefore assume the worst—that the amenders choose $\{O, N\}$ at the first division and O at the second division, which is the worst outcome for the defeaters. In this case, they could do nothing to prevent O's adoption at the second division since the passers would also vote for O.

But if in fact the amenders actually preferred N to O, the defeaters could ensure N's adoption by voting for it at the second division, so they should always vote for N at this division. Since the defeaters prefer A at the first division, they should vote for it because there is nothing more they can do to prevent their worst outcome, O, from being adopted. Consequently, they would have a single "best" sophisticated strategy, (a'), rather than the two sophisticated strategies, (a') and (b'), which they previously had when information was assumed to be complete.[18]

In this example, surprisingly, incomplete rather than complete information pares down the two secondarily desirable strategies of the defeaters, (a') and (b'), to a single best strategy. Moreover, the primarily (and secondarily) desirable strategies of the passers and amenders, (c') and (a'), respectively, remain unaffected since the sophisticated strategies of these voters allowed for the possibility that the defeaters would select either strategy (a') or (b').

We call voting *supersophisticated* if it reduces the sophisticated strategies of one or more voters to some smaller set when the assumption of complete information is relaxed. Supersophisticated strategies provide voters with insurance against uncertainty by allowing for the possible adoption of additional strategies on the part of the voters about whom their information on preference scales is incomplete. (The same strategies would be appropriate if there were information that these voters were truly indifferent—and presumably made choices randomly—with respect to some or all outcomes.) They also offer a safeguard against misinformation, whether it is intentionally distorted or not.[19] The precise

---

18. This procedure for finding "best" sophisticated strategies can be formalized, in the same way as that for finding sophisticated strategies, by making some additional assumptions. Specifically in the above case, one can establish that strategy (b') for the defeaters is better than any others by showing that, given that the passers choose their primarily desirable strategy (c'), strategy (b') for the defeaters is dominant for all possible strategy choices of the amenders—any one of which the defeaters must assume could be chosen, given their incomplete information about the amenders' preference scale—except strategy (d'). But while not dominated by strategy (b'), the defeaters' strategy (d') results in the social choice of outcome O, the defeaters' worst outcome—and strategy (b') does not—when the amenders choose either their strategy (a') or (d'); strategy (d') yields a better outcome for the defeaters (N rather than A) only when the amenders choose strategy (b'). Thus, since strategy (b') rules out the possibility that the defeaters' worst outcome will be selected in two contingencies in which strategy (d') ensures its selection, and strategy (d') does not rule out the selection of the defeaters' worst outcome under any circumstances that strategy (b') provides for it, strategy (b') would appear to be the preferable undominated strategy for the defeaters.

19. For an example of a voting situation in which a voter might profitably misrepresent his preference scale, see Kenneth J. Arrow, *Social Choice and Individual Values* (2d ed.; New Haven, Conn.: Yale University Press, 1963), n. 8, pp. 80–81. For other examples and references, see William H. Riker, "Voting and the Summation of Preferences: An Interpretive Bibliographical Review of Selected Developments during the Last Decade," *American Political Science Review*, 55 (Dec. 1961), pp. 904, 911; and R. Duncan Luce and Howard Raiffa, *Games and Decisions: A Critical Survey* (New York: John Wiley & Sons, 1957), pp. 357–63. Most of these alleged examples of "misrepresentation," however, simply refer to some profitable form of insincere (e.g., sophisticated) voting. They do not refer to voting

specification of information conditions related to supersophisticated voting strategies has not been investigated.

2.10. VOTING ON VOTING PROCEDURES

At the end of section 2.3 we showed that if voting were sincere, the order in which alternatives are voted upon under the successive procedure would result in different outcomes.[20] Specifically, under the successive procedure of Figure 2.1, the outcome selected was A, but under the successive procedure of Figure 2.5, the outcome selected was O. Given that each voter has complete information about the preference scales of all other voters, we suggested that each voter could anticipate the results of sincere voting under both procedures. Should a voter be given the opportunity to select which procedure to follow, he would choose the procedure that produced the better outcome, assuming that outcomes on other issues either did not matter or could not be anticipated.[21] We call voting *sincere-anticipatory* when voting on the substantive outcomes under each procedure is sincere, and voting on the procedures is based on an anticipation of the sincere outcomes produced by each procedure.

Because outcome A is preferred to outcome O by two of the three voters in our example (amenders and defeaters), we indicated that a majority would choose the procedure of Figure 2.1 over the procedure of Figure 2.5. The question we raised, but did not answer, in section 2.3 was: If there were *not* the opportunity to vote on procedures (i.e., the order in which alternatives would be voted upon), and the operative procedure were that of Figure 2.5, would the majority (amenders and defeaters) who do not favor the outcome selected by this procedure (O) have any recourse?

With a knowledge of sophisticated voting, we can now answer "yes" to this question. As we showed in section 2.7, the outcome under the successive procedure of Figure 2.5 would be A instead of O if voting were

---

where one announces in advance erroneous information about one's preference scale in order not only deliberately to mislead other voters but also to induce them to vote (perhaps insincerely) in a way that is profitable to oneself. This is a strategic calculation that we shall not consider here since our main interest is in showing how strategies are selected on the basis of the honest preferences of other voters and not how the outcomes implied by these strategies may in turn change voters' representations of their preferences. Cf. Tapas Majumdar, "Choice and Revealed Preference," *Econometrica*, 24 (Jan. 1956), pp. 71–73; Alan Gibbard, "Manipulation of Voting Schemes: A General Result," *Econometrica*, 41 (July 1973), pp. 587–601; and Richard Zeckhauser, "Voting Systems, Honest Preferences and Pareto Optimality," *American Political Science Review*, 67 (Sept. 1973), pp. 934–46.

20. For another example in which the procedure makes a difference, see T. C. Schelling, "What is Game Theory?" in *Contemporary Political Analysis*, ed. James C. Charlesworth (New York: Free Press, 1967), pp. 224–32.

21. When voting is sophisticated, Gerald H. Kramer has generalized Farquharson's results to cover sequential voting processes on more than one issue, given that each voter's utility function across the various issues is additive. Gerald H. Kramer, "Sophisticated Voting over Multidimensional Choice Spaces," *Journal of Mathematical Sociology*, 2 (July 1972), pp. 165–80.

sophisticated rather than sincere. This is the same outcome that would have been produced if there were a prior procedural vote and voting were sincere-anticipatory, given the majority's preference for the Figure 2.1 procedure (i.e., outcome *A*) over the Figure 2.5 procedure (i.e., outcome *O*). We generalize this result in the following theorem.

Procedure Theorem

For simple binary voting procedures of the kind illustrated earlier, if two different orderings in which outcomes are voted upon (Figures 2.1 and 2.5 in our example) produce two different outcomes under sincere voting, and these outcomes are anticipated on a procedural vote, then the outcome preferred on the procedural vote (outcome *A* in our example) could have been obtained under the "losing" ordering (Figure 2.5 in our example) if voting had been sophisticated.

Remark. In other words, if the would-be losers on a procedural vote (amenders and defeaters under Figure 2.5 procedure) recognize their disadvantageous position should they vote sincerely, they do not need the procedural vote to rectify it; the same outcome that would occur under sincere voting, preceded by a procedural vote, can be obtained immediately through sophisticated voting. Thus, a majority disadvantaged by a binary voting procedure when voting is sincere has a recourse: It can vote sophisticatedly and achieve the result favored by the majority, which is the same as that it would obtain by voting sincerely under a procedure that leads to the majority outcome.

The fact that sophisticated voting prevents the will of the majority from being thwarted is not inconsequential. (If there is no single majority, however, as in the case of the "cyclical majorities" discussed in section 2.11, the will of the majority will depend on the alternatives compared.) By providing a safeguard against arbitrary orderings of binary procedures that produce outcomes inimical to a majority when voting is sincere, sophisticated voting ensures the selection of a socially preferred outcome. Thus, the strategic calculations of sophisticated voting would seem to be defensible as not only optimal for individual voters but socially desirable as well, and so would voting systems that encourage the exchange of information about voters' preferences that makes sophisticated voting possible.

Given only the information that one binary procedure or ordering (say, that of Figure 2.1) produces an outcome different from that of another procedure or ordering (say, that of Figure 2.5), and the Figure 2.1 procedure is preferred to the Figure 2.5 procedure if voting is sincere-anticipatory, we shall now prove that the outcome of the preferred procedure under sincere voting is always the same as the outcome of the nonpreferred procedure under sophisticated voting. Our proof is independent of the preference scales of the voters.

Proof. Consider a "composite" binary voting procedure at whose first division a voter is given the opportunity to choose between two different binary procedures. Assume that voting under each of these two procedures is sincere, and under one procedure (the *A*-procedure, represented by Figure 2.1) outcome *A* is chosen and under the other procedure (the *O*-procedure, represented by Figure 2.5) outcome *O* is chosen.

Assume, as in our example, that the *A*-procedure defeats the *O*-procedure. Then we know that this is because a majority prefers outcome *A* to outcome *O*, which are the sincere outcomes of each procedure. We also know that since *O* is the outcome chosen under the *O*-procedure, it must be preferred to some third outcome *N* at the second division of this procedure (see Figure 2.5); otherwise, *N* would defeat *O* at this terminal division, and *O* could not be the outcome chosen under this procedure. Similarly, under the *A*-procedure (see Figure 2.1), since outcome *A* eventually emerges victorious under sincere voting, *A* must be preferred to *N* at the second division.

In summary, a majority prefers outcome *A* to outcome *O*, outcome *O* to outcome *N*, and outcome *A* to outcome *N*. This is the collective constraint on the individual preference scales of the voters that permits one procedure to be preferred to another if voting is sincere.

Under the *O*-procedure, assume that voting is sophisticated. If voting is sophisticated, we showed in the proof of the Binary Theorem (see section 2.8) that if one outcome is preferred by a majority at a terminal division, then the outcome not preferred is inadmissible. This division can then be deleted and the preferred outcome substituted in its place. Now since *O* is the preferred outcome at the second division of the *O*-procedure (see Figure 2.5), it can be substituted for this division; and since *A* is preferred to *O* at the first division, *O* is inadmissible. Thus, the outcome chosen under the *O*-procedure is *A* if voting is sophisticated.

Under the *A*-procedure (see Figure 2.1), assume that voting is sincere. Since outcome *A* is preferred to outcome *N*, *A* wins at the second division; and since *A* is preferred to both *O* and *N*, *A* also wins at the first division. Thus, the outcome chosen under the *A*-procedure is *A* if voting is sincere.

If voting is sincere-anticipatory, the composite binary procedure chooses the *A*-procedure over the *O*-procedure since outcome *A* is preferred by a majority to outcome *O*. If voting is sophisticated, however, the outcome of the *O*-procedure will also be *A*. Hence, the outcome of the preferred procedure under sincere voting is the same as the outcome of the nonpreferred procedure under sophisticated voting. This completes the proof.

The converse is not generally true: The outcome of sophisticated voting under the preferred procedure will not necessarily be the same as the outcome of sincere (or sophisticated) voting under the nonpreferred procedure. For example, assume that the preference scales of the three voters are as follows: passers—(*O, A, N*); amenders—(*A, N, O*); defeaters—(N, O, A). If voting is sincere, then the outcome under the successive procedure of Figure 2.1 is A and the outcome under the successive procedure of Figure 2.5 is N. Since A is preferred by a majority to N, on a procedural vote the Figure 2.1 procedure (i.e., outcome A) would be selected if voting were sincere-anticipatory.

From the Procedure Theorem, we know that outcome A would be selected under the (nonpreferred) Figure 2.5 procedure if voting were sophisticated. Under the (preferred) Figure 2.1 procedure, however, if voting were sophisticated, outcome O would be selected, which agrees with neither the sincere nor the sophisticated outcomes (N and A, respectively) under the nonpreferred Figure 2.5 procedure.

To summarize, when two different binary procedures lead to different sincere outcomes, only the sincere outcome under the preferred procedure will always match the sophisticated outcome under the nonpreferred procedure, whatever the preference scales of the voters consistent with the choice of the different outcomes are. On the other hand, if the sincere outcomes are the same under two different binary procedures, it follows by an argument similar to that given in the proof of the Procedure Theorem that the sophisticated outcomes under both procedures will match the sincere outcomes. Thus, sophisticated voting provides an alternative to sincere voting only if the sincere outcomes differ under two binary voting procedures.

The Procedure Theorem, whose proof is strictly applicable only to different orderings of the successive (i.e., disjoint) binary procedure, can be extended to different orderings of the amendment (i.e., overlapping) binary procedure; it would also appear to hold for more complicated binary procedures for choosing among more than three outcomes, but we shall not pursue this matter. Instead, it seems appropriate at this point to turn to some examples of parliamentary maneuvering in real voting bodies that illustrate some of the strategic notions of voting that we have so far discussed. Before doing so, however, we shall first describe the *paradox of voting*, which several of the empirical examples that we shall analyze concern.

2.11. THE PARADOX OF VOTING

The standard example of the paradox involves three voters, who rank a set of three alternatives (A, B, C) as shown in Table 2.5. The paradox arises from the fact that if every voter is assumed to have a *transitive* preference scale (i.e., if he prefers A to B and B to C, then he prefers A to C), the social ordering nevertheless is intransitive: A majority (voters 1 and 3) prefers A to B, a majority (voters 1 and 2) prefers B to C, but a majority (voters 2 and 3) prefers C to A rather than A to C. (Note that each alternative is ranked differently by every voter—first on one voter's preference scale, second on another's, and last on the third's—which is a general characteristic of preference scales that result in the paradox.)

TABLE 2.5 PARADOX OF VOTING

VOTER	PREFERENCE SCALE FOR ALTERNATIVES
1	(A, B, C)
2	(B, C, A)
3	$(C, A. B)$

This means that, given at least three alternatives, there may be no social choice that is decisive since every outcome that receives majority support in one comparison can be defeated by another majority in another comparison. For this reason, the majorities that prefer each outcome over some other in a series of pairwise comparisons are referred to as *cyclical majorities*, though the paradox is not dependent on a specific decision rule like simple majority, as Kenneth J. Arrow demonstrated over two decades ago in his famous General Possibility Theorem.[22]

Cyclical majorities may manifest themselves in other forms, including voting on party platforms.[23] Consider a voting procedure in which the voters do not have the opportunity to vote on single issues but instead must choose among candidates who take positions on two or more issues. For example, if there are two issues, assume that the alternatives (i.e., positions a candidate might take) on the first issue are A and A', and on the second are B and B'. If the preferences of three voters for the four possible platforms, comprising positions on the two issues, are as shown in Table 2.6, then there is no platform that can defeat all others in a series of pairwise comparisons. As shown by the arrows in Figure 2.8, which indicate majority preferences between pairs of platforms, the majorities are cyclical and social preferences are therefore intransitive.

What is particularly striking about this example is that if separate votes were taken on each of the two issues, A would be preferred to A' by voters 1 and 2 and B would be preferred to B' by voters 1 and 3 (compare the first preferences of the voters in Table 2.6). But despite the fact that majorities would prefer alternatives A and B were the issues voted on separately, platform $A'B'$ defeats platform AB since it is preferred by a

22. Arrow, *Social Choice and Individual Values*; the first edition of this classic work was published in 1951. For a game-theoretic formulation of the General Possibility Theorem in terms of voting in a cooperative game, see Robert Wilson, "The Game Theoretical Nature of Arrow's General Possibility Theorem," *Journal of Economic Theory*, 5 (Aug. 1972). pp. 14–20; see also Robert Wilson. "A Game-Theoretic Analysis of Social Choice " in *Social Choice*, ed. Bernhardt Lieberman (New York: Gordon and Breach, 1971), pp. 393–407.

23. Claude Hillinger, "Voting on Issues and on Platforms," *Behavioral Science*, 16 (Nov. 1971), pp. 564–66. See also Joseph B. Kadane, "On Division of the Question," *Public Choice*, 13 (Fall 1972), pp. 47–54, for an analysis of the effects of combining different alternatives.

TABLE 2.6 PLATFORM VOTING

VOTER	PREFERENCE SCALE FOR PLATFORMS HIGH PREFERENCE ⟶ LOW PREFERENCE			
1	*AB*	*AB'*	*A'B*	*A'B'*
2	*AB'*	*A'B'*	*AB*	*A'B*
3	*A'B*	*A'B'*	*AB*	*AB'*

FIGURE 2.8 CYCLICAL MAJORITIES FOR PLATFORM VOTING

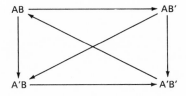

majority (voters 2 and 3). Thus, a platform whose alternatives, when considered separately, are both favored by a majority may be defeated by a platform containing alternatives that only minorities favor. A recognition that a majority platform may be constituted from minority positions is what Anthony Downs argued may make it rational for politicians to construct platforms that appeal to coalitions of minorities.[24]

The divergence between less-preferred individual alternatives and a more-preferred platform that combines them is not dependent on the existence of the paradox.[25] The paradox of voting in the preceding example, however, makes no platform invulnerable, which helps to explain the importance that politicans attach to anticipating an opponent's positions so that they can respond with a set that is more attractive to the voters. (In our previous terminology, if a strategy denotes positions on issues, no politician would have a straightforward strategy in the above example since no set of positions is unconditionally best.) Of course, many politicans try to avoid this problem by being intentionally vague about their positions in the first place, as Downs pointed out.[26]

24. Anthony Downs, *An Economic Theory of Democracy* (New York: Harper & Row, 1957), chap. 4.

25. For an example, see Hillinger, "Voting on Issues and Platforms," p. 565.

26. Downs, *Economic Theory of Democracy*, chaps. 8 and 9.

The paradox of voting demonstrates that a concept of rationality based on transitivity cannot be transferred directly from individuals to a collectivity via some decision rule like that of simple majority. Something queer happens on the way, something comes apart, that complicates the idea of a "social choice" based on this kind of internal consistency. In Chapter 4 we shall consider this problem further and one proposed means for avoiding the social incoherence of the paradox.

2.12. EMPIRICAL EXAMPLES

The first instance of the paradox that we shall examine concerns federal aid for school construction, which was considered by the United States House of Representatives in 1956 in terms of the following three alternatives:[27] (1) an original bill (O) for grants-in-aid for school construction; (2) the bill with the so-called Powell amendment (A), which provided that no federal money be spent in states that had segregated public schools (i.e., the southern states); and (3) no bill (N) (i.e., the status quo). As Riker reconstructed the situation, there were basically three groups of voters with the following preference scales:

1. The passers (mostly southern Democrats) favored school aid, but they were so repelled by the Powell amendment that they preferred no bill to the amended bill. Thus, their preference scale was (O, N, A).

2. The amenders (mostly northern Democrats) were pro-integration, but if the Powell amendment were not successful, would still prefer school aid to no action. Thus, their preference scale was (A, O, N).

3. The defeaters (mostly Republicans) preferred the status quo over either version of the bill, with the pro-integration bill apparently more palatable than the original bill. Thus, their preference scale was (N, A, O).

Using the amendment procedure of Figure 2.2, the roll-call votes on the Powell amendment, which passed, and on the amended bill, which failed, reveal that the northern Democrats needed (and got) the support of Republicans—whose first preference was defeat of any school-aid bill—on the Powell amendment. Riker's interpretation is that the Republicans, realizing that a majority probably favored the original bill, may have voted for the amendment so as to disattach southern Democratic support on the vote for final passage (of the amended bill).

Since the Republicans did not introduce the Powell amendment, however, it seems unfair to blame them for using it to defeat the school-aid bill. If there had been contrivance on their part, it would have arisen

27. William H. Riker, "Arrow's Theorem and Some Examples of the Paradox of Voting," in *Mathematical Applications in Political Science*, ed. John M. Claunch (Dallas: Arnold Foundation of Southern Methodist University, 1965), pp. 41–60.

from not voting sincerely to effect a better outcome. Yet in fact most Republicans preferred outcome *A* to *O*, which is consistent with their postulated preference scale.[28] Indeed, the sincere outcome under the amendment procedure of Figure 2.2 is *N*, which did in fact occur.

Since the sophisticated outcome under this procedure is *O*, it would seem that it was the northern Democrats, and not the Republicans, who had the opportunity to contrive an outcome that would have been pre-ferable for them over *N*—but they failed to exploit this opportunity. Specifically, had they voted for {*O*, *N*} instead of their sincere choice {*A*, *N*} on the amendment vote, {*O*, *N*} would have been selected. Then outcome *O* would necessarily have been the choice of the voting body on the final vote, since two of the three voters (southern Democrats and northern Democrats) would have preferred it to outcome *N*.

Given the preference scales of the different groups of voters, therefore, it was not contrivance on the part of the Republicans which sank the federal school-aid bill, but rather a failure to contrive (by voting sophisti-catedly) on the part of the northern Democrats. [Apparently, this possibility for contrivance was recognized by Congressman Dawson (an Illinois Democrat), who was alone among black Congressmen in voting against the Powell amendment.] In fairness to Riker, however, he does point out that if there were a paradox of voting, it may not have been a contrivance of the Republicans but "inherent in the distribution of opinion."[29] The fact that in 1957, a year after the Powell amendment succeeded (in effect, *A* defeated *O*) but final passage failed (*N* defeated *A*), the bill in its original form got majority support in the House (*O* defeated *N*) is convincing evidence that a paradox actually existed, since *O* defeats *N*, *N* defeats *A*, and *A* defeats *O*. But our analysis strongly suggests that the failure of either version of the school-aid bill to pass in 1956 was not a contrivance of the Republicans, at least in terms of the way in which they voted on the three alternatives. Furthermore, since they did not propose the amend-ment, it is hard to accuse them of splitting the Democratic opposition; the fault seems rather to lie with the Democrats themselves.

In his second example of the paradox, Riker presents more convincing evidence that an amendment was specifically proposed to divide the support of the voters favoring the original bill. In this case, which occurred

---

28. The proposition that the Republicans contrived the paradox by not voting directly according to their preferences could be supported only if their preference scale were (*N*, *O*, *A*) instead of (*N*, *A*, *O*)—that is, they actually preferred the original bill to the amended bill but voted nonetheless for the amended bill. Yet, if their preference scale were (*N*, *O*, *A*), then the outcome of both sincere and sophisticated voting would be *O*, which means that contrivance on their part would have been fruitless. Furthermore, not only is this outcome inconsistent with what actually happened, but given the preference scale (*N*, *O*, *A*) of the Republicans, the paradox of voting does not occur since *O* defeats *N*, *N* defeats *A*, but *O* defeats *A*. On the other hand, if the preference scale of the Republicans were (*N*, *A*, *O*), as postulated in the text, there is a paradox since *O* defeats *N*, *N* defeats *A*, and *A* defeats *O*, which is consistent with the facts in this case (see text).

29. Riker, "Arrow's Theorem and Some Examples of the Paradox of Voting," p. 56.

in the U.S. Senate in 1911, the original bill (O) was a proposal to amend the Constitution to provide for direct popular election of senators (which eventually became the Seventeenth Amendment). The amendment (A) provided that the guarantee against federal regulation of these elections in the original bill be deleted; by jeopardizing the ability of southern states to exclude blacks from the election process, it was specifically designed to antagonize senators from the South. The final alternative was the status quo, or no bill (N).

The preference scales of the southern Democrats, northern Democrats, and Republicans for $\{O, A, N\}$ duplicate those in the previous case, with the sincere outcome N once again triumphant due to the failure of the northern Democrats to vote sophisticatedly. If there were contrivance on the part of the Republicans in introducing an amendment that drove a wedge through Democratic support, there was certainly no contrivance on the part of the northern Democrats to thwart this strategy. Although the idealist might characterize the voting behavior of the northern Democrats in the two cases described earlier as "sincere," the realist might consider the epithet "naive" more fitting.[30] In cases like these, it is hard to believe that a knowledge of party positions, past voting, and so forth did not provide members with sufficient information on which to base predictions about outcomes—and, accordingly, vote sophisticatedly. Certainly in contemporary times, extensive debate and media coverage of controversial issues expose the preferences of many members. Perversely, however, this may inhibit sophisticated voting, for reasons given below.

When preferences are more fully exposed, members will vote sincerely if the sanctions from their party, constituents, and other groups for taking the "wrong" (i.e., insincere) public positions on roll-call votes override strategic considerations. Although the means used to achieve particular ends are irrelevant to our analysis, in principle the costs associated with the diabolical aspects of sophisticated voting could be incorporated in the valuation of the ends themselves (e.g., by distinguishing outcomes achieved by "devious" means from those achieved by "nondevious" means).

Another possible explanation for sincere voting might stem from the inability of members, who do have something to gain from sophisticated voting, to agree to coordinate their strategies. In situations wherein a party or coalition leader is skilled in parliamentary maneuver and persuasion, however, he may be able to convince a sufficiently large number of voters to take concerted action whose immediate effect is disadvantageous but whose expected ultimate effect is on balance beneficial. Such a person seems to have been Lyndon Johnson, who as majority leader of the U.S. Senate was instrumental in the following case.

---

30. When the possibility of sophisticated voting was pointed out to Jean-Charles de Borda (1733–99) under his method of vote counting, he replied, "My scheme is only intended for honest men." Duncan Black, *Theory of Committees and Elections* (Cambridge, England: Cambridge University Press, 1958), pp. 182, 238.

In 1955 members of the Senate faced the following three alternatives:[31] (1) an original bill (O) for an $18 billion highway program, which included the so-called Davis-Bacon clause that set up fair-pay standards for workers on federal construction projects; (2) the bill as amended (A), with the Davis-Bacon clause deleted; and (3) no bill (N) (the status quo). Since the Davis-Bacon clause was anathema to southern Democrats, the voters as before divided into three groups:

1. The northern Democrats, with preference scale (O, A, N).

2. The southern Democrats, with preference scale (A, N, O).

3. The Republicans, with preference scale (N, O, A).

Under the amendment procedure of Figure 2.2, the sincere outcome is N and the sophisticated outcome is A. The latter outcome was eventually adopted, for southern Democrats, led by Johnson, were able through parliamentary maneuvering to get the Davis-Bacon clause removed without a roll-call vote,[32] which is tantamount to getting northern Democrats to support them on {A, N} by voting sophisticatedly under the amendment procedure of Figure 2.2.[33] Then, on the vote pitting A against N, the northern Democrats joined the southern Democrats and supported A.

Farquharson quotes *Time* as reporting that the "Republicans flubbed their chance,"[34] presumably because their least-preferred outcome was adopted. Certainly this possibility exists since they do not have a straightforward strategy by the Separability Theorem: The top outcome for them of each subset at the first division (N) is ranked higher than the bottom outcome (A or O) of the other subset.

The situation is more complicated, however, if one assumes that the voters can communicate with each other for the purpose of coordinating joint strategies. Then A, the outcome adopted, is vulnerable to a coalition of northern Democrats and Republicans who support outcome O. Similarly, O is vulnerable to a coalition of southern Democrats and Republicans who support outcome N; and N is in turn vulnerable to a coalition of northern and southern Democrats who support outcome A. In other words, we have once again the paradox of voting since no coalition is an equilibrium choice of order two, and there is thus no outcome that cannot be defeated—including that which the Republicans most favored,

31. Farquharson, *Theory of Voting*, pp. 52–53.

32. *Congressional Quarterly Almanac*, vol. 11, 84th Congress, 1st Session (Washington, D.C.: Congressional Quarterly News Feature, 1955), p. 438.

33. Whereas contrivance of the paradox normally involves introduction by amendment of a third alternative, here an amendment was used in effect to suppress an alternative.

34. Farquharson, *Theory of Voting*, p. 53.

N. Thus, if the Republicans flubbed their chances, it was not because they had any surefire strategy for winning.

In the three instances of the paradox we have described, the sincere outcome was chosen twice and the sophisticated outcome once.[35] To a certain extent, these outcomes are an artifact of the voting procedure used. If the voting procedure had been that of the successive procedure of Figure 2.1, for example, both the sincere and sophisticated outcomes would have been different in the first two instances, though the same in the last instance.

What these examples share, however, whatever the sincere or sophisticated outcomes are, is that there is no decisive coalition. Given that groups of voters cooperate with each other, which seems a fairly reasonable assumption to make about members in both houses of Congress, some group of members in any coalition that forms can be tempted to defect and join another coalition. This means that the outcome that is supported by a particular coalition and eventually prevails is very much an artifact of the stage in the voting process when it forms. Assuming it takes some time to mount a challenge to a coalition after it has formed, it is evident that the later a coalition forms when there exists a paradox of voting, the more successful it will be in warding off such challenges.[36]

Paradoxes occur under some rather special conditions, and it is not entirely clear how frequently they crop up. Several theoretical studies have concluded that the greater the number of voters or alternatives— especially the latter—the greater the probability that the paradox will occur (i.e., no alternative will defeat all others), which is a relationship reinforced if voters do not exercise a common standard of judgment in deciding among alternatives.[37] Apart from its statistical frequency, the

---

35. For another example of a sophisticated outcome (from the Swedish riksdag), see Rustow, *The Politics of Compromise*, p. 194.

36. Duncan Black advances a similar argument that the later an alternative is introduced into the voting, the more likely it will be adopted (by a majority, but not necessarily a coalition whose members have agreed to cooperate). Black, *Theory of Committees and Elections*, pp. 39–45. But Farquharson shows that although this is true if voting is sincere, if voting is sophisticated an alternative fares better the earlier it is introduced. Farquharson, *Theory of Voting*, Appendix I, p. 62. See also Bernard Grofman, "Some Notes on Voting Schemes and the Will of the Majority," *Public Choice*, 7 (Fall 1969), pp. 65–80.

37. Richard G. Niemi and Herbert F. Weisberg, "A Mathematical Solution for the Probability of the Paradox of Voting," *Behavioral Science*, 13 (July 1968), pp. 317–23; Mark B. Garman and Morton I. Kamien, "The Paradox of Voting: Probability Calculations," *Behavioral Science*, 13 (July 1968), pp. 306–17; Richard G. Niemi, "Majority Decision-Making with Partial Unidimensionality," *American Political Science Review*, 63 (June 1969), pp. 488–97; Frank Demeyer and Charles Plott, "The Probability of a Cyclical Majority," *Econometrica*, 38 (March 1970), pp. 345–54; Robert M. May, "Some Mathematical Remarks on the Paradox of Voting," *Behavioral Science*, 16 (March 1971), pp. 143–51; Bernard Grofman, "A Note on Some Generalizations of the Paradox of Cyclical Majorities," *Public Choice*, 12 (Spring 1972), pp. 113–14; Peter C. Fishburn, "Voter Concordance, Simple Majorities, and Group Decision Methods," *Behavioral Science*, 18 (Sept. 1973), pp. 364–76; Peter C. Fishburn, "Single-Peaked Preferences and Probabilities of Cyclical Majorities," *Behavioral Science*, 19 (Jan. 1974), pp. 21–27; and Peter C. Fishburn, "Paradoxes of Voting," *American Political Science Review*, 68 (June 1974), pp. 537–46, which contains a discussion of several related "paradoxes" of voting.

more important and controversial an issue is, Riker has argued, the more likely some voters will be motivated to contrive a paradox that is not inherent in the distribution of preferences.[38]

The arbitrary social choices that may be foisted on members of a society because of the paradox have alarmed some analysts, though others have argued that it is of little moment.[39] For the purposes of our analysis it is only one—though perhaps the most striking—manifestation of voting situations vulnerable to strategic manipulation. In electoral systems, whose members often do not have the opportunities of sequential voting that members of committees and legislatures typically enjoy, there still may be the opportunity to alter a sincere outcome through sophisticated voting.

Consider, for example, the United States presidential election of 1912, in which there were three major parties/candidates. These alternatives, and

The effects of unequally weighted voters and conflict on the occurrence of the paradox are explored in Bo H. Bjurulf, "A Probabilistic Analysis of Voting Blocs and the Occurrence of the Paradox of Voting," in *Probability Models of Collective Decision Making*, ed. Richard G. Niemi and Herbert F. Weisberg (Columbus, Ohio: Charles E. Merrill Co., 1972), pp. 232–49; and the effects of uncertainty are analyzed in Kenneth A. Shepsle, "The Paradox of Voting and Uncertainty," in *Probability Models of Collective Decision Making*, pp. 252–70. For an elegant argument, based on an elementary geometric construction, that the paradox of voting is ubiquitous, see Howard Margolis, "What It Takes to Avoid Arrow's Paradox" (Paper delivered at the 1974 Annual Meeting of the American Political Science Association, Chicago, Aug. 30–Sept. 2); see also Charles R. Plott, "A Notion of Equilibrium and Its Possibility under Majority Rule," *American Economic Review*, 57 (Sept. 1967), pp. 787–806. Of related interest is Gerald H. Kramer's proof that in essentially quantitative kinds of social-choice situations—not the kind of "all-or-nothing" qualitative choice situations discussed in this chapter—cyclical majorities are ineradicable unless the (quantitative) preferences of voters are nearly coincidental. See Gerald H. Kramer, "On a Class of Equilibrium Conditions for Majority Rule," *Econometrica*, 41 (March 1973), pp. 285–97.

38. Riker, "Arrow's Theorem and Some Examples of the Paradox of Voting," p. 53. For accounts of paradoxes, some apparently contrived, that involved more than three alternatives and/or voters, see Michael J. Taylor, "Graph-Theoretical Approaches to the Theory of Social Choice," *Public Choice*, 4 (Spring 1968), n. 1, pp. 45–46, who analyzes voting in a committee of a university department; Richard G. Niemi, "The Occurrence of the Paradox of Voting in University Elections," *Public Choice*, 8 (Spring 1970), pp. 91–100, who analyzes voting in several elections held by a university faculty; John C. Blydenburgh, "The Closed Rule and the Paradox of Voting," *Journal of Politics*, 33 (Feb. 1971), pp. 57–71, who analyzes voting in the U.S. House of Representatives on amendments to the Revenue Acts of 1932 and 1938; and William H. Riker, "The Paradox of Voting and Congressional Rules for Voting on Amendments," *American Political Science Review*, 52 (June 1958), pp. 349–66, who analyzes voting on amendments to an appropriations bill for the Soil Conservation Service in the U.S. House of Representatives in 1952. In the last article, Riker suggests a procedural rule that would enable the House and Senate to discover the paradox and, at the prerogative of the presiding officer, prevent the arbitrary passage of a bill or resolution that happened to be successful only because of the order in which amendments were voted upon (up to four amending motions to a bill or resolution can be considered simultaneously in both houses of Congress); but see my comment on this rule in note 43 below. The most systematic investigations of the paradox in the Senate and House are Bruce D. Bowen, "Toward an Estimate of the Frequency of the Paradox of Voting in U.S. Senate Roll Call Votes," and Herbert F. Weisberg and Richard G. Niemi, "Probability Calculations for Cyclical Majorities in Congressional Voting," both in *Probability Models of Collective Decision Making*, pp. 181–203, 204–31.

39. For a penetrating assessment of different views, see Riker and Ordeshook, *Introduction to Positive Political Theory*, chap. 4.

the likely preference scales of most of the electorate favoring each, are:

1. Democratic Party/Woodrow Wilson—(Wilson, Taft, Roosevelt);

2. Republican Party/William Howard Taft—(Taft, Roosevelt, Wilson);

3. Progressive Party/Theodore Roosevelt—(Roosevelt, Taft, Wilson).

In this election, Wilson was conceded to be the leading candidate (he eventually got 42 percent of the popular vote) since Roosevelt, who had bolted the Republican Party to form the Progressive ("Bull Moose") Party after failing to get the Republican presidential nomination, was expected to divide the Republican vote (he eventually got 27 percent of the popular vote and Taft got 24 percent, with minor candidates splitting the remainder).

Because Democratic voters were most numerous, they had a straight-forward strategy (vote for Wilson) under the plurality procedure of Figure 2.3. Republican and Progressive voters each had two primarily desirable strategies (vote for Roosevelt, vote for Taft), which admits all three possible outcomes as sophisticated. (This example provides a concrete interpretation of the game discussed on pages 77–78, wherein outcome O corresponds to Wilson, outcome A to Taft, and outcome N to Roosevelt). If the Republican and Progressive voters could have combined on either Roosevelt or Taft, however, then they could have ensured the election of a candidate who, if not their favorite, was preferable to Wilson. (Carrying this reasoning one step further, if Wilson voters had suspected such a coalition might form that would support Roosevelt, then it would have been wise for them to have supported Taft over Roosevelt and, with the support of Taft voters, made Taft the coalition equilibrium choice, which provides an interpretation of the cooperative game discussed on pages 78–79.) But it was precisely because Taft and Roosevelt refused to cooperate, and their supporters apparently voted sincerely, that the election went to Wilson.

It is reasonable to suppose that sophisticated voting has become more prevalent in recent presidential elections because, with the advent of public opinion polls and the wide dissemination of poll results, more complete information has been available to the electorate. In the case of the 1948 United States presidential election, Anthony Downs notes that

> when the Progressive Party ran a candidate . . . , some voters who preferred the Progressive candidate to all others nevertheless voted for the Democratic candidate. They did so because they felt their favorite candidate had no chance at all, and the more people voted for him, the fewer would vote Democratic. If the Democratic vote fell low enough, then the Republicans—the least desirable group from the Progressive point of view—would win. Thus a vote for their favorite candidate ironically increased the probability that the one

they favored least would win. To avoid the latter outcome, they voted for the candidate ranking in the middle of their preference ordering.[40]

The fact that the Progressive candidate, Henry Wallace, got only 2 percent of the popular vote indicates that this irony was not acceptable to many Democratic voters who may have favored Wallace—at least as compared with the 50 percent who favored the Democratic candidate, Harry Truman.

More recently, George Wallace's candidacy in the 1968 presidential election appears to have suffered from the fact that he was given little chance to win. In the final two months of the campaign, from the beginning of September to Election Day at the beginning of November, polls indicate that he lost (to both major candidates) more than one-third of his early supporters, which would seem to indicate sophisticated voting on their part.[41]

In elections that involve more than two candidates, it is extremely dismaying for some people to see a strong "extremist" candidate triumph over two or more "moderate" candidates only because the moderates split the centrist vote. Although we may be quick to attribute this result to sincere voting, it is important to point out that such an outcome can arise from sophisticated voting as well. As we saw in the case of the 1912 presidential election, sophisticated voting alone could have produced any of the three outcomes, but cooperation was needed to ensure one in particular. If the moderates do not cooperate, even the sophisticated voter may be quite helpless to do anything.[42]

---

40. Anthony Downs, *Economic Theory of Democracy*, p. 47.

41. Richard M. Scammon and Ben J. Wattenberg, *The Real Majority* (New York: Berkley Medallion Books, 1971), p. 193. This case is a bit ambiguous because it appears that the early Wallace supporters who switched split into two groups on their second-choice preferences: some favored Humphrey, others Nixon. With the two major candidates neck-and-neck in the final days of the campaign, neither group therefore had a surefire (sophisticated) strategy for blocking the selection of his last-place choice. For another example of sophisticated voting under the plurality procedure, see Farquharson, *Theory of Voting*, pp. 40–42, on the Roman Senate; for a general theoretical treatment of plurality voting under risk, see Ralph Joyce, "Sophisticated Voting in Three-Candidate Contests: Simple-Plurality Voting Systems" (Paper delivered at the 1973 Annual Meeting of the Midwest Political Science Association, Chicago, May 3–5). For data on sophisticated voting in West German elections, see Stephen L. Fisher, "A Test of Anthony Downs' Wasted Vote Thesis: West German Evidence" (Paper delivered at the 1974 Annual Meeting of the Public Choice Society, New Haven, Conn., March 21–23); and for evidence of manipulative voting practices generally in ancient Greece and Rome, see E. S. Stavely, *Greek and Roman Voting and Elections* (Ithaca, N.Y.: Cornell University Press, 1972).

42. Outside the world of professional politics, it is interesting to observe that the world of letters is not immune from such strategic kinds of activities. In several formal and informal votes, the five-member committee that awarded the 1973 National Book Award for fiction could not, after apparently extensive sophisticated voting, decide upon a winner. (There is not sufficient information to reconstruct complete preference scales for the committee members.) Divided in the end on two books, the two factions agreed to cooperate by splitting the award between the two authors. See Eric Pace, "The National Book Award in Fiction: A Curious Case," *New York Times Book Review*, May 6, 1973, pp. 16–17.

In many elections today based on plurality voting, cooperation has been institutionalized in the form of a runoff election between the two top vote getters if no candidate in the plurality election receives a majority. This device forces supporters of the defeated candidates into agreement on either one of two alternatives, which prevents sophisticated voting in the runoff election. It is not eliminated entirely, however, but simply pushed back to the plurality election, complicating the strategic calculation somewhat since more than one choice is made in the two elections.

To sum up, it appears that members of not only small committees and medium-size legislatures, but large electorates as well, may engage in game-theoretic kinds of strategic voting. On some occasions there seems to be cooperation among voters (if only implicit), on others none. Our examples together suggest that one cannot always treat the choices of voters as a direct consequence of their preferences but must view them in the context of the preferences of other voters and relevant environmental constraints.

2.13. SUMMARY AND CONCLUSION

In this chapter we have looked at voting as a means for selecting alternatives that are in some sense socially preferred. We have shown that the social or collective choice that is arrived at will depend on the following environmental constraints: (1) the voting *procedure* used, including the order in which alternatives are voted upon; (2) the *information* available to individual voters about the preference scales of other voters; and (3) *communication* among voters, which allows them to coordinate their choices of strategies and form coalitions. An awareness only of the procedure is consistent with sincere voting, whereby individuals vote directly according to their preferences. Additional information about the preference scales of other voters is consistent with sophisticated voting (as well as sincere-anticipatory voting) in a noncooperative game, whereby individuals may not choose their most-preferred outcome at a particular division in order to prevent less-preferred outcomes from ultimately being selected. Finally, an ability to communicate with other voters is consistent with the selection of joint strategies in a cooperative game, which may yield still better outcomes than can be achieved if no cooperation is assumed.

The idea that these environmental constraints are consistent with different kinds of voting is fundamentally rooted to the concept of *rationality*. Having previously offered commonsensical reasons why voters would make certain choices, we shall indicate how these reasons constitute, in effect, different definitions of rational behavior.

At the level of environmental constraint (1), a voter is rational if he *chooses his most-preferred outcome*. In this case, we assume that a voter's knowledge of the social choice process extends only to the voting procedure, so he can do no better than select his most-preferred outcome at each division (i.e., vote sincerely).

At the level of environmental constraint (2), a voter is rational if he selects strategies, insofar as possible, that *block the adoption of his least-preferred outcomes, assuming that other voters will do likewise.* If a voter has knowledge of the preference scales of other voters, then he can anticipate what strategies they will *not* select, and thereby narrow down his own strategies to those which are most desirable. Since the choices of other voters may block the selection of a "most-preferred" outcome at any division, a voter's most sensible course, therefore, is to reduce the set of outcomes to those that are "not least preferred" by voting sophisticatedly. Supersophisticated voting, by relaxing the assumption of complete information, offers one criterion for choosing the "most-preferred" from the "not least-preferred" set, but information conditions associated with this concept have not been systematically developed.

At the level of environmental constraint (3), a voter is rational if he joins a coalition that can enforce an agreement, binding on its members, which *provides an outcome better than that which any other coalition can provide.* Clearly, if any member of a coalition can be tempted to defect to a coalition that can guarantee a better outcome, other voters will recognize this possibility and the first coalition will not form. In this case, as in the previous case, a voter chooses from outcomes that cannot be blocked, but now blockage can occur due to the selection of coordinated strategies by coalitions of voters.

In sum, the concept of rationality can be defined at various levels, depending on the environmental constraints assumed to be operative. At all levels the focus is exclusively on outcomes, which we have assumed are the only things valued by voters. Under environmental contraints (2) and (3), rationality is the force that fuels strategic calculations on the part of voters, who are linked to each other through a mutual perception of each other's rationality. Only when voters cannot perceive the effect of their individual choices on other voters' choices, and ultimately on the collective choice of the voting body—as when they are familiar only with the voting procedure—is a voting situation not a game.

Games may be cooperative or noncooperative, depending on whether voters are able to communicate and coordinate joint strategies that lead to enforceable outcomes. In noncooperative games, where information is complete but there is no communication among voters, the sophisticated outcome will depend on the voting procedure. We showed that binary procedures are always determinate, leading to single outcomes, but under other procedures sophisticated voting may not reduce the strategies of voters to those associated with only a single outcome. In cooperative games, where information is complete and there is communication among voters, the formation of coalitions does not depend on the voting procedure but only on the structure of voters' preference scales. This may render certain (equilibrium) outcomes incapable of being blocked, even by a coalition of voters. On the other hand, if these scales are such as to create a paradox of voting, binding and enforceable agreements are rendered nugatory, perhaps belying the name "cooperative."

Does the possible existence of the paradox doom any social-choice process to an inescapable arbitrariness? In one sense it does, because there is no procedure or type of voting so far considered that can surmount the arbitrariness of the paradox—except, perhaps, one that demonstrates its existence, but this does not lead to a recommended course of action.[43] Certainly the outcomes in our legislative examples of the paradox had nothing to justify them as the most-preferred alternatives; they were simply the products of sincere and sophisticated voting under a particular procedure. Whereas in two of the three legislative examples sincere voting resulted in maintenance of the status quo, in the third case a parliamentary maneuver that had the effect of sophisticated voting led to a result different from the status quo. It seems that a judgment on whether the strategies that produced these outcomes are clever contrivances or brilliant maneuvers must ultimately rest on a value judgment about the desirability of the outcomes themselves. In the absence of an unambiguous first choice of voters in these cases, it is hard to think of any other grounds for labeling one outcome as socially preferred.

When no paradox of voting exists, is there any best procedure or best way of voting? In the presidential election of 1912, voting was apparently sincere and Wilson, the leading vote-getter, won under the plurality procedure. (We can safely ignore the effects of the Electoral College—at least until Chapter 7—since the popular-vote winner in this election was the electoral-vote winner.) But given the postulated preference scales of voters in this example, if voting had been sophisticated, the outcome would have been indeterminate (i.e., any of the three candidates might have won). In a series of pairwise contests, however, Taft would have defeated the other two candidates, making him the so-called Condorcet winner; indeed, if coalitions could have been formed, Wilson supporters would have voted for Taft. Yet, if voters were sincere and there had been a runoff election between the two top vote getters (Wilson and Roosevelt), Roosevelt would have been elected.[44] What then, if anything, makes Wilson—the historical winner—the rightful winner?

From this example it is apparent that even if there exists no paradox, different combinations of voting procedures and types of voting can produce a rather rich menu of outcomes. For those who have a low tolerance for ambiguity, this richness may be quite unpalatable. It may, however, be an accurate reflection of the general incoherence of social-

---

43. Riker has recommended that further action in such cases be at the discretion (with restrictions) of the presiding officer, but this simply substitutes one person's arbitrariness for the arbitrariness of a fixed procedure for voting on alternatives. Riker, "The Paradox of Voting and Congressional Rules for Voting on Amendments," pp. 364–66. The value of exposing the paradox is questioned in Roger H. Marz, Thomas W. Casstevens, and Harold T. Casstevens II, "The Hunting of the Paradox," *Public Choice*, 15 (Summer 1973), pp. 97–102, where details on the number of votes required to reveal its existence are also given.

44. For a similar analysis of this election, see Riker and Ordeshook, *Introduction to Positive Political Theory*, n. 13, p. 98, who incorrectly identify Roosevelt as the Condorcet winner.

choice processes. As Riker and Ordeshook observe, there is a qualitative difference between individual and social decisions:

> Social decisions are not the same kind of thing as individual decisions, even though the former are constructed from the latter. As a consequence of that difference, social decisions are sometimes arbitrary in a way personal decisions are not: personal decisions follow from persons' tastes, but social decisions do not follow from the taste of society simply because it is never clear what the taste of society is.[45]

Our task in this chapter has been to try to show how different environmental constraints mediate the selection of outcomes for voters with given preference scales. Untangling the effect of these different constraints has enabled us to explicate different concepts of rationality that take account of the preferences of other voters and the resultant social preference implied by these individual preferences. The very existence of a coherent social preference or social choice, however, is called into doubt by the paradox of voting. The lack of social coherence suggested by the paradox helps to explain not only the lack of agreed-upon criteria for deciding what voting procedures and types of voting are "best" but also why strategies that exploit this incoherence were considered—if not always followed—by actors in several real-life voting situations.

---

45. Riker and Ordeshook, *Introduction to Positive Political Theory*, pp. 114–15.

3

QUANTITATIVE VOTING GAMES

3.1. INTRODUCTION

In this chapter we shall analyze the general properties of voting systems—and the game-theoretic implications of one in particular—that allow voters to express their intensities of preference for alternatives. In these systems, in which the outcome is sensitive not only to how individuals rank-order alternatives but also to the "distance" that separates the ranks, voting games take on a quantitative dimension. Unlike the games discussed in Chapter 2, quantitative voting games take account of the fact that in summing preferences, one person may very much prefer one alternative over another whereas another person may have only a slight preference for one alternative over another. Furthermore, the games we shall describe in this chapter do not limit either voters or the voting body to the selection of only one alternative.

Voting systems that allow individuals to express intensities of preference have been proposed as a means for alleviating the plight of minorities that may feel very intensely about an issue. By giving the members of such minorities the opportunity to indicate the strength of their feelings directly through voting, it has been argued that minorities can be more equitably represented.[1]

The concept of representation has been much discussed and debated in the literature of political science. We shall not try to review or evaluate the different meanings that have been given to this concept but instead shall focus on a specific voting scheme that allows the expression of minority (as well as majority) viewpoints in an elected body in rough proportion

This chapter is based largely on Steven J. Brams, "The APSA and Minority Representation," *PS*, 3 (Summer 1970), pp. 321–35; the permission of the American Political Science Association to adapt material from this article is gratefully acknowledged.

1. For a general discussion of this question in democratic theory, see Willmore Kendall and George W. Carey, "The 'Intensity' Problem and Democratic Theory," *American Political Science Review*, 62 (March 1968), pp. 5–24; Douglas W. Rae and Michael Taylor, *The Analysis of Political Cleavages* (New Haven, Conn.: Yale University Press, 1970), chap. 3; and Alvin Rabushka and Kenneth Shepsle, *Politics in Plural Societies: A Theory of Democratic Instability* (Columbus, Ohio: Charles E. Merrill Publishing Co., 1972), chaps. 2 and 3.

to the number of supporters of these viewpoints in a larger electorate. Although there are many different systems of proportional representation in use today—some brief references to these are given in subsequent footnotes—the system we shall analyze has manifest game-theoretic properties, which is the main reason why we have singled it out for detailed examination.[2]

From the perspective of minority and majority groups of voters, representation will be considered in terms of the operation of voting rules that allow the preferences of a group to be translated into the election of representatives who would be *capable* of expressing the group's preferences in an elected body. Whether these representatives, upon election, *should* faithfully attempt to reflect the views of their supporters (the "delegate model"), or instead exercise their own, independent judgments on matters before the body (the "free agent model"), is a question we shall not consider.[3]

We shall begin our analysis by postulating two abstract requirements that it seems desirable that a voting system of proportional representation should meet. After showing that the particular voting system we shall study meets these requirements, we shall then investigate the logical interrelatedness of the requirements. Next, the effects that the size of an elected body, and the information available to groups seeking representation on it, have on the selection of strategies will be analyzed and illustrated by several examples. Finally, the strategies used by players in two empirical cases will be discussed.

3.2. REQUIREMENTS FOR A VOTING SYSTEM OF PROPORTIONAL REPRESENTATION

A voting system based on proportional representation should provide:

1. Individual members of the electorate with the capability to express their *intensities* of preference (i.e., their degrees of preference measured on an interval scale) for particular candidates to an elected voting body;[4]

---

2. For an interesting analysis of another voting system that also possesses such properties, see Alvin K. Klevorick and Gerald H. Kramer, "Social Choice on Pollution Management: The Genossenschaften," *Journal of Public Economics,* 2 (1973), pp. 101–46.

3. On the role of the representative, see, among other sources, Heinz Eulau, "Changing Views of Representation," in *Contemporary Political Science: Toward Empirical Theory,* ed. Ithiel de Sola Pool (New York: McGraw-Hill Book Co., 1967), pp. 53–85; and *Representation,* ed. Hanna Fenichel Pitkin (New York: Atherton Press, 1969). For a brief summary of empirical studies on representative attitudes and behavior, see William H. Riker and Lloyd S. Shapley, "Weighted Voting: A Mathematical Analysis for Instrumental Judgments," in *Representation: Nomos X,* ed. J. Roland Pennock and John W. Chapman (New York: Atherton Press, 1968), p. 215.

4. This requirement is not satisfied by "list systems" of proportional representation in which the individual voter has no opportunity to express preferences for *particular* candidates

2. An "organized" group in the electorate with the capability to *ensure* its representation on an elected voting body in rough proportion to its numbers in the electorate.

Briefly, the justification for the first requirement is that no person should be forced to cast the same number of votes (typically, one) for candidates to an elected body for whom he may have a differential set of preferences (i.e., not feel indifferent toward). Neither should he be forced by the voting rules to give so many votes to his first-choice candidate, so many fewer to his second-choice candidate, and so on, as is true, for example, in the Borda scheme.[5]

The second requirement is designed to thwart the circumstances in which a bare majority (of, say, 51 percent of the electorate), or perhaps only a plurality, can swamp all opposition in an election. This situation occurs in voting systems that require members of the electorate to vote for candidates for *all* seats in an elected body, for under such systems the majority (or plurality) will always prevail: It will be able to win all elected positions and minorities will therefore be unrepresented. Even in plurality-majority systems where representatives are elected from different (usually single-member) constituencies—in some of which, presumably, minority viewpoints are dominant—the empirical evidence supporting the "cube law" gives a large group with supporters distributed across all constituencies an advantage in winning seats in an elected body far out of proportion to its aggregate size in the electorate. If the majority has a two-to-one edge over a minority, for example, it will tend to have an eight-

---

of a party. Allowing the expression of intensities in a one-shot affair (e.g., election) is equivalent to what Buchanan and Tullock call "logrolling"—wherein a voter in a one-man, one-vote situation trades his votes on measures toward which his preferences are weak for those of other voters on measures toward which his preferences are strong—over a continuing sequence of votes. See James M. Buchanan and Gordon Tullock, *The Calculus of Consent: Logical Foundations of Constitutional Democracy* (Ann Arbor, Mich.: University of Michigan Press, 1962), chap. 10; in Chapter 4, some game-theoretic implications of logrolling will be developed in detail. It also may be argued that nonparticipation in an election is one means, albeit a very limited one, an individual has for expressing his intensities of preference—or lack thereof.

5. For a discussion and critique, see Duncan Black, *The Theory of Committees and Elections* (Cambridge, England: Cambridge University Press, 1958), pp. 59–66, 157–59, 180–83; also, Alfred De Grazia, "Mathematical Derivation of an Election System," *ISIS*, 44 (June 1953), pp. 42–51. The Hare system involving the single transferable vote also requires that the voter rank all candidates; if his first-choice candidate does not receive a "quota," the lowest-ranking candidate is dropped, and his second choices are allocated to the remaining candidates, with the process terminating when all seats are filled by candidates who have received a quota. Thus, second choices may be used to express intensities of preference in a limited way, but only for those voting for candidates dropped because they rank lowest in first choices.

to-one edge in an elected body whose representatives are chosen by the separate constituencies of the electorate.[6]

3.3. RULES THAT MEET THE REQUIREMENTS

What voting rules meet the two requirements and obviate the difficulties we discussed in section 3.2? We shall propose a set of rules with a specific example, show that they satisfy the requirements, and then demonstrate how the requirements themselves are interdependent.

Assume there is a single minority position among the electorate favored by one-third of the voters and a majority position favored by the remaining two-thirds. If there are 300 voters, for example, assume that 100 hold a minority viewpoint against the majority viewpoint supported by 200. Assume further that a simple majority is required to elect a 6-member governing body from this 300-member electorate. Now if the majority puts up one slate of six, and the minority another slate of six, and each voter casts six votes, one for each of six elected positions, the two-thirds majority slate will win all six seats and the one-third majority none. If, however, we allow each voter to allocate his six votes among one to six candidates, he would then be able to express his intensities of preference for particular candidates, which satisfies the first requirement. At one extreme he might cast his six votes for one candidate, or at the other extreme he might cast his six votes for six candidates. Such a system of voting in which voters in the electorate in multimember constituencies are allowed to cumulate their votes on fewer than the number of candidates to be elected is called *cumulative voting*.

(It would be a practical question whether a voter should be allowed to cast fractional votes—for example, $1\frac{1}{2}$ votes for each of four candidates. The answer to this question depends on how finely one believes an individual should be allowed to quantify his preference intensities. One rule of thumb might be that fractional allocations would be disallowed except when an individual desires to distribute his votes *equally* among a set of candidates whose number does not exactly divide his total number of votes—as in the above example in which four chosen candidates do not exactly divide an individual's six votes.)

Such a voting scheme, which allows members to allocate their votes according to the intensities of their *individual* preferences, also meets the second requirement—but only when a group is capable of *collectively*

6. The theoretical generality and empirical applicability of this law have recently been challenged. See Edward R. Tufte, "The Relationship between Seats and Votes in Two-Party Systems," *American Political Science Review*, 67 (June 1973), pp. 540–54; David Sankoff and Koula Mellos, "The Swing Ratio and Game Theory," *American Political Science Review*, 66 (June 1972), pp. 551–54; David Sankoff, "Party Strategy and the Relationship between Seats and Votes" (Paper delivered at the 1973 Annual Meeting of the Public Choice Society, College Park, Md.); and Rein Taagepera, "Seats and Votes: A Generalization of the Cube Law of Elections," *Social Science Research*, 2 (Sept. 1973), pp. 257–75.

disciplining its members on how to cast their votes. This is why we used the term *"organized" group* in stating the second requirement: If the members of the group do not follow the (presumably intelligent) voting instructions of their leaders (to be described below), then they will *not* be able to ensure their proportional representation.

To carry further our example in which each voter has six votes to cast for as many as six candidates, if each of the 100 voters in the minority casts three votes each for only two candidates, these voters could ensure the election of these two candidates no matter what strategy the 200 voters in the majority follow. For each of these two minority candidates would get a total of 300 (100 × 3) votes, and the two-thirds majority, with a total of 1,200 (200 × 6) votes to allocate, could at best match this sum for its choices by instructing its supporters to distribute their votes equally among four candidates (1,200/4 = 300). If the two-thirds majority instructed its supporters to distribute their votes equally among five candidates (1,200/5 = 240), it would not match the vote totals of the two minority candidates (300) but could still ensure the election of four (of its five) candidates—and possibly get its fifth candidate elected if the minority put up three candidates and instructed its supporters to distribute their votes equally among the three (giving each 600/3 = 200 votes).

Against these strategies of either the majority (support five candidates) or the minority (support two candidates), it is easy to show that neither side could improve its position. To elect five (instead of four) candidates with 301 votes each, the majority would need 1,505, instead of 1,200, votes, holding constant the 600 votes of the minority; similarly, for the minority to elect three (instead of two) candidates with 241 votes each, it would need 723 instead of 600 votes, holding constant the 1,200 votes of the majority.

It is evident that the optimal strategy for the leaders of both the majority and minority is to instruct their members to allocate their votes as equally as possible among a certain number of candidates. And that number of candidates which they support for the elected body should be proportionally about equal to the number of their supporters in the electorate (if known). Any deviation from this strategy—for example, by putting up a full slate of candidates and not instructing supporters to vote for only some on this slate—offers the other side an opportunity to capture more than its proportional "share" of the seats. [7]

To illustrate this point, if the majority puts up a full slate of six candidates in our example and is not careful to instruct its supporters to cast one vote apiece for each of the six, then instead of the 1,200 votes being equally distributed among the six (giving each of the six 200 votes), some

7. This was apparently first recognized by James Garth Marshall in a pamphlet entitled "Majorities and Minorities: Their Relative Rights" (1853). See Duncan Black, "Lewis Carroll and the Theory of Games," *American Economic Review: Papers and Proceedings of the Eighty-first Annual Meeting of the AEA*, 59 (May 1969), p. 207.

of the six (say, three) will have more than 200 votes and some (also three) less. Now the minority could counter with a strategy of supporting three candidates (instead of its proportional "share" of two), and if its supporters followed closely their leaders' instructions for distributing their 600 votes evenly among their three candidates so as to give each 200 votes, these minority candidates would end up with fewer votes than the top three majority candidates (who we assumed got more than 200 votes each), but more than the bottom three majority candidates (who we assumed got less than 200 votes each). By pursuing an informed counterstrategy, in other words, the minority could end up with representation equal to that of the majority in the elected body (i.e., three members) even though out-numbered two to one in the electorate.

This example is not meant to be advice to a minority on how to stage a take-over, but rather an illustration of the principle that cumulative voting gives a premium to "organization"—either on the part of the majority or minority (or minorities)[8] By "organization" we mean that if each side knows its approximate number of supporters, it can get them to cast their votes as it prescribes—and this prescription should always be to distribute their votes as evenly as possible among a certain number of candidates. Against a rational opponent, the number of candidates each side should support should be about equal to its proportional strength in the electorate in order to ensure its approximate proportional represen-tation. It is sensible to support more than this number only when there is reason to believe that the other side, as in the preceding example, will not pursue an optimal strategy—or the "other side" comprises several opponents not acting in concert.

In the case of the example we have been discussing, the optimal strategies are also applicable when each voter is restricted to casting one vote for one candidate. (This restriction would, in effect, limit the ex-pression of intensities of each voter to one candidate.) In this case, the optimal strategies for each side would be to instruct some of its supporters to vote for certain of its candidates and some of its supporters for others, with the objective still being to produce as even a distribution of votes as possible among the candidates chosen for support. Clearly, these in-structions specifying which supporters should vote for which candidates would have to be more explicit than those given to all voters in the previous example to distribute their six votes equally among all those candidates chosen for support.

3.4. INTERDEPENDENCE OF THE REQUIREMENTS

It is important to point out that the two requirements we have postulated are inconsistent when an *individual's* intensities of preference are not

---

8. Because only a small minority can assure itself of the election of at least one seat, one possible effect of cumulative voting may be to encourage the formation of several minority, or splinter, factions, whose influence would perhaps be less a function of their numerical size than their holding the balance of power between major factions in an elected body.

compatible with the optimal strategies of organized *groups* supporting candidates for elected positions. But if we restrict a voter to giving to a complete slate of candidates only one vote each (i.e., not allow him to express his intensities of preference by allocating more votes to some candidates than to others and thereby not satisfying the first requirement), the optimal strategies for ensuring proportional representation cannot in general be pursued (the second requirement is not satisfied).

This is so because a restriction on the expression of intensities (first requirement not satisfied) has the effect of ensuring that some voters will not be represented (not get their candidates elected), as Duncan Black (following Lewis Carroll) has shown.[9] But if some voters are not represented, the group of which they are a part cannot achieve proportional representation (second requirement not satisfied). By converting this contrapositive statement between the two requirements into an equivalent conditional statement, we can say that the second requirement in general implies the first requirement.[10] In other words, the second requirement is in general a sufficient condition for the first requirement, and the first requirement is in general a necessary condition for the second.[11]

Thus, the two requirements are not logically independent: If the second requirement, then in general the first. On the other hand, we previously showed that allowing *individual* intensities to be registered (first requirement satisfied), subject to following an optimal *group* strategy, ensures proportional representation of the group (second requirement satisfied) and, by extension, all of its members. But as the qualifying phrase "subject to following an optimal *group* strategy" implies, satisfying the first requirement for individuals may not be compatible with satisfying the second requirement for groups—that is, the interests of an individual may not coincide with the interests of any group in the electorate. Yet, since we cannot dispense with the first requirement without forfeiting the second, we have no choice but to retain the first and accept this possible incompatibility if we want groups to be able to ensure their approximate proportional representation on an elected body. How "approximate" will

---

9. See Duncan Black, "The Central Argument in Lewis Carroll's *The Principles of Parliamentary Representation*," *Papers on Non-Market Decision Making* (now *Public Choice*), 3 (Fall 1967), pp. 1–17; and Black, "Lewis Carroll and the Theory of Games," pp. 206–10.

10. See John G. Kemeny, J. Laurie Snell, and Gerald L. Thompson, *Introduction to Finite Mathematics*, 2d ed. (Englewood Cliffs, N.J.: Prentice-Hall, 1966), pp. 41–42, for definitions of contrapositive and conditional statements.

11. We say "in general," because under special circumstances voting rules that do not allow for the full expression of intensities may still be able to ensure proportional representation. As we indicated in the text in the case of our example, if each voter can cast one vote for one candidate, the majority and minority can instruct their supporters to distribute their votes equally among five and two candidates, respectively, and the majority can ensure the election of four of its five candidates and the minority both of its two candidates. The minority could not accomplish this feat, however, if each voter had, say, three votes and could give no more than one vote to a candidate, for the minority could give at most $\frac{300}{3} = 100$ votes to each of (three) candidates, but the majority could give $\frac{600}{5} = 120$ votes to each of five candidates and thus capture five of the six seats.

be a function of the number of representatives, as we shall show in the next section.

3.5. THE EFFECTS OF SIZE

In order to illustrate with numerical examples some effects which the size of the elected body has on the strategies available to the minority and majority under cumulative voting, we start by asking: If the minority wishes to assure itself of one-third representation on the body, and the majority two-thirds representation, what is the minimal support each group must receive from the electorate in order to achieve these goals?

First consider an elected body of three members. If the majority wishes to assure itself of two of the three seats, it can do so only if over 50 percent of the electorate allocate their three votes to two majority candidates. For no matter what strategy the minority pursues, with less than 50 percent of the remaining votes it cannot give two candidates more than the $\frac{50}{2} = 25$ percent of the votes received by the two majority candidates. It can at best ensure the election of only one candidate with less than 50 percent support from the electorate.

In fact, it is necessary only that the minority give one candidate more than 25 percent of the total number of votes to assure itself of one seat, because the majority with less than 75 percent of the total number of remaining votes cannot give each of three majority candidates the greater than 25 percent received by the one minority candidate. In a three-member body, therefore, the majority can assure itself of two of the three seats if more than 50 percent (say, 51 percent) of the voters allocate their three votes equally to two majority candidates, and the minority can assure itself of one of the three seats if more than 25 percent (say, 26 percent) of the voters cast their three votes for one minority candidate.

The majority can assure itself of more than two seats only if more than 75 percent (say, 76 percent) of the voters allocate their three votes equally among three majority candidates. In this case the minority, with less than 25 percent of the votes, cannot match the more than $\frac{75}{3} = 25$ percent of the votes the majority is able to distribute among its three candidates. Put another way, if fewer than 25 percent (say, 24 percent) of the voters support a minority candidate, that candidate will not be elected if the majority pursues an optimal strategy of instructing its supporters to divide their three votes equally among three candidates. In other words, as many as 24 percent of the voters may be *unrepresented* in a three-member body if a majority of more than 75 percent pursues an optimal strategy.[12] In

---

12. This figure is a maximum value, in contradistinction to Black's (Carroll's) "mathematical expectation of the percentage of the voters represented" (or unrepresented), which is a function of both the percentages of the electorate unrepresented (based on the difference between the minimum percentages necessary to elect $n - 1$ versus n members, for all n seats on the body) and the probability distribution of group preferences. See Black, "The Central Argument in Lewis Carroll's *The Principles of Parliamentary Representation*," pp. 10ff.

TABLE 3.1   REPRESENTATION FIGURES FOR ELECTED BODIES

NO. OF MEMBERS IN BODY	MINIMAL % OF ELECTORATE NECESSARY TO ENSURE REPRESENTATION OF			MAXIMAL % OF ELECTORATE THAT CAN BE UNREPRESENTED	MAXIMAL % OVERREPRE-SENTATION OF THOSE VOTING FOR SUCCESSFUL CANDIDATES
	$\frac{1}{3}$	$\frac{2}{3}$	$\frac{3}{3}$		
3	26	51	76	24	32
6	29	58	86	14	16
9	31	61	91	9	10
12	31	62	93	7	8
30	33	65	97	3	3
99	34	67	100	0	0

still different terms, the maximum number of voters that can be unrepresented (24 percent), as a proportion of the minimum number that can assure itself of all seats (76 percent), may be viewed as the maximum amount by which the represented voters are *overrepresented*, which in this case is 0.32. In a three-member body, therefore, there may be as much as a 32 percent inflation of the votes of members of the electorate who support successful candidates. Furthermore, these successful candidates need not be those supported by the postulated 76 percent majority: If a 50+ percent majority elects two candidates, and a 25+ percent minority elects one candidate, another (smaller) 25− percent minority (or minorities) will go unrepresented—and so contribute to the overrepresentation of the portion of the electorate that casts votes for successful candidates.

Of course, even more might be unrepresented—and fewer overrepresented to a greater degree—if the three successful candidates receive less than a total of 75 percent of all votes cast due to the existence of more than two competing slates. For the purposes of the discussion that follows, the "minimal" and "maximal" figures given in Table 3.1 are based on the assumption that there is a single majority and a single minority operating under the rules of cumulative voting.

The figures we calculated for a three-member body are given in the top row of Table 3.1. Applying the same reasoning as described previously to elected bodies of 6, 9, 12, 30, and 99 members (all these figures are exactly divisible by three in order that we could obtain integer values for $\frac{1}{3}$, $\frac{2}{3}$, and $\frac{3}{3}$ representation in these bodies), we have calculated for these larger bodies the minimal percentages of the electorate necessary to ensure representation of a $\frac{1}{3}$ minority and of $\frac{2}{3}$ and $\frac{3}{3}$ majorities, as well as the maximal percentage of the electorate that can be *unrepresented* and the maximal percentage by which the votes of those electing successful candidates can be *overrepresented.*

The minimal percentages clearly reveal that as the elected body gets larger and larger, both the minority and majorities need the support of a greater percentage of the electorate in order to ensure their representation at a particular level. When the elected body doubles in size from three to six members, for example, a majority needs the support of 58 percent of the electorate (instead of the previous 51 percent) in order to assure itself of two-thirds representation with four out of six seats; and when the size of the body doubles again to twelve members, the majority needs 61 percent support from the electorate to guarantee the same two-thirds advantage with eight out of twelve seats.

The electoral support that a group needs to ensure the same proportion of seats rises because the group's votes must be distributed evenly over more seats as the body increases in size. This means that each seat is necessarily "held" by proportionally fewer votes, so an opposition group needs proportionally fewer supporters to capture an additional seat. Because each of the seats which give a group a particular proportion are therefore more vulnerable in a larger body, the group needs a proportionally greater *total* number of votes to ensure its representation at a particular level in a larger body.

It is evident from the figures given in Table 3.1 that the minimal percentages necessary for a group to assure itself of one-third, two-thirds, and three-thirds representation approach the values of $\frac{1}{3}$, $\frac{2}{3}$, and $\frac{3}{3}$ as the body increases in size—but less rapidly as the body gets larger. When the body increases from three to six members, for example, the minimal percentage necessary to ensure a two-thirds majority increases by 7 points (from 51 to 58 percent); from six to nine members, the percentage rises 3 points (from 58 to 61 percent), and from nine to twelve members only 1 point (from 61 to 62 percent). The fact that this *marginal* increase in "representativeness" of the body decreases as more members are added to the body can also be seen from the figures in Table 3.1 giving the maximal percentage of the electorate that can be unrepresented.[13] These percentages are the complements of the minimal percentages of the electorate necessary to ensure a majority of three-thirds representation (all seats) on the elected body. As the size of the body doubles from three to six members, the maximal size of the minority (or minorities) that can be unrepresented decreases from 24 percent to 14 percent (a 10 percent drop), and from 14 to 7 percent (a 7 percent drop) when the body again doubles in size from six to twelve members. Similarly, the overrepresentation figures decrease most rapidly when the size of the body increases from three to six to nine members, and less rapidly when the size of the

---

13. Cf. these theoretical results with Douglas W. Rae's similar empirical findings for Western democracies. Douglas W. Rae, *The Political Consequences of Electoral Laws* (New Haven, Conn.: Yale University Press, 1967), pp. 117ff.

body increases from twelve to thirty to ninety-nine members. Inequities caused by unrepresentativeness are therefore attenuated by the greatest amount as bodies go from small (say, three to six members) to moderate (say, nine to twelve members) size, and by lesser amounts as they become still larger.

We conclude that exact proportional representation can always be approximated to a finer and finer degree by increasing the total number of seats (and votes of an individual) up to the number of voters in the electorate. At this extreme, exact proportional representation of a group—or even an individual—can be achieved because the elected body is as large as the electorate! In practical situations involving a decision about how large an elected body should be, however, gains in "representativeness" would have to be weighed against the possible sacrifices in the cost and unwieldiness of a larger voting body. As we have seen, the "representativeness" of a body, which increases at a decreasing rate with more members, is quite high for bodies of even moderate size. Moreover, since groups (or parties) often have only a rough idea of their degree of support among the electorate prior to an election—and therefore cannot make very exact calculations for the purpose of pursuing an optimal strategy, anyway— it would appear that the number of seats (and votes of an individual) need not be unduly large to be responsive to the cleavages that divide the electorate into different groups.

3.6. INFORMATION AND THE CHOICE OF STRATEGIES

In section 3.3 we indicated by way of example that in a six-candidate race, in which each voter has six votes, the optimal strategy of a two-thirds majority would be to support five candidates, the optimal strategy of the one-third minority would be to support two candidates. If the majority supported only four candidates, however, it would not decrease its payoff of four seats, given that the minority chooses its optimal strategy. What makes the optimal strategy of the majority (support five candidates) superior to the strategy of supporting only four candidates—which yields the same payoff of four seats when the minority pursues its optimal strategy of supporting two candidates—is that the optimal strategy could provide the majority with a fifth seat if the minority pursued a nonoptimal strategy of supporting three (instead of two) candidates. In game-theoretic terms, the strategy of supporting five candidates *dominates* the strategy of supporting four candidates for the majority (see section 1.4), because the former is at least as good as the latter if the minority chooses its optimal strategy, possibly better if it chooses some other strategy.

Dominant strategies may not be superior if information is incomplete. For although a dominant strategy may allow one to exploit an opponent's failure to choose his optimal strategy—and thereby offer one the chance of capturing an additional seat—if a group's estimate of its electoral strength falls below expectations, this strategy may not only fail to give it a chance at an extra seat but also may cost it a seat that could have been ensured by choosing a more conservative strategy.

To illustrate this point, consider again the example in which a two-thirds majority can assure itself of four out of six seats on an elected body by supporting five candidates. Now let us assume that the majority over-estimated its strength and in reality only 59 (instead of 67) percent of the electorate followed its instructions to distribute their votes equally among its five candidates. Then each of these candidates would receive $\frac{59}{5}$ = 11.8 percent of all votes cast.

On the other hand, if the minority had supported three candidates, each would have received $\frac{41}{3}$ = 13.6 percent of all votes and won three seats on the elected body, with the remaining three seats going to three of the five majority candidates with 11.8 percent of the votes each. Thus, the 59 percent majority would have ended up with only 50 percent representation. However, if the majority had chosen the "safer" strategy of supporting four (instead of five) candidates, each would have received $\frac{59}{4}$ = 14.8 percent of all votes (versus 13.8 percent for the three minority candidates), and the majority would have won four seats (instead of three), which would have given it two-thirds representation.

Obviously, then, the confidence a group has in its estimate of its electoral strength and the likely strategy of its opponent should play a part in its choice of a strategy. Although we shall suggest no formal scheme for incorporating subjective probabilities into such an estimate,[14] we give in Table 3.2 the minimal percentages at which one can pursue a *safe* strategy of ensuring one-third and two-thirds representation, pursue a *dominant* strategy of capturing an extra seat without sacrificing an assured one-third or two-thirds representation, and pursue a *new-safe* strategy that ensures the one-third and two-thirds representation plus the extra seat.[15] For example, from Table 3.2 we see that a group in a 6-member body with

---

14. For a procedure for incorporating subjective probabilities in decision making under uncertainty, using Bayesian methods, see Howard Raiffa, *Decision Analysis: Introductory Lectures on Choices under Uncertainty* (Reading, Mass.: Addison-Wesley, 1968); Bruce W. Morgan, *An Introduction to Bayesian Statistical Decision Processes* (Englewood Cliffs, N.J.: Prentice-Hall, 1968); and at an advanced level, with direct relevance to game-theoretic strategies, J. C. Harsanyi, "Games with Incomplete Information Played by 'Bayesian' Players," Part I, *Management Science*, 14 (Nov. 1967), pp. 159–89; Part II, *Management Science*, 14 (Jan. 1968), pp. 320–34; Part III, *Management Science*, 14 (March 1968), pp. 486–502.

15. In general, a group supported by n voters (out of N in the electorate) can pursue a safe strategy that ensures the election of k candidates for m seats ($k \leq m$) whenever $n/k > (N - n)/(n - k + 1)$; a group can pursue a dominant strategy of running ($k + 1$) candidates [$(k + 1) \leq m$], and ensure the election of k, whenever $n/(k + 1) > (N - n)/(m - k + 1)$; and a group can pursue a new-safe strategy that ensures the election of ($k + 1$) candidates [$(k + 1) \leq m$] whenever $n/(k + 1) > (N - n)/(m - k)$. For a given n, the maximum value of k which satisfies each of these inequalities determines the optimal strategy of a group, except when the *same* maximum value satisfies the inequalities for both the safe and dominant strategies; in this case, the dominant strategy of running an additional candidate—i.e., ($k + 1$) instead of k—is to be preferred (subject to the qualifications related to incomplete information given in the text). Of course, when the same maximum

TABLE 3.2 MINIMAL PERCENTAGES OF ELECTORAL SUPPORT NECESSARY FOR ADOPTION OF STRATEGIES

STRATEGIES

No. of members in body	Safe: Ensure $\frac{1}{3}$	Dominant: Ensure $\frac{1}{3}$, Pursue $\frac{1}{3}+1$	New safe: Ensure $\frac{1}{3}+1$	Safe: Ensure $\frac{2}{3}$	Dominant: Ensure $\frac{2}{3}$, Pursue $\frac{2}{3}+1$	New safe: Ensure $\frac{2}{3}+1$
3	26 ·····15····· 41	·····10····· 51		51 ·····10····· 61	·····15····· 76	
6	29 ·····9····· 38	·····5····· 43		58 ·····5····· 63	·····9····· 72	
9	31 ·····6····· 37	·····4····· 41		61 ·····3····· 64	·····7····· 71	
12	31 ·····5····· 36	·····3····· 39		62 ·····3····· 65	·····5····· 70	
30	33 ·····2····· 35	·····1····· 36		65 ·····1····· 66	·····3····· 69	
99	34 ·····1····· 35	·····0····· 35		67 ·····0····· 67	·····1····· 68	

a 58 to 62 percent majority can assure itself of two-thirds representation (four seats) by supporting four candidates, with a 63 to 71 percent majority can assure itself of four seats by supporting five candidates—and capture the fifth seat if the minority should pursue a nonoptimal strategy of supporting more than two candidates—and assuredly win the fifth seat, no matter what strategy the minority adopts, if its majority should reach the 72 percent mark or greater. (From Table 3.1 we know in addition that a majority can assure itself of a sixth seat if, with 86 percent or more of the total vote, its supporters distributed their votes equally among six candidates.)

As we showed with our earlier example, however, when a group's estimated electoral support is based on incomplete information, it may be prudent for it to stick to a safe strategy even when its estimated strength indicates that it should pursue a dominant strategy. Either strategy is "optimal" in the sense of assuring the group the same guaranteed minimum number of seats, but the advantage the dominant strategy offers for winning an extra seat may be more than offset by the risk that, should the group's actual electoral support fall below the dominant threshold, it would jeopardize the guaranteed minimum that a safe strategy ensures.

value of k satisfies the inequalities for both the dominant and new-safe strategies, there is no need to choose between these two strategies since both involve running the same number of candidates—i.e., $(k+1)$.

Indeed, a group's risk aversion (or proneness) near a dominant threshold might be determined from whether it chooses to buy extra insurance for its guaranteed minimum with a safe strategy, or chooses to forsake this insurance and attempts to exploit the possible weakness of an opponent with a dominant strategy.

The difference in thresholds between safe and dominant, and dominant and new-safe, strategies are given along the dotted lines in Table 3.2. As the size of the body increases, these incremental values decrease, since pursuing an extra seat (dominant strategy), and ensuring an extra seat (new-safe strategy), take a smaller "expenditure" of votes when the votes must be divided among more candidates who are each worth proportionally fewer votes. Curiously, however, the one-third minority must increase its safe threshold by a greater amount to get to the dominant threshold (by 15 percentage points for the three-member body) than increase its dominant threshold to get to the new-safe threshold (by 10 percentage points for the three-member body); the situation is reversed for the two-thirds majority.

The reason is that the step to a dominant strategy for the one-third minority involves spreading a smaller initial percentage of the votes relatively further, and thus increasing it by more, than for the two-thirds majority. On the other hand, the step from the dominant to the new-safe strategy for the one-third minority involves raising the percentage at the dominant level to the new-safe level for relatively fewer seats, and thus increasing it by less, than for the two-thirds majority. We would suspect, therefore, that majorities near an estimated safe threshold would more often pursue dominant strategies (i.e., run one more candidate than their guaranteed minimum) than minorities, since dominant strategies involve a smaller departure from safe strategies for majorities; near a dominant threshold, however, we would suspect minorities would more often be successful with new-safe strategies (i.e., win the extra seat) than majorities, since new-safe strategies involve a smaller departure from dominant strategies for minorities.

Finally, note that in the ranges where it is rational for the minority to pursue a dominant strategy (41 to 50 percent for the three-member body), it is rational for the complementary majority to pursue a safe strategy (51 to 60 percent), and vice versa. In other words, a dominant strategy for one group is incompatible with a dominant strategy for the other group: At most one group can pursue an optimal (dominant) strategy of seeking an extra seat beyond its guaranteed minimum without risk of losing one of its guaranteed seats. The optimal strategy of the other group must therefore be safe;[16] since neither group has anything to gain by changing its strategy unilaterally, these strategies are in equilibrium.

---

16. In the special case in which there is the possibility of a tie, neither group may have a dominant strategy. For example, if each person in the electorate is allowed to cast a total of

Given that the number of seats is fixed so that the more seats one side wins, the fewer go to the other, the voting situation we have described can be conceptualized as a two-person constant-sum game. If the payoff is defined as the number of seats won by one player (a positive number) and lost by the other (a negative number)—instead of the number won by each—then the game is zero-sum. Since seats are neither created nor destroyed, to win seats necessarily entails winning them from the other player. This renders collusion useless because there are no joint strategies that can simultaneously improve the positions of both players.

In cumulative voting, the payoff is determined by the electoral rules that award seats to the candidates who receive the greatest numbers of votes. The strategies are the sets of ways that each player (group) has of partitioning his votes among candidates. Since the solution to a two-person cumulative voting game is in pure strategies, as we have shown, the payoff corresponding to these strategies is a saddlepoint. Thus, the optimal strategies that assure two players (groups) of proportional representation under cumulative voting are simply the minimax/maximin strategies in a two-person zero-sum game.[17] In adopting an optimal strategy, a player minimizes the maximum number of seats the other player can win, and maximizes the minimum number of seats he can win. Whatever strategy the other player adopts, each player, by choosing the best of the worst possible outcomes that the other player can inflict upon him, assures himself of a guaranteed minimum that maximizes his security level. Since the optimal strategies of both players are in equilibrium, each player is assured that he could not have done better—at least if the other player acted similarly.

Of course, this is true only if the game is one of complete information. If a group does not have complete information on the number of its supporters and those of the other side, it may be well advised not to select its optimal game-theoretic strategy since, as we have illustrated, what is "optimal" cannot be ascertained with confidence. Accordingly, it might be better in the face of uncertainty to forego the possibility of winning an extra seat for the security of almost certainly capturing those one chooses to contest. This and other considerations are relevant to the empirical examples that follow.

---

five votes for up to five candidates for five seats, a one-third minority could assure itself of one seat (20 percent representation), and a two-thirds majority of three seats (60 percent representation), but neither side could assure itself of greater representation. Indeed, if the one-third minority distributed its votes equally between two (instead of one) candidates, and the two-thirds majority among four (instead of three) other candidates, all six candidates —for the five seats—would receive the same number of votes! (In reality, especially for a large electorate, the possibility of such a deadlock is so remote that the electoral rules usually do not even prescribe how it would be resolved.)

17. This was first demonstrated by Gerald J. Glasser in "Game Theory and Cumulative Voting for Corporate Directors," *Management Science*, 5 (Jan. 1959), pp. 151–56.

3.7. EMPIRICAL EXAMPLES

Our first example illustrates some of the hazards of a majority group's choosing a risky strategy. It concerns a fight for control of the board of directors at the 1883 meeting of the Sharpsville Railroad Company.[18] As is rather common in elections of corporate directors in the United States, cumulative voting was used to choose the directors. Two factions controlled all the shares voted at the meeting, with each faction having a number of votes equal to the number of shares it controlled times the number of directors to be elected (six).

The majority group, with 53 percent of the votes, chose a strategy of dividing its votes equally among a full slate of six candidates, and the minority group, with 47 percent of the shares, chose a strategy of dividing its votes among a slate of four. The minority group easily captured the four directorships it contested, with each of its candidates getting 11.8 percent of the vote to the 8.8 percent for each of the majority group's candidates. In fact, if the minority group had pursued the bolder strategy of contesting five directorships, it could have won all five by giving 9.4 percent of all votes to each of its candidates.

The minimax strategies, on the other hand, are for the majority group to pursue the dominant strategy of contesting four directorships, the minority group to pursue the safe strategy of contesting three. Adoption of these strategies by each group results in each group's winning three directorships. Thus, even the minority group took a chance in contesting four directorships, because if the majority group had followed their dominant (minimax) strategy, they could have won four of the six directorships against this nonoptimal strategy of the minority group.

Why did the factions behave so foolishly in this election? Although it is hard to understand what motivated the majority faction to spread its votes thinly across all the directorships—and thereby in effect hand majority control of the board to the minority faction—the motivations of the minority faction are somewhat clearer: It apparently had a priori knowledge that the majority faction would act as it did. Despite its knowledge, however, it did not fully exploit this intelligence, as we have indicated, although it did capture majority control of the board of directors, which may have been its goal.

It is evident from this example that in a two-person zero-sum game, knowledge that one player will not adopt a minimax strategy may motivate the other player to abandon his own minimax strategy and try to capitalize on the first player's failure to "play it safe." In these cases, game theory offers no general prescriptions on how best to exploit an opponent's imprudence.

In our second example, the players were prudent in the extreme, with their strategy choices exhibiting a marked aversion to risk. The players

---

18. Glasser, "Game Theory and Cumulative Voting for Corporate Directors," pp. 154–55.

in this example were the members of the Democratic and Republican parties that nominate candidates to run for the office of representative to the Illinois General Assembly. This legislature is unique among state legislatures in the United States in basing elections for representative on cumulative voting, as described by Sawyer and MacRae:

> As it is applied in Illinois, three representatives are elected from each district, and each voter has three votes, which he may distribute 3–0, 2–1, $1\frac{1}{2}$–$1\frac{1}{2}$, or 1–1–1, among the candidates. Each party may nominate for the general election one, two, or three candidates, and the *number* to be nominated is decided upon and announced prior to the primary. This decision is made more or less independently by separate three-man committees elected by members of each party, with the voters in the primary election then determining *who* the candidates shall be.
>
> The committee's decision is made under uncertainty as to the percentage of the vote that the party will receive, and often, though not always, the number of candidates that the other party will nominate. . . . The behavior of the two nominating committees—one for each party—may be viewed as a two-person game in which the payoff is the number of candidates elected.[19]

Associated with the three ranges of vote percentages shown in Table 3.3 for the expected minority/majority parties in each election district are the following (optimal) minimax strategies and payoffs (the number to the left of the slash refers to the minority party, the number to the right the majority party):

1. 0 to $25-$ /$75+$ to 100: run 1/3; 0/3 elected;

2. $25+$ to $40-$ /$60+$ to $75-$: run 1/3; 1/2 elected;

3. $40+$ to $50-$ /$50+$ to $60-$: run 2/2; 1/2 elected.

The only minimax strategy not heretofore discussed is that of the minority's running one candidate when its expected vote percentage is in the 0 to $25-$ range. Clearly, this strategy dominates the strategy of running no candidates and is therefore optimal.

19. Jack Sawyer and Duncan MacRae, Jr., "Game Theory and Cumulative Voting in Illinois: 1902–1954," *American Political Science Review,* 56 (Dec. 1962), p. 937; italics in original. Since 1954, legislative districts have been reapportioned, but cumulative voting has been retained in the election of representatives to the Illinois General Assembly. See George S. Blair, *Cumulative Voting: An Effective Electoral Device in Illinois Politics* (Urbana, Ill.: University of Illinois Press, 1960); and George S. Blair, "Cumulative Voting: An Effective Electoral Device for Fair and Minority Representation," *Annals of the New York Academy of Sciences (Democratic Representation and Apportionment: Quantitative Methods, Measures, and Criteria,* ed. L. Papayanopoulos), 219 (New York: New York Academy of Sciences, 1973), pp. 20–26.

**TABLE 3.3   DISTRIBUTION OF ILLINOIS GENERAL ASSEMBLY
ELECTIONS, 1902–1954, BY STRATEGIES OF MINORITY/MAJORITY
PARTY AND PERCENT VOTE[a]**

STRATEGIES OF MINORITY/ MAJORITY PARTY	PERCENT VOTE FOR MINORITY/MAJORITY PARTY			
	0 to 25 − / 75 + to 100	*25 + to 40 − / 60 + to 75 −*	*40 + to 50 − / 50 + to 60 −*	PERCENT TOTAL
Run 1/3	20[b]	4[b]	1	4
Run 2/2	3	12	68[b]	41
Run 1/2	77	83	32	56
No. of elections	122	516	715	1,353

[a] Adapted from Sawyer and MacRae, "Game Theory and Cumulative Voting in Illinois 1902–1954," Table II, p. 941. Percentages were combined from this table and may not add up to one hundred due to rounding. These percentages were checked against original figures kindly supplied by Jack Sawyer (personal communication, March 20. 1974).

[b] Rows give minimax strategies for the ranges of vote percentages in each column.

To what extent have the nominations of the two major parties in Illinois's 51 election districts reflected—at least after the fact—the application of the minimax criterion? We give the results in Table 3.3 for 1,353 of the 1,377 biennial elections (27 in each of Illinois's 51 districts) from 1902 to 1954, excluding only those that involve strategies other than the three given in Table 3.3 (e.g., because of the bid of a strong third party).[20] What is astonishing about the nomination strategies and election results is that in more than one-half of the elections (56 percent), the minority/majority parties chose the strategy pair "run 1/2," which can *never* be optimal for both parties whatever the outcome of the election.[21] As can be seen from Table 3.3, this nonoptimal strategy was most often adopted by the minority/majority parties when the actual vote fell in the

20. For a brief discussion of strategies in cumulative voting when there are more than two groups in what may be modeled as essentially *n*-person zero-sum games, see Glasser, "Game Theory and Cumulative Voting for Corporate Directors," pp. 155–56.

21. Sawyer and MacRae miss this point when they say that the "expected number" of candidates (which they imply to be the optimal number "based upon a combination of probability and game theory principles") that the parties nominate is 1/2 in the 25 + to 40 − /60 + to 75 − range. This erroneous conclusion seems to be based on their mistaken belief that a party can have more than one minimax strategy, which is never the case in cumulative voting involving two players. Sawyer and MacRae, "Game Theory and Cumulative Voting in Illinois: 1902–1954," pp. 939–40. As we have shown for an expected vote in this range, the minimax safe strategy of the minority party is to run one candidate, and the minimax dominant strategy of the majority party is to run three candidates. If both parties adopt these optimal strategies, the minority party will elect its one candidate and the majority party will elect two of its three candidates.

range 25+ to 40−/60+ to 75− for the minority and majority parties, respectively.

Why in these cases did the expected majority party generally eschew its dominant minimax strategy of running three candidates? With the foresight that it was the strong favorite (based on previous election results in the district and other information), the answer would appear to be that majority party leaders were inclined to play it safe just in case their expected percentage might fall below the expected range, as suggested earlier. Of course, such fears are groundless if the expected minority party nominates only one candidate, but in 12 percent of the elections that fell in the 25+ to 40−/60+ to 75− range (see Table 3.3), the minority party did pursue the (a posteriori) risky strategy of running two candidates.

Even more compelling evidence that the majority plays it safe can be found from election results that fell in the 0 to 25−/75+ to 100 range. Provided that the majority party can accurately estimate the outcome in such instances, it would seem utter folly not to run three candidates and get them all elected. In this case, unlike the previous case, the expected majority party can be assured of sweeping the district, rather than simply exploiting a possible mistake on the part of the expected minority party. Despite this fact, the expected (overwhelming) majority party put up only two candidates in 80 percent of these lopsided elections (see Table 3.3).

As a possible explanation for this self-abnegating behavior on the part of the majority party, Sawyer and MacRae suggest that majority parties may be reluctant to try to demolish all opposition in a district. It seems that the majority party often views the third seat as less valuable if it can be assured of a majority of two seats in a district, especially if there is some uncertainty that incumbents might not be reelected when a third candidate is nominated. As a consequence, bipartisan agreements may be reached which, in effect, cede one seat to the minority party in such districts.[22]

These considerations, of course, are not relevant to game theory—at least the theory of two-person zero-sum games that are assumed to be one-shot affairs in which cooperation is of no value. On the other hand, such self-denying considerations apparently played much less of a part in the most competitive elections. In elections whose outcomes fell in the 40+ to 50−/50+ to 60− range, the expected minority/majority parties pursued their optimal minimax strategies 68 percent of the time, as compared with 4 and 20 percent adoption rates in the 25+ to 40−/60+ to 75− and 0 to 25−/75+ to 100 ranges, respectively (see Table 3.3). This would seem to mean that parties in the most competitive election districts either have more complete information about their likely prospects or are less accommodating to each other. Since there seems to be no reason why parties would have better information in these districts, it seems

---

22. Sawyer and MacRae, "Game Theory and Cumulative Voting in Illinois: 1902–1954," pp. 939, 941, 945.

plausible to assume that competition in these districts takes on more of a zero-sum character. This may be related to the idea that significance is attached to being the majority party in a district, although evidence on this point is lacking.

We should not let the peculiarities of competitive districts obscure the fact that in only 39 percent of all elections did both parties adopt their minimax strategies.[23] In almost all cases, deviations from the minimax strategy were in the direction of conservatism—playing it safe—which is especially surprising in view of the fact that many analysts consider minimax strategies themselves to be "conservative," in the sense that they anticipate the worst.

To be more precise, in only 5 percent of all elections did the expected minority party adopt a strategy of nominating two candidates, which in light of the results of the election was "risky" (nonminimax). Yet in none of these elections did the risky strategy cost the minority a seat because of the counterbalancing conservative bias of the majority party. Similarly, in the very few elections (less than 1 percent) in which the returns proved that the majority party had adopted a risky strategy of nominating three candidates, the minority party never took advantage by running more than one candidate.

This low propensity of the parties to adopt risky strategies contrasts sharply with their high propensity to adopt strategies more conservative than minimax, which was true of one of the parties in 60 percent of all elections. Of these failures on the side of conservatism, 21 percent cost one party an extra seat, which means that the results of 12 percent (0.60×0.21) of all elections could have been changed if one party had changed its strategy to that of its minimax strategy.

How significant is this failure rate? It seems hard to make a judgment on this question without knowledge about the accuracy of preelection predictions of the actual outcome. Obviously, we cannot maintain that the parties acted irrationally if a skilled game-theoretician could not have done better on the basis of the (incomplete) information available at the time. What the game-theoretician can say, however, is that given complete information, adoption of the "run 1/2" strategy is never optimal for one of the two parties in a zero-sum game. We suspect that shrewd politicians also recognize this fact but choose to ignore it because of an "understanding" that they might have with politicians of the other party. This is not to say that politicians act irrationally but rather that other considerations may displace the minimax logic. In other words, players in a game may be rational with respect to goals other than that of minimaxing their payoffs.

---

23. Contrast this with the figure of 69 percent given by Sawyer and MacRae, whom we previously pointed out misidentified the minimax strategy in the 25 + to 40 − /60 + to 75 − range. Sawyer and MacRae, "Game Theory and Cumulative Voting in Illinois: 1902–1954," p. 945.

Our data from the Illinois General Assembly indicate that minimax strategies hold up best in competitive elections when both parties anticipate a close outcome. Although such an outcome might have been expected by both contestants in the fight for control of the Sharpsville Railroad Company, what contravened the minimax logic in this case was one player's advance knowledge of the other player's (nonminimax) strategy. In such situations, there is no reason to rely on minimax strategies if one player suspects that he can probably do better.

3.8. SUMMARY AND CONCLUSION

In Chapter 2 we showed that if communication is permitted and there is no paradox of voting, a coalition can form that ensures an outcome better for all of its members than any other coalition can ensure. By contrast, we showed in this chapter that when a set of voters does not constitute a group sufficiently large to guarantee adoption of outcomes preferred by its members, its members can achieve, via cumulative voting, proportional representation on a voting body that itself decides on the selection of outcomes. In shifting the focus of our analysis from the behavior of individuals to that of groups reflecting particular viewpoints, we also shifted attention from the selection of single alternatives to that of multiple alternatives (i.e., representatives on an elected body).

More specifically, we showed that through cumulative voting a group can ensure its approximate proportional representation on an elected body by marshaling its supporters to concentrate their votes on a set of candidates roughly proportional to the group's size in the electorate. After demonstrating that proportional representation can in general be achieved only if individual voters are able to express their intensities of preference, we showed that allowing individuals to express their intensities of preference may be incompatible with requiring them to follow a group strategy that assures them of proportional representation. This is just another way of expressing the age-old conflict between the interests (or rights) of the individual and the interests (or rights) of a group. In a sense there is justice in this conflict, because when the incompatibility between the policies of a group and the preferences of its members is great, the group will be ineffective in acting as a cohesive bloc—as perhaps it should be! Recognition of this individual/group trade-off in the context of voting is important, for it permits us to assess its quantitative implications in an empirical setting rather than in the ethereal realm of speculative philosophy.

The quantitative effects that the size of a voting body has on the ability of groups to achieve proportional representation were shown to decrease quite rapidly as the body increases from three to ninety-nine members. At least for minority and majority groups that comprise one-third and two-thirds of the electorate, proportional representation can be closely approximated for elected bodies with about ten or so members.

If there are only two groups, we showed that their optimal strategies can be conceptualized as minimax strategies in a two-person zero-sum game. Given complete information, one player in such a game always has a "dominant" strategy and the other player a "safe" strategy; these strategies guarantee each player proportional representation regardless of the strategy choice of the other player.

In our first empirical example illustrating the strategic aspects of cumulative voting, we looked at a fight for control of the board of directors in a corporate election. This example demonstrated some of the risks involved in departures by players from their minimax strategies. Despite the fact that both players in this example apparently had complete information, knowledge that the majority player would not choose his minimax strategy was exploited by the minority player, who abandoned his own minimax strategy—though, in retrospect, not to the degree he might have—and won majority control of the board of directors.

Our second example illustrating the effects of cumulative voting was of elections of representatives to the Illinois General Assembly. In these election games, the players did not have complete information about their own size and that of the opposition prior to the elections. Perhaps as a consequence, their strategies were anything but risky; by and large, the players erred on the side of extreme restraint, at least as gauged by the election results. There is evidence, however, that not all the games played in these elections were conceived by the players to be zero-sum, especially those in noncompetitive districts. This makes it difficult to determine the extent to which deviations from minimax strategies are attributable to incomplete information in a zero-sum game, or to the nonzero-sum character of the elections themselves in which bargaining and negotiation processes may have exerted a significant effect.

Politics is both a practical and a theoretical science, and in this book we stress its theoretical side. At the same time, however, we should not ignore the practical aspects of politics that relate to using knowledge to achieve ends considered desirable. If we are interested in not only the possible but also the desirable in politics, normative considerations, which relate to what should or ought to be done, naturally complement theoretical and empirical studies.[24] Thinking back to the subject matter of Chapter 2, for example, one might argue that voting sophisticatedly, bargaining with other voters, and joining a coalition not only occur but may also be useful. One can hardly condemn "strategic" voting if it furthers goals one considers important and, moreover, prevents the adoption of outcomes not favored by a majority.

With respect to the analysis of this chapter, it is perhaps unnecessary to point out that the representation of a minority in a voting body in

---

24. William H. Riker and Peter C. Ordeshook, *An Introduction to Positive Political Theory* (Englewood Cliffs, N.J.: Prentice-Hall, 1973), p. 6.

rough proportion to its electoral strength is no guarantee that its views will be expressed effectively. Obviously, a minority on an elected body can be systematically outvoted by the majority. Indeed, it would be most unfortunate if the electoral disenfranchisement in majority-plurality systems that cumulative voting and other systems of proportional representation are designed to work against were reinstituted by a tyrannical majority, or coalition of minorities, on an elected body.

In situations where a majority may offer concessions to a minority by proposing to include its representatives on a single compromise slate, it seems that taking the choice of what minority members will be supported in an election out of the hands of even a benevolent majority, and putting this choice into the hands of the leaders of the minority through the operation of voting rules that can ensure their approximate proportional representation, is a step toward more equitable representation. For it is precisely the voting rules to which an opposition group might appeal as a last resort by running an opposition slate that would give it the bargaining power necessary to negotiate the composition of a compromise slate in line with its proportional electoral support. Hopefully, our theoretical and empirical analysis has helped clarify the nature of the voting rules that satisfy the general requirements for a system of proportional representation—both of minorities and majorities.

4

VOTE-TRADING GAMES

4.1. INTRODUCTION

In Chapter 2 we analyzed a variety of voting situations whose outcomes are subject to strategic manipulation by individual voters and coalitions of voters. The most pathological, as well as most studied, of these situations occurs when every outcome is dominated—there is no outcome that can defeat all others in a series of pairwise comparisons—which is called the paradox of voting. These situations are one manifestation of Arrow's General Possibility—sometimes called Impossibility—Theorem that condemns all methods of summing the preferences of voters, which meet certain plausible conditions, to an ineradicable arbitrariness.[1] Expressed another way, the social outcome of any summation procedure that satisfies these conditions must necessarily be insensitive to individuals' preferences or be imposed (e.g., by a dictator), since the conditions preclude any kind of "social consensus."

Perplexed and dismayed by the negativeness of this result, many theorists have attacked the reasonableness of certain of Arrow's conditions. This attack has led to various refinements in Arrow's statement of the original conditions, new axiomatizations of social choice, and many new theorems.[2] But except for one minor flaw in Arrow's original proof (discussed in the second edition of *Social Choice and Individual Values*), Arrow's basic impossibility result has remained intact for a generation. If the correctness of Arrow's theorem has proved logically unassailable, however, there nevertheless remains a good deal of controversy over its interpretation and its significance for the study of social processes.

---

1. Kenneth J. Arrow, *Social Choice and Individual Values*, 2d ed. (New Haven, Conn.: Yale University Press, 1963).

2. For recent reviews and syntheses of the literature related to the General Possibility Theorem, see Peter C. Fishburn, *The Theory of Social Choice* (Princeton, N.J.: Princeton University Press, 1972); Prasanta K. Pattanaik, *Voting and Collective Choice: Some Aspects of the Theory of Collective Decision Making* (New York: Cambridge University Press, 1971); and Amartyak Sen, *Collective Choice and Social Welfare* (San Francisco: Holden-Day, 1970).

We cannot possibly do justice to the various sides of this controversy in a short amount of space. Anyway, since most of the issues that have been raised are not game-theoretic in nature, they do not properly fall within the compass of this book.

One proposed "solution" to the paradox of voting does raise questions of a game-theoretic nature, however, and it is these that we shall pursue in this chapter. First, however, to provide some background on the controversy related to this proposed solution, which is based in part on a fundamental confusion over the meaning of one of Arrow's conditions, we shall try to clarify what the General Possibility Theorem precludes.

This discussion will serve as a preface to our main analysis of vote trading, which would seem to provide one means, albeit limited, for voters to register their intensities of preference across a set of issues in cases where the voting procedure itself allows voters to express only approval or disapproval for individual issues. We shall demonstrate that, contrary to the writings of several theorists, vote trading does not provide an escape from the paradox of voting; indeed, the very conditions that permit vote trading also ensure that no set of outcomes dominates all others and is therefore stable. Furthermore, we shall show that under particular circumstances, vote trading may lead to a kind of *n*-person Prisoner's Dilemma situation wherein everybody, as a result of making individually rational trades, ends up worse off than before trading.

4.2. WHAT THE GENERAL POSSIBILITY THEOREM PRECLUDES

Since the conditions in Arrow's proof of the General Possibility Theorem are all necessary, and together sufficient, to ensure the occurrence of the voting paradox, one way to avoid the paradox is to drop one of the conditions. The most controversial, and some have argued the most unrealistic, of Arrow's conditions is the so-called independence from irrelevant alternatives condition. Unfortunately, practically all the *examples* that have been used to illustrate the meaning of this condition, including one given by Arrow himself, misinterpret it, as Charles R. Plott has recently demonstrated.[3] Here is Arrow's example:

> The reasonableness of this condition can be seen by consideration of the possible results in a method of choice which does not satisfy Condition 3 [independence from irrelevant alternatives], the rank-

3. Charles R. Plott, "Recent Results in the Theory of Voting," in *Frontiers of Quantitative Economics*, ed. M. D. Intriligator (Amsterdam: North-Holland Publishing Co., 1971), pp. 109–29; Charles R. Plott, "Ethics, Social Choice Theory and the Theory of Economic Policy," *Journal of Mathematical Sociology*, 2 (July 1972), pp. 181–208; and Charles R. Plott, "Rationality and Relevance in Social Choice Theory" (Social Science Working Paper Number 5, California Institute of Technology, August 1971).

order method of voting frequently used in clubs. With a finite number of candidates, let each individual rank all the candidates, i.e., designate his first-choice candidate, second-choice candidate, etc. Let preassigned weights be given to the first, second, etc., choices, the higher weight to the higher choice, and then let the candidate with the highest weighted sum of votes be elected. In particular, suppose that there are three voters and four candidates, x, y, z, and w. Let the weights for the first, second, third, and fourth choices be 4, 3, 2, and 1, respectively. Suppose that individuals 1 and 2 rank the candidates in order x, y, z, and w, while individual 3 ranks them in order z, w, x, and y. Under the given electoral system, x is chosen. Then, certainly, if y is deleted from the ranks of the candidates, the system applied to the remaining candidates should yield the same result, especially since, in this case, y is inferior to x according to the tastes of every individual; but, if y is in fact deleted, the indicated electoral system would yield a tie between x and z.[4]

We shall not try to give a precise mathematical statement of the independence condition, but in words it can be stated as follows: If S is a subset of the set of available alternatives, and the preference scales of individuals change with respect to some alternatives not in S, then the social ordering of alternatives in S does not change. Clearly, what this condition does *not* assume is that the set of alternatives changes; rather, only individual preferences are allowed to vary. In Arrow's example, however, the set of alternatives changes from $\{w, x, y, z\}$ to $\{w, x, z\}$, whereas individual preferences are held fixed.[5] In fact, rather than violate the independence condition, this example satisfies it. Given the initial preferences and the set of alternatives $\{w, x, z\}$, the social choice is $\{x, z\}$. Now if individuals 1 and 2 continue to rank the three alternatives in the order x, z, w, and individual 3 ranks them in the order z, w, x, the choice under the rank-order method remains $\{x, z\}$ since z and x both get 7 votes, w gets 4 votes. This is true regardless of how the individuals feel about alternative y, because the condition assumes that the postulated set of alternatives $\{w, x, z\}$ from which a choice is made cannot change. Unavailable alternatives not included in this "feasible" set are assumed to have no effect on the social-choice process, which Plott argues is a rather innocuous assumption to make, even on ethical grounds: "As a question of 'ethics' the axiom simply says 'best' depends upon preferences over candidates for 'best' and not on unavailable alternatives."[6]

---

4. Arrow, *Social Choice and Individual Values*, p. 27.

5. The literature is filled with other examples where this mistake is made, even by perceptive scholars. See, for example, William H. Riker and Peter C. Ordeshook, *An Introduction to Positive Political Theory* (Englewood Cliffs N.J.: Prentice-Hall, 1973), pp. 88–90, 109–14, a generally outstanding work that we have referred to frequently throughout this book.

6. Plott, "Recent Results in the Theory of Voting," p. 116.

Arrow's example does violate what Plott calls a (collective) "rationality" condition, since it says that over one set of alternatives x is preferred to z, but over another set of alternatives (the set of alternatives, but not individuals' preferences, is allowed to vary under this condition) neither candidate is preferred to the other (i.e., the social choice is one of indifference between the two candidates). Since there is no binary relation, satisfying certain properties (e.g., transitivity), which can "rationalize" these variable social choices over different sets of alternatives, the rationality condition is not satisfied.

Should we be unduly upset by this fact? Plott thinks not, arguing that the idea of "better" inherent in binary relations is excessively narrow and "uninteresting." He offers examples of social-choice processes that violate various rationality conditions yet do not depend on the sequence in which alternatives are voted upon (i.e., are independent of the "path" of choice over two-element sets), which Arrow and others have argued justifies the rationality conditions. Thus, to devise social-choice processes that are path independent—unlike several voting procedures discussed in Chapter 2 wherein the outcome is sensitive to the order in which alternatives are voted upon—does not require that we impose the rationality condition: Although rationality implies path independence, path independence does not imply rationality.[7]

Since the rationality and independence-of-path conditions are not equivalent, there exist social-choice processes that satisfy all of Arrow's conditions except rationality and still possess the independence-of-path property. Hence, they are not subject to the General Possibility Theorem and would appear to yield "reasonable" outcomes that are not simply artifacts of the voting procedure or social-choice process used, at least if voting is sincere.

What if voting is not sincere? We shall not investigate this question in its full generality but rather shall focus on one form of insincere voting—vote trading—which, in systems that permit only qualitative choices among a set of alternatives, allows members to express their intensities of preference. Systems that allow the expression of intensities, like cumulative voting or the rank-order method, do not in general satisfy the rationality conditions, as we showed in the case of Arrow's example. Yet, as we shall demonstrate, vote trading may not only offer no escape from the arbitrariness of the paradox of voting guaranteed by the General Possibility Theorem, but it may lead one unwittingly into it. Before developing an example where this is indeed the case, we shall first survey the recent literature that portrays vote trading as by and large socially beneficial and compare it with an older literature in which a more skeptical view is taken about the value of vote trading.

---

7. Plott, "Rationality and Relevance in Social Choice Theory," pp. 27–29, and Charles R. Plott, "Path Independence, Rationality, and Social Choice," *Econometrica*, 41 (Nov. 1973), pp. 1075–91.

4.3. JUDGMENTS ABOUT VOTE TRADING[8]

The conventional judgment on vote trading in legislatures is one of severe disapproval. In the United States, at least, the usual idiom is "logrolling," and this word has always had pejorative connotations.[9] The common sense of the language has been reinforced by the judgment of scholars. E. E. Schattschneider, whose study of the writing of the Smoot-Hawley tariff of 1930 is the most detailed examination of a single event of vote trading, concluded his book with the remark: "To manage pressures [i.e., for vote trading] is to govern; to let pressures run wild is to abdicate."[10] Furthermore, it seems likely that the widespread scholarly support for concentrated national leadership—as embodied, for example, in such writings as the report of the Committee on Political Parties of the American Political Science Association, a report to which Schattschneider contributed substantially—was engendered in part at least by a desire to minimize vote trading.[11]

Despite the long-standing unanimity of scholarly and popular judgment, however, it has recently been argued by a number of scholars that vote trading is socially desirable because it allows the expression of degrees of intensity of preference. James M. Buchanan and Gordon Tullock argue that the voter or legislator can improve his welfare "if he accepts a decision contrary to his desire in an area where his preferences are weak in exchange for a decision in his favor in an area where his feelings are stronger."

---

8. The remainder of this chapter is based largely on William H. Riker and Steven J. Brams, "The Paradox of Vote Trading," *American Political Science Review*, 67 (Dec. 1973), pp. 1235–47; the permission of the American Political Science Association to adapt material from this article is gratefully acknowledged.

9. *A Dictionary of Americanisms on Historical Principles* (Chicago: University of Chicago Press, 1951), which is probably the most authoritative collection, cites many political examples from 1812 (Ninian Edwards) to 1949 (*Time* magazine), in every one of which the word is clearly used in a perjorative sense. In the quotation from Edwards, logrolling is specifically identified with intrigue; and the climate of intrigue is conveyed by almost all later quotations. The Oxford Dictionary (i.e., *A New English Dictionary on Historical Principles* [Oxford, England: Clarendon Press, 1908]) cites political usage from both sides of the Atlantic from 1823 (*Niles Weekly*) on to the end of the nineteenth century, and its quotations also are uniformly pejorative. John Russell Bartlett's *Dictionary of Americanisms* (New York: Bartlett and Welford, 1848) contains a lengthy and righteously indignant explication of the political sense of "logrolling." None of these sources indicates any pejorative connotation at all for the word when it refers to clearing land or building houses. It seems to have acquired the sense of intrigue only when it was transferred from the bartering of labor to the bartering of votes, which demonstrates that the bartering of votes has regularly been socially disapproved.

10. E. E. Schattschneider, *Politics, Pressures and the Tariff* (New York: Prentice-Hall, 1935), p. 293. Schattschneider interpreted the process of tariff writing as an "attempt to set up a beneficently discriminatory set of privileges"—beneficent in the sense that a few particular industries deemed worthy would benefit. This attempt failed, he argued, because the legislation as actually written protected so "indiscriminately . . . as to destroy the logic and sense of the policy" (p. 283). One of our purposes here is to explain the conditions for such failure.

11. Committee on Political Parties of the American Political Science Association, *Toward a More Responsible Party System* (New York: Rinehart, 1950).

Hence they say: "Bargains among voters can, therefore, be mutually beneficial."[12] The authors recognize that when bargains are concluded "in which the single voter does not participate, . . . he will have to bear part of the costs of action taken."[13] But they nevertheless conclude that these external costs are on the average less than the benefits obtained so that, for the society as a whole, vote trading or logrolling is a desirable kind of event.

Following this argument, a small literature on the subject has developed with the same tone. Tullock elaborated his earlier position.[14] James S. Coleman has argued that logrolling is a device both to avoid the Arrow paradox and to arrive at optimal allocations.[15] A number of writers has pointed out, however, that vote trading does not eliminate these paradoxes.[16] But these same writers and others have taken pains to argue that it does improve allocations in the direction of Pareto optimality.[17]

We thus have an intellectual confrontation between an older (and popular) tradition and new developments in social science. To try to reconcile this conflict, we shall define the relationship between salience and utility in the next section, where we shall also describe the assumptions of our model of vote trading. Then we shall offer an example that demonstrates how, through vote trades which are all individually rational (defined in section 4.4) for pairs of members of a voting body, every member may end up worse off than before. This is what we shall call the *paradox of vote trading*. It will then be shown that if the original trades are not considered binding, it is rational for pairs of members to violate their original vote-trading agreements and consummate new ones, which may result in indefinite cycling and even a return to each member's original positions. Furthermore, we shall demonstrate that the formation of coalitions that exclude some members offers no guarantee that these trading cycles can be broken, which underscores the lack of a stable equilibrium. Next, several real-world conditions that may retard the incidence of vote trading and its disequilibrating effects will be indicated. The significance of the vote-trading paradox, in light of some empirical

---

12. James M. Buchanan and Gordon Tullock, *The Calculus of Consent* (Ann Arbor, Mich.: University of Michigan Press, 1962), p. 145.

13. Buchanan and Tullock, *Calculus of Consent*, p. 145.

14. Gordon Tullock, "A Simple Algebraic Logrolling Model," *American Economic Review*, 60 (June 1970), pp. 419–26.

15. James S. Coleman, "The Possibility of a Social Welfare Function," *American Economic Review*, 56 (Dec. 1966), pp. 1105–22.

16. Dennis C. Mueller, "The Possibility of a Social Welfare Function: Comment," *American Economic Review*, 57 (Dec. 1967), pp. 1304–11; and Robert Wilson, "An Axiomatic Model of Logrolling," *American Economic Review*, 59 (June 1969), pp. 331–41.

17. See Wilson, "An Axiomatic Model of Logrolling"; Edwin T. Haefele, "Coalitions, Minority Representation, and Vote-Trading Probabilities," *Public Choice*, 8 (Spring 1970), pp. 75–90; and Dennis C. Mueller, Geoffrey C. Philpotts, and Jaroslav Vanek, "The Social Gains from Exchanging Votes," *Public Choice*, 13 (Fall 1972), pp. 55–79.

examples, will then be assessed. Finally, we shall indicate the relationship of vote trading to Arrow's independence-from-irrelevant-alternatives condition and conclude by suggesting how the basic model of vote trading presented might be refined and extended.

4.4. DEFINITIONS AND ASSUMPTIONS

In the analysis that follows, we assume that each member of a hypothetical voting body can rank all roll calls in terms of their salience to him. By *salience* we mean the difference it makes to a member to be in the majority (i.e., win) versus the minority (i.e., lose). More precisely, if $u_i(M)$ is the utility that member i associates with being in the majority (M) on a particular roll call—which may not be positive if he votes insincerely, as we shall presently show—and $u_i(N)$ is the utility he associates with being in the minority (N), then we define the salience of that roll call to him as

$$s_i = u_i(M) - u_i(N) \cdot$$

For the two roll calls shown in Table 4.1, the second is more salient to member i than the first (since $s_i = 4$ on roll call 2, $s_i = 2$ on roll call 1), whereas the first is more salient to member j than the second (since $s_j = 3$ on roll call 1, $s_j = 1$ on roll call 2).

Note that even though $s_i > s_j$ for member i on roll call 2, and member j on roll call 1, we make no comparisons of salience *between* members. Such interpersonal comparisons of salience, and the utilities associated with salience, we rule out in the absence of any universal standard (e.g., based on money, power, or prestige) that can be presumed for all members in their evaluation of social choices.[18]

In our subsequent analysis, we shall continue the development of the example in Table 4.1 to illustrate our definition of vote trading. First, however, we present the assumptions of our model:

1. A voting body of (initially) three members, $V = \{i, j, k\}$.

2. A set of (initially) six roll-call votes, $R = \{1, 2, 3, 4, 5, 6\}$, on each of which a member has a preference for either one of two *positions* (for or against): If two of the three members agree on a roll call (by voting either for or against), this is the majority position, which we denote by M; we denote the minority position of the third member by N.

3. There are no roll calls on which all three members agree. Winning majorities are thus assumed to be always *minimal winning*, which means that if either member of the two-member majority changes his position on a roll call, this changes the majority outcome on that roll call. Thus, the votes of the majority members on all roll calls are *critical*.

---

18. For a detailed discussion of this important but often misunderstood aspect of utility theory, see R. Duncan Luce and Howard Raiffa, *Games and Decisions: Introduction and Critical Survey* (New York: John Wiley & Sons, 1957), chap. 2.

TABLE 4.1   UTILITY OF POSITIONS AND SALIENCE OF ROLL CALLS FOR TWO MEMBERS

	UTILITY AND SALIENCE					
	Roll call 1			*Roll call 2*		
MEMBERS	u(M)	u(N)	s	u(M)	u(N)	s
i	1	−1	2	2	−2	4
j	1	−2	3	1	0	1

4. Each member of the voting body can rank all roll calls in terms of the importance, or *salience*, of the outcome to him. As we shall show, the differential salience members attach to the outcomes of different roll calls is the condition that makes vote trades possible.

5. Each member possesses *complete information* about the positions and salience rankings of all other members and can *communicate* with them.

6. The *goal* of each member is, through vote trading, to maximize his utility across all roll calls by being in the majority on the roll calls whose outcomes are most salient to him.

Clearly, if members value the outcomes on roll calls differently (assumption 4), then it follows from the goal assumption (assumption 6) that a rational member would be willing to give up a preferred position on one roll call to secure a preferred position on another roll call whose outcome was more salient to him. The mechanism by which he accomplishes this we call a *vote trade*. The possibility of a vote trade depends on the presence of two conditions for any pair of members i and j who vote on roll calls 1 and 2:

(a) Member i is in the majority on roll call 1, the minority on roll call 2.

(b) Member j is in the majority on roll call 2, the minority on roll call 1.

A trade occurs when member i (majority position holder) gives his support to member j (minority position holder) on roll call 1 in return for the support of member j (majority position holder) on roll call 2 (where member i is in the minority).

It is precisely the asymmetry of the positions of the two members on the two roll calls that permits an exchange of support. By assumption 3, such an exchange will change the majority position on each roll call because the switcher (member i on roll call 1, member j on roll call 2) casts a critical vote when majorities are minimal winning. Under this circumstance, the minority position for which a member (i on roll call 1, j on roll call 2) receives support becomes the majority position when this support is given.

A third condition is necessary to ensure that a vote trade is *individually rational* for each member in a pair:

(c) The minority position for which a member receives support changes an outcome more salient to him than the outcome of the roll call on which he switches his vote.

In other words, a trade is individually rational for a pair of members when it not only changes the outcomes of the two roll calls on which they exchange votes but also assures each member a majority position on the roll call whose outcome is more salient to him. As we shall show later, however, the trades of other pairs of members may upset these calculations in such a way that the trades which produce positive gains for each pair of trading members may lower the collective benefits for all members of the voting body. Before discussing the interdependence of vote trades, though, we must distinguish between "sincere" and "insincere" voting.

4.5. SINCERE AND INSINCERE VOTING

By *sincere voting* we mean, as in Chapter 2 (see section 2.3), that a member of a voting body votes directly according to his preferences, which in this chapter translates to mean that a member takes the position on a roll call (for or against) that offers him the greatest utility. This is to be contrasted with *insincere voting*, which is defined as voting against one's interest— that is, for one's less-preferred, rather than one's more-preferred, position on a roll call—when there is an incentive to do so. Insincere voting includes sophisticated voting, discussed in Chapter 2, but it also includes voting in cooperative situations where voters can communicate with other voters and make agreements. Surely one of the main reasons for insincere voting is logrolling, which we analyze in this chapter.

Whatever one thinks of the propriety of this kind of behavior, it may be rational with respect to the goal postulated earlier (see assumption 6). Consider again the situation of two members i and j whose (sincere) positions on roll calls 1 and 2 are such that member i is in the majority on roll call 1, which we denote by $M(1)$, and member j is in the majority on roll call 2, which we denote by $M(2)$. Similarly, let $N(1)$ and $N(2)$ indicate the minority positions on these roll calls, which we assume are the positions, respectively, of members j and i.

Now let us assume, as was the case in Table 4.1, that roll call 2 is the roll call whose outcome is more salient to member i, roll call 1 is the roll call whose outcome is more salient to member j. This situation is summarized in Table 4.2, with the utilities associated with the positions of each member, originally given in Table 4.1, also shown. ("Total utility" is simply the algebraic sum of the utilities of each member across the two roll calls.) Given the goal of each member to be in the majority on roll calls most salient to him (assumption 6), member i would prefer to be in

TABLE 4.2   POSITIONS AND UTILITIES BEFORE TRADING

SALIENCE

| MEMBERS | High | | Low | | TOTAL |
	Position	Utility	Position	Utility	UTILITY
i	$N(2)$	-2	$M(1)$	1	-1
j	$N(1)$	-2	$M(2)$	1	-1

the majority on roll call 2 and member j on roll call 1. We shall now show that each member can realize his goal through a vote trade.

To effect such a trade, each member will have to vote insincerely on the roll call whose outcome is of less salience to him. We label the insincere positions of the members on these roll calls as M_d to signify support of a majority with which one disagrees (d). Assuming that the positions N and M_d are valued as equally bad—that is, on any roll call $u(M_d) = u(N)$ for each member—the utilities of the new positions of members after the trade, originally given in Table 4.1, are shown in Table 4.3. The value of the trade can be seen by comparing the total utilities of the members before (Table 4.2) and after (Table 4.3) trading, which increase for both from -1 to $+1$. These numbers are purely illustrative, for individually rational vote trades as we have defined them will always be profitable for the participating members, given the equivalence we have assumed between the N and M_d positions.

This equivalence assumption is justified by the fact that in the case of both positions N and M_d, a member's sincere position differs from the prevailing one, expressed in one case by *public nonsupport* of the majority position (i.e., voting with the minority, N), and in the other by *public support* of a majority position with which one privately disagrees (M_d). Although one might regard the hypocrisy of the latter position as distasteful, if not unconscionable, it would seem that whether one's disagreement is public or private, the results are the same: One's preferred position on a roll call is thwarted by the majority, even if it is a majority one publicly supports.

For this reason we consider these positions equivalent in terms of both the cardinal utilities associated with each and an ordinal ranking of preferences. Indeed, implicit in the goal assumption (assumption 6) is that a member's majorities on roll calls whose outcomes are less salient to him are negotiable, and in particular on these roll calls he will be willing to sacrifice preferred positions—and form majorities inimical to his interests—if he can prevail on roll calls whose outcomes are more salient to him. We assume that these more salient outcomes override any ethical

TABLE 4.3 POSITIONS AND UTILITIES AFTER TRADING

| | SALIENCE | | | | |
| | High | | Low | | TOTAL |
MEMBERS	*Position*	*Utility*	*Position*	*Utility*	UTILITY
i	$M(2)$	2	$M_d(1)$	-1	1
j	$M(1)$	1	$M_d(2)$	0	1

qualms associated with insincere voting—that is, the utilities are a true expression of preferences.

So far so good for the vote-trading members in our previous example: They both come out better after the trade. Their good fortune is not costless to all members, however, for the reversal of the majorities on the two roll calls must be at somebody's expense—namely, the third member of the voting body.

To see this, consider the original (sincere) positions of members i and j on the two roll calls given in Table 4.2. By assumption 2, a third member k must be in the majority on both roll calls in order that two members support the majority on each roll call and one member the minority. Thus, the (sincere) positions of a third member k on roll calls 1 and 2 will necessarily be $M(1)$ and $M(2)$.

Now recall that after the vote trade of members i and j, the old minority positions became the new majority positions because of the switch of member i on roll call 1, and member j on roll call 2, to positions with which they disagree (see Table 4.3). Because of the critical nature of members' votes on roll calls, these switches produce new majorities on the roll calls. These new majorities, that members i and j now together constitute on the two roll calls 1 and 2, recast the previous majority positions of member k on these roll calls into minority positions, $N(1)$ and $N(2)$. Thus, the external costs of the favorable trade between members i and j fall on the nonparticipant, member k. This follows from the fact that the number of majority and minority positions must remain constant by assumption 2 (in a three-member body, a 2:1 ratio of M's and N's must be preserved on all roll calls under simple majority rule), so all switches in positions produced by the vote-trading members on each roll call will perforce generate an equal number of involuntary position changes (from majority to minority) for the nontrading member. If we assume that member k associates the same utilities with the majority and minority positions on the two roll calls as member i (see Table 4.1), then the vote trade between members i and j drops his total utility from $+3$ to -3, as shown in Table 4.4.

TABLE 4.4 POSITIONS AND UTILITIES OF LEFT-OUT MEMBER
BEFORE AND AFTER TRADING

	SALIENCE				
	High		Low		TOTAL
MEMBER k	Position	Utility	Position	Utility	UTILITY
Before trading	M(2)	2	M(1)	1	3
After trading	N(2)	−2	N(1)	−1	−3

If the gains from a vote trade can be wiped out by external costs (defined in the following paragraph), the enthusiasm of logrolling's proponents may be misplaced. Indeed, it is quite possible to imagine a system in which everyone gains from individual trades—and so has a positive incentive to logroll—and yet everyone also loses if all such trades are carried out. This is the paradox of logrolling, which we shall illustrate in section 4.6.

External costs are the costs one bears as a result of the actions of others. They are to be contrasted with internal costs, which result from one's own action. The noise that comes from a pneumatic drill is an internal cost with respect to the driller because it is part of what he must suffer to earn his pay. But to the neighbor at the construction site, the noise is an external cost. He must bear it, not in order to get something else, but merely because he is there.

Debate is never ending about the existence of some external costs. (For example, does the noise of the drill really bother the neighbor or does he see a chance, by complaining of the noise—even though it does not bother him—to blackmail the contractor?) But it seems impossible to doubt the existence of external costs in vote trading. Consider the position of some member who is not a party to a trade: Since the trade brings about a different winner on a roll call, the nontrader bears an external cost if he were originally in the majority, or he receives an external gain if he were originally in the minority. In our example, member k is entirely passive throughout the entire transaction, but two of his positions are reversed from majority to minority.

External costs have mostly been ignored in the writings of those who extol the benefits of vote trading.[19] Yet they must by assumption 3 on

---

19. An exception is Coleman, "The Possibility of a Social Welfare Function," p. 1121, who notes in passing that the inclusion of such costs requires a "much more extensive calculation," but he does not try to develop it. In a more recent article, Coleman provides an explicit calculation and concludes that vote trading will be generally profitable only under conditions of "absolutely free and frictionless political exchange" that preclude arbitrary decision rules (e.g., majority rule). See James S. Coleman, "Beyond Pareto Optimality," in

critical votes be present: Since trading changes the outcomes, there must always be innocent bystanders who suffer. Although the size of their losses in particular cases will depend on the amounts of utility associated with their salience rankings, it is possible to show that, in particular cases, suffering is general and universal, which is the paradox of logrolling.

In summary, it would appear that the optimal joint strategy of a pair of vote trading members is to vote insincerely on roll calls whose outcomes are less salient and sincerely on the roll calls whose outcomes are more salient. The trade puts both members on the side of the majority on both roll calls, one of which (the roll call whose outcome is more salient) each member agrees with, the other of which he disagrees with. The external cost which this trade imposes on member k in our example is to change his previous majority positions on both roll calls into minority positions.

4.6. INITIAL TRADES AND THE PARADOX OF VOTE TRADING

We now turn to a simple example to demonstrate that vote trades that are individually rational for pairs of members in a voting body may leave all members worse off after the trades. We assume that roll calls are voted upon serially so that future roll calls are not necessarily anticipated when current roll calls are considered. Note that this assumption about the voting process is quite different from the assumption made in the previously cited works of proponents of vote trading, where it is typically laid down that all roll calls are simultaneously before the legislature. In such a system, support can be traded back and forth in all possible ways, but in the present system support can be traded only with respect to the subset of roll calls currently before the legislature or anticipated to come before it.[20] The rationale for this more restrictive assumption is simply that it accords with a crucial feature of the real world.

In order for each of the three members of our postulated voting body, i, j, and k, to be able to trade with the two other members, the trading must occur over a minimum of six roll calls. This is so because there is a total of three different pairs of members in the three-member voting body— (i, j), (j, k), and (i, k). Since a trade for each pair involves two different roll calls, there must be at least six different roll calls if all three pairs are to engage in vote trading. In general, on any single round of trading there can at most be as many trades as half the number of roll calls,

---

Philosophy, Science, and Method: Essays in Honor of Ernest Nagel, ed. Sidney Morgenbesser, Patrick Suppes, and Morton White (New York: St. Martin's Press, 1969), pp. 415–39; also, James S. Coleman, *The Mathematics of Collective Action* (Chicago: Aldine Publishing Co., 1973), where computer simulation models of vote trading are developed.

20. What comes before a legislature may very well be a product of vote trades in committees, where typically support is exchanged among members on the provisions of a single bill rather than on wholly different bills.

TABLE 4.5   POSITIONS OF MEMBERS BEFORE INITIAL TRADES ON
SIX ROLL CALLS

HIGH SALIENCE ⟶ LOW SALIENCE

MEMBER	MAJORITY AND MINORITY POSITIONS OF MEMBERS BEFORE TRADES						TRADES	SWITCHES	KEY
i	$M(1)$	$M(2)$	$N(3)$	$N(4)$	$M(5)$	$M(6)$	with k	to $M_d(6)$	———
							with j	to $M_d(5)$	·······
j	$M(3)$	$M(6)$	$N(5)$	$N(2)$	$M(1)$	$M(4)$	with i	to $M_d(4)$	······
							with k	to $M_d(1)$	– – – –
k	$M(5)$	$M(4)$	$N(1)$	$N(6)$	$M(3)$	$M(2)$	with j	to $M_d(2)$	– – – – –
							with i	to $M_d(3)$	———

since on any trade support is exchanged between members on two different roll calls that we assume cannot serve as the basis for more than one trade.

Given an *initial* set of trades among pairs of members, let us ignore for the moment *subsequent* possible trades that these initial trades may set up. As we shall show later, however, an indefinitely large number of possible trades may follow from the initial trades, so the number of roll calls on which members vote serves to limit only the number of initial trading possibilities. Moreover, a member may have a choice of more than one trading partner on two roll calls, but this and other possible complications of the simple example that we shall elaborate in this section will be reserved for our later discussion.

To illustrate the paradox of vote trading wherein each member is worse off after trading, consider a voting body whose members' positions on six roll calls, indexed by the numbers 1 through 6 in parentheses, are indicated in Table 4.5. For each member we assume that the salience of the outcome of each roll call decreases from left to right in each row of the table. Thus, for member *i* the outcome of roll call 1 is most salient to him, the outcome of roll call 6 least salient; on both roll calls, his position is the majority position.

Vote trades can readily be identified for each pair of members in Table 4.5. Thus, member *i* can trade with member *k* on roll calls 3 and 6. That is, member *i* votes against his interest on his least salient roll call, 6, and gains thereby from member *k*'s switch on *i*'s third most salient roll call, 3. Similarly, member *k* votes against his interest on his next to least salient roll call, 3, and gains thereby on his roll call of next higher salience, 6. This trade is indicated in Table 4.5 by solid lines connecting the positions

TABLE 4.6 POSITIONS OF MEMBERS AFTER INITIAL TRADES ON SIX ROLL CALLS

MEMBER	HIGH SALIENCE⟶LOW SALIENCE MAJORITY AND MINORITY POSITIONS AFTER INITIAL TRADES	TRADES	SWITCHES	KEY
i	$N(1)$ $N(2)$ $M(3)$ $M(4)$ $M_d(5)$ $M_d(6)$	with j with k	to $M(6)$ to $M(5)$	——— ·······
j	$N(3)$ $N(6)$ $M(5)$ $M(2)$ $M_d(1)$ $M_d(4)$	with k with i	to $M(4)$ to $M(1)$	— — — ———
k	$N(5)$ $N(4)$ $M(1)$ $M(6)$ $M_d(3)$ $M_d(2)$	with i with j	to $M(2)$ to $M(3)$	······· — — —

of members i and k; other trades between members i and j and between members j and k are indicated by dotted and dashed lines, respectively.

After the three trades indicated in Table 4.5 have been made, the situation will be as shown in Table 4.6 (ignore for now the subsequent trades indicated in this table), where the two most salient M's for each member have been replaced by N's, the medium salient N's by M's, and the two least salient M's by M_d's (majority positions with which members disagree). As a consequence of the trades, the winning positions on all six roll calls are reversed.[21]

This is a dismal result for all three members, for on the debit side each member not only loses a majority position on his two most salient roll calls but also acquires a majority position, contrary to his interests, on his two least salient roll calls. For thus losing on his two most and his two least salient roll calls, on the credit side he gains only on his two medium salient roll calls. Clearly, since the amount of utility by which he worsens his position is greater than the amount by which he betters his position, he is in sum hurt by the set of trades.

Any set of cardinal utilities consistent with the salience rankings in our example can be used to illustrate this conclusion. For example, assume for each member that the utility he associates with voting with the majority (M) on the roll call whose outcome is most salient to him is 6, on the next most salient is 5, and so on to the least salient, whose utility is 1. Assume

21. Note that for each member, the M to N changes and N to M changes from Table 4.5 to Table 4.6 arise from other members' switches, not from a member's changing the positions which he supports; the latter changes occur only on roll calls on which a member's position changes from M to M_d.

further that the utility each member associates with voting with the minority (N) is 0 on all roll calls. Then the salience for each member,

$$s = u(M) - u(N),$$

will be $6 - 0 = 6$ on the roll call whose outcome is most salient to him, $5 - 0 = 5$ on the next most salient, and so forth, which is consistent with the ordinal salience rankings.

Before trading, M's on each member's two most salient roll calls give him $6 + 5 = 11$ utiles, N's on his two next most salient roll calls give him $0 + 0 = 0$ utiles, and M's on his two least salient roll calls give him $2 + 1 = 3$ utiles, for a total utility of 14. After trading, assuming $U(M_d) = u(N) = 0$, N's on each member's two most salient roll calls give him $0 + 0 = 0$ utiles, M's on his two next most salient roll calls give him $4 + 3 = 7$ utiles, and M_d's on his two least salient roll calls give him $0 + 0 = 0$ utiles, for a total utility of 7. Thus, for the particular cardinal utilities, consistent with the ordinal salience rankings, which we have assumed in this example, trading halves each member's total utility from 14 to 7. The paradox is confirmed: All members are absolutely worse off from trading.

It might be thought that this result could be avoided by a different set of initial trades. But it is apparent in light of the situation as given in Table 4.5 that such trades are not possible. Member i, for example, wants to gain on roll calls 3 and 4 by getting someone to switch on these roll calls. He cannot trade with member j on roll call 3, however, because this would force j to switch on his most salient roll call, which is in clear violation of condition (c) in section 4.4 of individual rationality. Therefore, member i *must* trade with member k on roll call 3, as in Table 4.5. Similarly, member i cannot trade with member k on roll call 4, because this would force k to vote against his interest on a more salient roll call (i.e., 4) in order to gain his interest on a less salient one (i.e., 1 or 6). Again, therefore, member i *must* trade with member j on roll call 4, exactly as in Table 4.5. By a similar argument, each of the other two members in this table is constrained to exactly those trades indicated in it.

4.7. SUBSEQUENT TRADES AND THE INSTABILITY OF VOTE TRADING

The example in Table 4.5 shows that the paradox is inherent in the ordinal relationships and is not dependent on particular cardinal utility numbers, which we used simply to provide one illustration of salience and utility values in this example. Following the trades indicated in Table 4.5, however, the situation depicted in Table 4.6 is not in equilibrium. The positions generated by the initial set of trades are themselves unstable and generate the new set of trades shown in Table 4.6. Except for the fact that the majorities a member trades are now ones that he disagrees with (M_d's), conditions (a), (b), and (c), which ensure an individually rational vote trade, are all met for the trading pairs indicated in this table.

M_d's, however, can be traded like M's. In fact, there is an added incentive to switch an M_d on a roll call of lower salience than an M, because one switches from a majority one disagrees with back to a majority one agrees with. To illustrate, in the initial trading in our example, it was rational for member i to support member k on roll call 6 by switching from $M(6)$ (see Table 4.5) to $M_d(6)$ (see Table 4.6). But, as indicated in Table 4.6, after this initial trade it is now rational for member i to switch back to $M(6)$ so that he can support member j on roll call 6 in return for member k's support on roll call 1.

What happens after members make these subsequent trades? It is easy to show that when members trade their M_d's back for M's, their positions after the subsequent trades will bring the trading back to where it started (i.e., the positions of members given in Table 4.5). The reversion of members after two rounds of trading back to their original positions, with nothing gained and nothing lost, is an immediate consequence of the two switches in position made by members' voting insincerely on the initial round but sincerely once again on the subsequent round. Since each member has only one sincere position on each roll call, the initially postulated sincere positions of members, and the sincere positions after the subsequent trades, must be the same.

Voting that involves first a switch to a position contrary to one's preference on a roll call and then a reverse switch back may seem more devious than sincere. But this is only the beginning. For after a reverse shift, when everybody assumes once again his original sincere position, there is no reason to believe that pairs of members will not plunge themselves once again into a new round of trades which are individually rational. Indeed, the inexorable logic of individual rationality dictates that if a member is willing to doublecross an old trading partner in order to improve his position through another trade, he would be willing to triplecross his new trading partner, and so on. The consequence of all members' acting in this self-serving fashion is that trading continues indefinitely—in theory, at least, if not in practice. The lack of a stable equilibrium that could break this cycle at any point means that there is no guarantee that even the improved benefits to all, as before the initial or after the subsequent trades in our example, will halt the process. In practice, only when roll calls are actually voted upon will the process of vote trading be terminated.

The instability of the positions of members after the initial trades is underscored by the fact that a member still has an incentive to switch his M_d positions on roll calls back to his original M positions even if the other members refuse to trade. For even without the cooperation of the other members, the switches themselves will increase each member's utility. To illustrate in terms of the cardinal utilities we gave earlier, any single member in Table 4.6 would gain $1 - 0 = 1$ utile by switching from M_d

to M on his least salient roll call, and $2 - 0 = 2$ utiles on his next least salient roll call, for a total utility gain of 3.[22] In fact, in *all* cases of vote trading, the generation of M_d's on the initial trades are by their nature positions that members would like to get rid of—and can, by reneging on vote trading agreements. They are, therefore, unstable.

4.8. THE CONSEQUENCES OF REFUSING TO TRADE

In theory, it would seem, one possible way in which members might transcend the upsetting logic of individual rationality would be to make irrevocable trading agreements. Even if members have no ethical compunctions about reneging on pairwise agreements that turn out to be no longer individually rational to honor, it is possible that the penalties (e.g., reprisals by other members) for acting deceitfully may in the long run outweigh any benefits that accrue from trading. If this is the case, it would be rational for all members to honor trading agreements made on the initial round of trading.

To assume that the individual rationality assumption is operative for only one round of trading, however, seems in some ways artificial. Although the blatant disregard of earlier agreements may seem duplicitous behavior that members could not get away with over the long haul, there is not much basis for believing this will bind them to their initial vote-trading agreements, especially when these agreements lead to collective disaster for all, as our example illustrating the paradox demonstrated.[23] This is the worst possible situation when all members trade; with apparently nothing to lose, any behavior, however unscrupulous, would seem justified on the part of members who are trying to extricate themselves from this unfortunate situation.

As unfortunate as the everybody-worse-off situation is after the initial round of vote trading, members would do still worse by refusing to trade. For example, if in Table 4.5 member i refused to trade with members j

---

22. If all three members switched back from M_d's to M's on their two least salient roll calls in Table 4.6, they would thereby effect three pairwise vote trades, and the total utility gain to each member would be 7. This follows from the fact that the total utility of each member before trading—and after the subsequent trades, when there is a reversion to the original positions—is 14, but after the initial trades only 7, as we showed earlier. Hence, by reclaiming their initial positions before the initial vote trading, each member gains the difference of $14 - 7 = 7$ utiles.

23. Even in the case where the rules or norms of a legislature absolutely proscribe the breaking of trading agreements, this would not prevent members from calculating the probable behavior of others, others' calculations about their own calculations, and so forth. This mental juggling of anticipated trades may well carry members beyond the initial round of vote trading into subsequent rounds—at least in the course of their thinking, if not in the course of their actual trading. Even though members would not be able to cancel old trades, then, the irrevocable trades they do make will not necessarily be what we have called "initial trades." Trading cycles (and possible paradoxes), in other words, might be generated by some artful combination of mental constructions and physical trades—even when agreements are considered binding and trades cannot be undone.

and k, but they were willing to trade with each other, the positions of member i on the six roll calls after this trade would be:

$$N(1) \quad N(2) \quad N(3) \quad N(4) \quad M(5) \quad M(6).$$

These positions are clearly inferior to the positions of member i if he had traded with members j and k (see Table 4.6),

$$N(1) \quad N(2) \quad M(3) \quad M(4) \quad M_d(5) \quad M_d(6),$$

for the majorities he agrees with on roll calls 3 and 4 are more salient to him than the majorities he would have been left with on roll calls 5 and 6 had he not traded. In terms of the cardinal utilities given earlier, member i's total utility, summed across the six roll calls from most to least salient, would be

$$0 + 0 + 0 + 0 + 2 + 1 = 3$$

if he refused to trade,

$$0 + 0 + 4 + 3 + 0 + 0 = 7$$

if he agreed to trade.

Similarly, it is possible to show that it would not be in the interest of either member j or member k to refuse to trade if the other two members did trade. Since it is individually rational for the other pair always to trade, there is no way to freeze the desirable status quo positions shown in Table 4.5. This means that, despite the unfavorable consequences for all members set loose by the trades depicted in Table 4.5, it is irrational for any single member to refuse to trade.

Conceivably, the members might escape the paradox of Table 4.6 by all agreeing not to make the trades depicted in Table 4.5. For example, it might be said that if the members were foresighted, they would refuse to make trades at all, lest the trades lead to the realization of the paradox. Self-restraint is not easy, however. Suppose each member makes no trade himself, but he anticipates a trade between the pair that does not include him. That is, he himself behaves with self-restraint; but, with a conservative appreciation of what others can do to harm him, anticipates the worst from them.

This situation is analogous to an n-person Prisoner's Dilemma, which is simply an extension to n players of the two-person Prisoner's Dilemma discussed in section 1.8. If any member suspects that the others may not abide by their agreements not to trade, then his optimal strategy is to trade. For, as we have shown, the consequences of refusing to trade are more undesirable than the consequences of trading: Anticipating the worst (i.e., that the other pair will trade), a member maximizes his security level—the minimum amount he can guarantee for himself—by trading. In this way, it is rational to trade not only to maximize one's utility if other members do not trade but also to avoid results that could be even more unpalatable than those of the paradox of Table 4.6 when other members do trade. Trading is, therefore, the unconditionally best strategy for members. Thus, with every member forced to trade in order to behave rationally, each perversely may come to a worse end than if he did not trade and always voted sincerely.

To be sure, the logic of metagame theory (see sections 1.8 and 1.9) may enable members to extricate themselves from the paradox. In fact, as we argued in Chapter 1, the selection of the cooperative solution by players in certain two-person nonzero-sum games in which communication is allowed seems to provide prima facie evidence that players actually choose strategies on the basis of metastrategic calculations. But these calculations, and the conditional strategies on which they are based, offer a less persuasive resolution to Prisoner's Dilemma-type situations in n-person games, where contingent calculations become far more complex. As Russell Hardin has shown, this problem is equivalent to the problem of collective action in the theory of collective (or public) goods.[24] As applied to vote-trading situations, it means that every pair of members that can benefit from trading has no incentive to cooperate with other members whose positions they might adversely affect by trading. This makes it impossible for members of a voting body to reach a collective agreement not to trade unless there are sanctions (e.g., group norms) penalizing such actions.[25]

4.9. THE CONSEQUENCES OF FORMING COALITIONS

In the previous section we showed that the refusal of member i to trade, given that members j and k trade with each other, leads to a more undesirable outcome for member i than if he traded. Just as member i comes out worse off by refusing to trade, it follows from the assumption of individual rationality that members j and k come out better off.

Exactly how much better off, with member i out of the picture, is worth considering. In the case of member j, for example, his trade with member k changes his positions from those given in Table 4.5 to

$$M(3) \quad M(6) \quad N(5) \quad M(2) \quad M_d(1) \quad M(4)$$

---

24. Russell Hardin, "Collective Action as an Agreeable n-Prisoners' Dilemma," *Behavioral Science*, 16 (Sept. 1971), pp. 472–81. A *collective good* is one that, when supplied to some members of a collectivity, cannot be withheld from other members of that collectivity. Thus, a vote trade that changes the majorities on two roll calls is a collective good or bad, depending on one's preferences, to all members of the voting body.

25. For an interesting example in which effective sanctions were not (and probably could not) be imposed, and the attempt to concert action therefore failed, see J. Ronnie Davis and Neil A. Palomba, "The National Farmers Organization and the Prisoner's Dilemma: A Game Theory Prediction of Failure," *Social Science Quarterly*, 50 (Dec. 1969), pp. 742–48. For other examples and a general analysis of the "free rider" problem in the theory of collective goods, see Mancur Olson, Jr., *The Logic of Collective Action: Public Goods and the Theory of Groups* (Cambridge, Mass.: Harvard University Press, 1965); Norman Frohlich, Joe A. Oppenheimer, and Oran R. Young, *Political Leadership and Collective Goods* (Princeton, N.J.: Princeton University Press, 1971); and Jeffrey Richelson, "A Note on Collective Goods and the Theory of Political Entrepreneurship," *Public Choice*, 16 (Fall 1973), pp. 73–75.

after the pair exchanges support on roll calls 1 and 2. Had member i not refused to trade with members j and k, member j's positions after the three trades by all pairs of members on roll calls 3 and 6, 4 and 5, and 1 and 2 would have been (see Table 4.6)

$$N(3) \quad N(6) \quad M(5) \quad M(2) \quad M_d(1) \quad M_d(4).$$

While member j does pick up a majority that he agrees with on his third most salient roll call (i.e., 5) in exchange for voting insincerely on his least salient roll call (i.e., 4), he forfeits majorities on his two most salient roll calls (i.e., 3 and 6) when member i "cooperates" (by trading) rather than refuses to trade. In terms of the cardinal utilities given earlier, member j's total utility summed across the six roll calls from most to least salient is

$$6 + 5 + 0 + 3 + 0 + 1 = 15$$

when member i refuses to trade,

$$0 + 0 + 4 + 3 + 0 + 0 = 7$$

when member i does trade with the other two members. This difference, of course, is mostly due to the external costs inflicted upon member j by the vote trade between members i and k to which he is not a party.

The apparent lesson that follows from this example, which applies no matter which single member (i, j, or k) refuses to trade, is that two members can do better by agreeing to trade with each other but not the third member—that is, by forming a coalition whose members pledge to trade only with each other. Thereby they can obtain the benefits of trading and avoid the external costs of others' trades. Of course, this solution to the paradox simply denies that conditions (a), (b), and (c) in section 4.4, which define an individually rational trade, are operative for all members, but it is one way to cut the Gordian knot. It is comparable to cooperative solutions to Prisoner's Dilemma in which, with preplay communication allowed and the suspects able to make binding agreements, they agree never to confess and thereby thwart the prosecutor. In the absence of enforceable agreements, however, there may be instabilities in such solutions simply because they are not individually rational.[26] So it is here.

Referring again to Table 4.5, suppose members j and k coalesce on roll calls 1 and 2. Member i is excluded, but he may propose that he and member j coalesce instead on roll calls 4 and 5. Once this alternative is proposed, member j prefers the coalition with member i on roll calls 4 and 5 to the coalition with member k on roll calls 1 and 2 simply because

---

26. This illustrates the possible conflict between individual rationality and collective rationality. (The positions of members are collectively rational, or Pareto optimal, if there are no vote trades that would leave no voter worse off and at least one voter better off.) Without an enforceable agreement that guarantees the sanctity of the collectively rational positions (i.e., the status quo positions in Table 4.5 before the initial trades), it is in the interest of every member to break the agreement. Particularly if each member assumes that every other member is individually rational, all members together may arrive at positions which are not collectively rational (i.e., the positions in Table 4.6 before the subsequent trades).

he (member j) gains more from an improvement of his position on his more salient roll call, 5, than he does from an improvement on his less salient roll call, 2, and he loses less from voting insincerely on his least salient roll call, 4, than from so voting on his next to least salient roll call, 1. (Put somewhat differently, in member j's salience ranking in Table 4.5, the roll calls on which he trades with member k are "embedded" in the roll calls on which he trades with member i and hence are clearly inferior.) Thus, member j prefers coalition (i, j) to coalition (j, k), and of course member i—who is in the former but not the latter—also prefers coalition (i, j) to coalition (j, k). One can then write

$$(i, j) \, D \, (j, k),$$

where D means that the former coalition dominates the latter in that both members of the former prefer it to the latter.

By a similar argument it can be shown that coalition (j, k) dominates coalition (i, k). Suppose coalition (i, k) forms over roll calls 3 and 6. Then coalition (j, k) involving roll calls 1 and 2 is preferred by member k because he gains in coalition (j, k) on roll call 1, which is more salient for him than roll call 6, and loses in coalition (j, k) on roll call 2, which is less salient for him than roll call 3; i.e., roll calls $(3, 6)$ are embedded in roll calls $(1, 2)$ in member k's salience ranking. Hence

$$(j, k) \, D \, (i, k).$$

Finally, coalition (i, k), which involves roll calls 3 and 6, is preferred to coalition (i, j), which involves roll calls 4 and 5, simply because roll call 3 is of greater salience to member i than is roll call 4, and roll call 6 is of lesser salience to him than is roll call 5; i.e., roll calls $(4, 5)$ are embedded in roll calls $(3, 6)$ in member i's salience ranking. Hence

$$(i, k) \, D \, (i, j).$$

Putting these statements together, we obtain an intransitive preference ordering of all two-member (majority) coalitions:

$$(i, j) \, D \, (j, k) \, D \, (i, k) \, D \, (i, j).$$

Thus, when vote trading is unrestricted, there may be no single (majority) coalition which is dominant. In the above example, there is always one member of every two-member coalition who can be tempted by the offer of the third (left-out) member to defect and form a new coalition from which both derive greater total utility. (In game-theoretic terms, there is no imputation, or set of payoffs for members of a coalition that meet certain conditions to be described in Chapter 6, that dominates all other imputations with respect to all possible coalitions in the voting body.) When domination is intransitive, the system is unstable because the third (left-out) member has not only an incentive to break up an existing coalition but also always can do so by offering an alternative coalition that dominates an existing one.

TABLE 4.7 PREFERENCES OF MEMBERS FOR TRADES

MEMBER	PREFERENCES FOR TRADES		
	Highest	*Medium*	*Lowest*
i	(3, 6)	(4, 5)	(1, 2)
j	(4, 5)	(1, 2)	(3, 6)
k	(1, 2)	(3, 6)	(4, 5)

It can be shown that vote trading, even in the absence of the paradox of vote trading, is a product of the same conditions that lead to the paradox of voting.[27] In Table 4.7 we have given the preferences of each member for the possible trades that might be made. Thus, for member i his highest preference is for the trade on roll calls 3 and 6 with member k, because roll call 3 is the most salient roll call on which he can improve his position and roll call 6 is the least salient roll call on which he can vote insincerely. His medium preference is for the trade on roll calls 4 and 5 with member j, because 5 is the next most salient roll call on which he can improve his position and 4 is the next least salient roll call on which he can vote insincerely. His lowest preference is for the trade on roll calls 1 and 2 (between members j and k) on which he is left out. If, now, the members vote on which trades they wish to occur, the outcome will be an intransitive social arrangement:

(3, 6) is socially preferred to (4, 5) by members i and k,

(4, 5) is socially preferred to (1, 2) by members i and j,

(1, 2) is socially preferred to (3, 6) by members j and k,

so the arrangement is a cycle,

$$(3, 6) \rightarrow (4, 5) \rightarrow (1, 2) \rightarrow (3, 6),$$

where the arrow indicates a majority preference (i.e., by two of the three members). In sum, we have shown that no trade is preferred to all other

---

27. See Peter Bernholtz, "Logrolling, Arrow Paradox, and Cyclical Majorities," *Public Choice*, 15 (Summer 1973), pp. 87–95; Peter Bernholtz, "Logrolling, Arrow-Paradox and Decision Rules—A Generalization," *Kyklos*, 27 (Fasc. 1, 1974), pp. 41–62; and David H. Koehler, "Vote Trading and the Voting Paradox: A Proof of Logical Equivalence," *American Political Science Review* (forthcoming). Furthermore, the paradox of vote trading is not dependent on the majority rule assumption, as shown in Eric M. Uslaner and J. Ronnie Davis, "The Paradox of Vote Trading: Effects of Decision Rules and Voting Strategies on Externalities," *American Political Science Review* (forthcoming), where necessary and sufficient conditions for the occurrence of the paradox are also given.

trades, which is equivalent to our earlier demonstration that no coalition can defeat all others. Since there is no preferred set of trades on which to base a stable coalition, no stable coalition is possible. The paradox of vote trading cannot, therefore, be simply solved by waving it away with a coalition. Rather, it is inherent in the nature of the legislative process and, given an appropriate distribution of preferences and external costs, cannot be avoided.

This lack of a stable equilibrium is significant, for it shatters any hope one might harbor that a dominant collective preference can be ensured through vote trading.[28] Such instability extends our previous conclusions in section 4.7 about the cyclic nature of vote trading for individual members of a voting body to proper subsets of members, whose formation as coalitions that attempt to exclude vote trades with other members offers no guarantee that trading cycles can be banished once and for all. There seems, therefore, to be a fundamental disequilibrium in vote trading which is motivated by the assumption of individual rationality.

4.10. CONDITIONS LIMITING VOTE TRADING

It is evident that not all the assumptions of the model we have presented will accurately reflect characteristics of vote trading in the real world. In this section we consider some of the most important characteristics of real-world voting bodies that may limit the applicability of the model.

Incompleteness of Information

Not only are the positions of members on upcoming roll-call votes often not known by other members, but the salience rankings by members of future roll calls may be equally mysterious (even to themselves). In a temporal world, moreover, only a finite number of issues can be juggled at once. In a session of a typical national legislature, there are probably no more than, say, two thousand roll calls, which gives a theoretical maximum of one thousand trades on a single round. But, of course, many roll calls are trivial so that no trade is worthwhile; many others come up

---

28. For the case in which the preferences of members for outcomes on roll calls can be expressed in terms of cardinal utilities, R. E. Park has shown that there is in general no stable equilibrium if there is a majority that can, through vote trading, improve the payoffs to all its members over those without vote trading. Furthermore, he has shown that if there is a stable equilibrium with vote trading, it must be exactly the same as the outcome without vote trading. See R. E. Park, "The Possibility of a Social Welfare Function: Comment," *American Economic Review*, 57 (Dec. 1967), pp. 1301–1305. But see James S. Coleman, "The Possibility of a Social Welfare Function: Reply," *American Economic Review*, 57 (Dec. 1967), pp. 1311–17, who argues that Park's assumptions are unrealistically restrictive. For other "negative" results on vote trading, see John A. Ferejohn, "Sour Notes on the Theory of Vote Trading" (Social Science Working Paper Number 41, California Institute of Technology, June 1974).

for decision so quickly that the elaborate arrangements of logrolling are not possible; for many others it may be difficult for a member to discover another one with whom conditions (a), (b), and (c) for an individually rational vote trade can be satisfied; and, finally, since roll calls come up serially so that many future roll calls cannot be anticipated and so that past roll calls are irrevocably settled, only a small subset is available for trading at any one moment. In fact, it would seem, the practical maximum of trades may be at most one-tenth of the theoretical maximum. Consequently, in the typical legislature (with, say, one hundred members) in the typical session it may well be that the average member can expect to be involved in at most one trade, especially if trading agreements are considered binding. In general, the impediments to vote trading which we have indicated would probably be least serious in small voting bodies where there is a free exchange of views among members and future issues can more readily be anticipated.

Binding Agreements

Vote trading beyond the initial round (or some mentally constructed subsequent round), whether it involves all members or only subsets of members, can occur only if members do not consider their previous agreements binding. The willingness of members to break agreements, we have shown, sets in motion successive rounds of vote trading that may lead members back to their original positions and the resurrection of earlier outcomes.

In voting bodies whose members—for whatever reasons—are scrupulous about sticking to their pledges of support, the paradox in which members are collectively worse off may occur. Should this paradox occur, the unwillingness of members to trade further, which is a violation of the individual rationality assumption, will provide no escape. On the other hand, if agreements are not considered binding, in practice the paradox may also occur whenever actual balloting on a roll call terminates a round of vote trading that generates the paradox. Thus, although binding agreements in real voting bodies preclude indefinite cycling, such agreements probably have little effect on the occurrence of the paradox.

Party Discipline

With party discipline, by which we mean the sanctioning of trading agreements only among members of some subset of the voting body (e.g., a political party), the tools for control are stronger. When trading across party lines is forbidden, a majority of members who agree to trade only with each other (e.g., by forming a party) may not only be able to forestall vote-trading cycles, but they may also circumvent the paradox.

To understand why this is so, recall in our example in the last section that no majority of two members was able to resist the challenge of another majority that included the third (left-out) member. If members of the original majority form a party that categorically forbids trading with the third member, however, then no challenge to this majority is possible. This renders individually rational trades with nonparty members impossible, which in turn enables the original majority to avoid the paradox that unrestricted vote trading may produce (as in our example) after one round of trading.

Thus, just as one breaks a Prisoner's Dilemma by cooperation (members of the Mafia agree never to confess and captains enforce the agreement), so one can break the paradox of vote trading by agreeing not to trade across party lines. Since it is the absence of discipline—the freedom, instead, to trade with anyone—that renders vote-trading cycles and the paradox possible in the first place, we would expect to find most instances of these phenomena in voting bodies with weak or nonexistent party discipline. By contrast, in a legislature with just two disciplined parties, logrolling is probably extremely rare. This is undoubtedly one reason why so many American writers from Woodrow Wilson onward have admired the British Parliament, especially when it has approached the two-party situation in which conditions (a) and (b) for a vote trade cannot be satisfied. They should also have admired state legislatures like those of New York and Pennsylvania, which party bosses have usually run with an iron hand, because logrolling is equally impossible in them. But it has never been fashionable to admire boss rule.

Of course, once vote trading is banished from the legislature, political compromise goes on someplace else politically antecedent to the legislature. Thus, in state legislatures and city councils with disciplined parties, it is in the majority caucus or in the mind of the boss that the compromise takes place. In England, the Cabinet serves as one place of compromise and very likely something like vote trading goes on there; since the Cabinet situation is unstructured in comparison with the situation in the Parliament, however, it is probably hard to identify the trades and compromises that do occur.

Agreement on Which Roll Calls Are Most Salient

In our previous vote-trading example, each of the three members of the voting body rated as the two roll calls whose outcomes were most salient to him ones different from those chosen by the other two members. It was this asymmetry of choices by members of the most (as well as the least) salient roll calls that permitted them to support each other on different roll calls.

This is not the case when members give the same or similar salience rankings to all roll calls. In the extreme case when the salience rankings of all members coincide, some members may be in the majority and some in the minority on any particular roll call. But the members in the minority are not going to get support from the majority members by offering to vote insincerely on some lower salience roll call (where they are in the majority), for this roll call will also be of lower salience to the majority members. If the majority members, therefore, have no incentive to trade, the conditions for a vote trade clearly are not met.

Practically, it would seem, there will be more vote-trading possibilities the more varied are members' primary concerns (i.e., saliences for different roll calls). As a corollary, we might add that the more these primary concerns are rooted in relatively privatizable interests (e.g., rural versus urban), the more likely representatives from the different interests will support each other through vote trading. On the other hand, when the furtherance of members' interests depends on the provision of some public good from which none can be excluded (e.g., national defense), then vote-trading possibilities will be reduced because the effects of the public good will be felt by all.

Consistent Majorities

Vote trading will be inhibited by the existence of a subset of members whose sincere positions are those of the majority on most or all roll calls. One can visualize this circumstance in the extreme by imagining a voting body in which the positions of a simple majority of members are M's on all roll calls and the positions of all other members (i.e., the minority) are N's. In terms of our three-member voting body, the rows of our previous tables would contain M's on all roll calls for two members (say, i and j) and N's on all roll calls for one member (say, k). Since no member has any M's preceding N's in his salience ranking, each would have no less salient majorities to trade for more salient minorities. In this extreme case, the salience rankings by members on roll calls would have no bearing on the possibility of trades, as they did in the limitation described above.

Although the existence of a subset of members who consistently take the majority position may reflect the discipline of party-directed voting, party discipline as such is not what retards vote trading in this case. Rather, it is a *structural feature of positions*, just as it was a *structural feature of salience rankings* (as we saw in the above case of agreement on salience rankings), which limits the vote-trading possibilities. By contrast, incompleteness of information, binding agreements, and party discipline all relate to restrictions on permissible trades, which may be regarded as *substantive features* of the voting body; they include its environment (e.g., incompleteness of information) and its norms and procedures (e.g., binding agreements and party discipline).

4.11. EMPIRICAL EXAMPLES OF THE PARADOX OF VOTE TRADING

The popular distrust of logrolling probably stems from an intuitive, incomplete, but, nonetheless, sure comprehension of the vote-trading paradox, or at least of the external costs on which it is based. To make a judgment on the validity of this popular distrust requires, therefore, that one estimate the likelihood that the paradox arises in real legislatures. If it arises frequently, then surely the popular distrust has a reasonable basis. If it arises only rarely, however, then one probably ought to conclude that logrolling is a socially useful technique to approach collectively rational states of the world. Even though there seems no systematic way to investigate real-world frequencies of the paradox, one can specify types of circumstances in which it is likely to occur and then estimate the likelihood of these circumstances.

The crucial element of the paradox is external costs, which of course depend on the specific trade. It is always the case, however, that more people gain than lose on each switch, for the gainers (including the critical member who switches) necessarily constitute a majority of members. Consequently, in a system of trades—and usually for each participant as well—more instances of gain occur than do instances of loss. But gains and losses are not necessarily equal in magnitude, so the paradox occurs when, in a system of trades, members generally lose more when they lose than they gain when they gain. The question then is: Under what real-world circumstances might these large losses be expected to occur?

Manifestly, if there is only one trade, the paradox is impossible because the gainers on that trade cannot suffer external costs on another, which does not exist. Consequently, if trades are occasional and isolated and not part of a system of interrelated bargains, losses are probably less than gains for many members of the legislature. This kind of sporadic logrolling is often found in American legislatures, where party discipline is not usually very strong, but it is not the kind so earnestly recommended by the several scholars cited in the beginning of this chapter. Rather, the kind they recommend is that in which members engage in a constant round of bargaining in order to arrive at optimal outcomes over the whole set of issues in the legislative session. Unfortunately, it is exactly this kind of logrolling that probably increases the likelihood of the vote-trading paradox.

If one can separate out his reformist concerns, this is what Schattschneider seems to have perceived about the writing of the Smoot-Hawley tariff. Each member who joined the tariff combination for a price on some issue (usually a tariff on something manufactured in his district) gained an advantage for himself and his constituents. Such a large number joined, however, that protection was made indiscriminate—nearly everybody got something. Thereby, international trade was discouraged and disrupted— to the disadvantage of everybody in the society. To the extent that the Smoot-Hawley tariff deepened and extended the Great Depression in the 1930s, even the gainers probably suffered more than they gained.

One can see the same process at work in, for example, the Rivers and Harbors bills of the twentieth century when they had ceased to have the character of vital internal improvements that they had in the nineteenth century. A majority gains from the pork in such barrels, else the bills would not pass. But there are also losses, external costs. Once taxes are paid, the net gains of the gainers are probably pretty small. And if the projects are for the most part economically useless (e.g., locks on a canal that only a few pleasure boats use), then opportunity costs probably exceed the gainers' gains; if the projects are positively harmful (e.g., dikes on a river that would otherwise support wet lands and a higher water table), then their absolute losses almost certainly exceed the gainers' gains.

Still other possible cases come readily to mind. As a long-term example, there is the system of income tax exemptions and deductions, which are so generally distributed that they actually provide savings for only an unidentifiable few and at the same time probably assess very high costs against nearly everybody in the form of distortions of the market for many goods. For another and even longer-term example, there is the proliferation of army bases throughout the country, each providing a small economic benefit to its neighborhood, but by their inefficiency considerably increasing in sum the military cost for everybody.

As these examples suggest, the bills and circumstances likely to occasion paradoxes are those which bring together an interrelated set of issues, many of which are of interest to only a few legislators. Tax bills, internal improvement bills, and redistributions of income, to cite but a few cases, seem especially prone to providing the kind of situation in which the paradox can easily arise. Since these kinds of bills and business occupy a good part of any legislature's time, it must be concluded that the paradox of vote trading is a real fact of political life. One suspects that, because of it, logrolling leads more often away from, rather than toward, collectively rational social allocations.

With a bow in the direction of reformers, both those who recommend and those who excoriate logrolling, it might be asked whether or not occurrences of the paradox can be prevented. The answer is, probably not, short of highly unpalatable restrictions. Legislatures do occasionally adopt self-denying ordinances, like the so-called closed rule in the U.S. House of Representatives, a rule that prohibits amendments from the floor. Although this perhaps prevents the grosser and more thoughtless types of logrolling, anticipation of it probably forces the trading back off the floor into committee. Given all the rational pressures for trading, it is not really to be expected that a policy of self-denial will work. Stronger measures, like a system of responsible and disciplined parties, may well make trading impossible in the legislature, but they certainly cannot prevent it in the executive branch or in the party caucus.

It seems, then, that so long as we have the open kind of legislature in which any pair of members can trade, trading cannot be eradicated. Nevertheless, speaking normatively, the paradoxical consequences are unpalatable and one probably ought to try—by such devices as closed rules and popular condemnation—to discourage the kinds of logrolling that generate paradoxes.

4.12. SUMMARY AND CONCLUSION

At the beginning of this chapter, we gave an example in which a rationality (but not independence-from-irrelevant-alternatives) condition underlying Arrow's General Possibility Theorem was violated, which seemed to provide an escape from the paradox of voting (as do some path-independent procedures, also). In particular, voting procedures that allow members to express their intensities of preference were shown to violate this condition. Vote trading, by allowing members to vote insincerely on less salient roll calls in exchange for support from other members on more salient roll calls, *appeared* to be one device that permitted the expression of intensities and, consequently, avoidance of the paradox.

Yet we showed that without restrictions on vote trading, the paradox of vote trading cannot be avoided for at least some preference scales of voters. The example used to illustrate the analysis also demonstrated that when vote trades are possible, it is *always* individually rational, whatever round of vote trading is reached, for pairs of members to continue trading. For even if the vote-trading paradox looms ahead, as it did in our example, it is difficult for members to extricate themselves from the *n*-person Prisoner's Dilemma created by the paradox. Moreover, the formation of coalitions, whose members attempt to realize positive benefits through vote trading and prevent other members from partaking of them, may be vulnerable to challenges by other coalitions, which leads to indefinite cycling. These cycles, we showed in our example, could be traced to an underlying paradox of voting, which in general characterizes all vote-trading situations.

In short, there may be no stable equilibrium that is collectively rational (i.e., Pareto optimal) either for members who act singly or for members who attempt to form a coalition that restricts vote trading. This instability engenders trading cycles, sometimes vicious, that may lead to the paradox of vote trading.

Because the paradox of voting underlies all vote-trading situations, vote trading clearly does not violate the rationality condition since this condition is necessary for the occurrence of the paradox. Certainly if one interprets vote trading as simply one way in which support develops around some issues and not others, its effect is no different in principle from that which occurs when provisions are added to, and subtracted from, bills that gain some supporters and lose others. In the end, of course, when members of a voting body must decide on whether they will vote for or against a bill, they cannot express their intensities of preference. (This is why it was necessary to assume only that voters could rank-order the salience of roll calls, but not that they could attach cardinal utilities to each outcome.) In the absence of a mechanism that permits voters to register their intensities of preference in the actual voting process, their choices necessarily remain qualitative and the rationality conditions are not violated by vote trading.[29]

---

29. Arrow makes a similar argument in responding to critics in the second edition of *Social Choice and Individual Values*, p. 109.

We discussed several conditions in real voting bodies that would inhibit the free trading of votes, including incomplete information that members have about each other's salience rankings and an inability to anticipate the future agenda of issues. A general agreement by members on which roll calls are most salient, which may be reinforced by party discipline and may take the form of some members' voting consistently with the majority, would also tend to retard vote trading. The fact that vote-trading agreements are considered binding by members may dampen the cyclical effects of endless vote trading, but it provides no guarantee that the paradox of vote trading can be avoided.

Several examples of issues in which interests are relatively circumscribed, and where support can therefore more readily be exchanged, were suggested as those most likely to engender vote trading and, under certain conditions, the paradox. Devices that limit vote trading on such issues have in some cases been institutionalized—for example, the closed rule in the U.S. House of Representatives—which may discourage rampant trading and its most deleterious effects. These restrictions, however, usually do not prevent vote trading entirely but simply push it back to more private settings (e.g., committees), where exchanges occur in a smaller market among fewer members.

Our knowledge of the effects of vote trading is still scant, and it seems useful to conclude this chapter by indicating three areas in which further research would be helpful in extending the basic model presented in this chapter and refining some of the conclusions drawn from it. First, the results of our analysis for three-member voting bodies need to be generalized to bodies of any size.[30] Our three-member examples can, of course, be embedded in any legislature; and, of course, the three members can be three factions. But whereas the existence of the vote-trading paradox, vote-trading cycles, and unstable coalitions clearly extend to larger voting bodies, the general conditions under which these phenomena occur need to be further specified. As the number of members and the number of roll calls increase, the number of possible structural configurations of member positions and salience rankings increase enormously. A complete enumeration of these configurations and possible trades may be facilitated by use of the computer, as previous studies have demonstrated (see the Haefele and Mueller *et al.* articles cited in note 17 above and Coleman's book cited in note 19 above), but in the end theoretical simplifications undoubtedly will be required to generalize about different classes of vote-trading situations and behavior.

Second, the assumption that winning majorities on roll calls are minimal winning (with respect to a given decision rule) seems unduly restrictive if the model is to have general applicability to real voting bodies. If majorities are not minimal winning, then one might postulate trades among critical collective actors (e.g., several single members) who

---

30. In part this has been done by Peter Bernholtz, and Eric M. Uslaner and J. Ronnie Davis, in their articles cited in note 27 above.

are capable in concert of changing outcomes on roll calls. Alternatively, one might define the "criticalness" of members as their probability of changing outcomes in concert with others, from which one might develop a probabilistic notion of vote trading.

Third, in cases where members have the opportunity to choose from among several possible trades (about which they may be indifferent), one would need to extend the concept of rationality to take account of vote-trading possibilities on subsequent roll calls, which their earlier trades may affect. This would require members to anticipate trades, based on the sincere or insincere voting of other members, beyond the immediate round. The failure of members to trace out the possible consequences of subsequent rounds of trading when they—as well as, perhaps, their potential trading partners—are faced with more than one trading possibility on any particular round could mean that they will not make the best "long-run" choice on that round. In our simple examples, there were no alternative trading partners and it was always individually rational for members to trade, but this concept of rationality needs to be extended in order to distinguish between the putative states of instant gratification (i.e., making the single best trade or coalition choice on a round) and extended bliss (i.e., making the trade or coalition choice that will ensure optimal future trades or coalition choices, given knowledge of how others are likely to vote). These short- and long-term ends may very well be in conflict; they certainly involve patent game-theoretic considerations not incorporated in our vote-trading model as it now stands.

5

VOTING POWER

5.1. INTRODUCTION

In Chapters 2 through 4 we analyzed the optimal strategies of individual voters and cohesive blocs of voters in several different games. Prior to becoming a player in a game and selecting an optimal strategy, an actor might try to determine whether a game is worth playing at all, given that he has a choice of whether or not to become a participant. If a game is corrupt, immoral, or unethical, a prospective player may very well choose not to participate.

But what is "corrupt," "immoral," or "unethical" to one actor may not be so to another. Since there are no agreed-upon societal standards for evaluating games in these terms, it is useful to consider more restricted concepts of value. These concepts relate to what a player can get out of a game, which we shall assume depends solely on the outcome.

We commonly attribute "influence" or "power" to actors who can control outcomes, although other definitions of influence and power stress the effect that actors can have on each other.[1] For the purpose of defining the power of players in voting games, however, an outcome-oriented measure is preferable to an actor-oriented measure. In large voting bodies or even the electorate, where the influence of each person on every other person is for all practical purposes negligible, an actor-oriented measure would suggest that no one has any power. In fact, if each person has one vote, each person would have an equal chance to influence the outcome, which seems a more reasonable way to view power in voting situations.[2]

This view is not compatible with defining the voting power of an actor to be proportional to the number of votes he casts, because votes per se may have no bearing on outcomes. For example, in a three-member committee (a, b, c), where a has 4 votes, b 2 votes, and c 1 vote, members b and c are powerless if the decision rule is simple majority (4 out of 7).

---

1. For a good collection of readings on the concept of power, see *Political Power: A Reader in Theory and Research*, ed. Roderick Bell, David V. Edwards, and R. Harrison Wagner (New York: Free Press, 1969).

2. John F. Banzhaf III, "Weighted Voting Doesn't Work: A Mathematical Analysis," *Rutgers Law Review*, 19 (Winter 1965), n. 31, pp. 329–30.

Since the fact that members b and c together control $\frac{3}{7}$ of the votes is irrelevant to the selection of outcomes by this committee, we call these members *dummies*. Member a is a *dictator*, on the other hand, since his votes by themselves are sufficient to determine the outcome and only coalitions in which he is a member are winning. Note that there can be only one dictator in a voting body, whose existence renders all other members dummies, but there may be dummies without there being a dictator.

The votes cast by a member of a voting body are relevant in the selection of outcomes only in the context of the number of votes cast by other members and the decision rule of a voting body. All measures of voting power that we shall develop in this chapter utilize this information, albeit in different ways. After defining and illustrating these measures, we shall consider their relevance to the question of what prospects a player can entertain from participation in a game, as well as their relevance in the comparison of the power positions of players in several real voting games.

5.2. THE SHAPLEY-SHUBIK INDEX OF VOTING POWER

Consider a three-person voting body (or committee) whose members have 1, 49, and 50 votes. Assume that a simple majority of 51 votes is needed to win passage of measures. What would be a reasonable indicator of the voting power of each member of this committee?

If we consider these voters as players in a voting game of complete information, the reasoning may go something like this:

1. For the 1-vote player: (a) If I join the 49-vote player, our coalition (49, 1) will not be winning; (b) if I join the 50-vote player, our coalition (50, 1) will be winning.

2. For 49-vote player: (a) If I join the 1-vote player, our coalition (1, 49) will not be winning; (b) if I join the 50-vote player, our coalition (50, 49) will be winning.

3. For 50-vote player: (a) If I join the 1-vote player, our coalition (1, 50) will be winning; (b) if I join the 49-vote player, our coalition (49, 50) will be winning.

From the perspective of which of the three players joins what other player, there are thus six possible ways in which a two-member coalition can form, and for each of these only one way in which the third (remaining) player can join. If there is no a priori information on the order in which each member will join a coalition, it is convenient to assume that all orders will be equally likely; each will therefore occur with probability equal to $\frac{1}{6}$.

Assume that the value of coalition S is 0 if it is losing and 1 if it is winning. Then we define the *characteristic function v* of a voting game as follows:

$v(S) = 0$ if S is a losing coalition;

$v(S) = 1$ if S is a winning coalition.

The characteristic function assigns a value to every subset (or coalition) of players; this value is the minimum amount that the coalition can ensure for itself whatever other coalition(s) forms.[3] Voting games in which there are only two values, 0 and 1—associated with losing and winning, respectively—are called *simple games*.[4] The simple game we have described is also called a *weighted majority game*, since it is a game defined by players of specified weights and a decision rule (e.g., simple majority) that distinguishes between coalitions of winning and losing size.

The value of the characteristic function of this game is 1 for two of the two-member coalitions—(1, 50) and (49, 50)—and 0 for the remaining two-member coalition—(1, 49). It is 0 for all of the three one-member coalitions and 1 for the *grand coalition*, which is the coalition containing all three members. For the coalition consisting of no members—denoted by ϕ (the Greek letter "phi"), which signifies the empty or null set—its value is assumed to be 0. In sum, for all subsets of members, the characteristic function for the previously described game is:

$$v(\phi) = 0;$$

$$v(1) = v(49) = v(50) = 0;$$

$$v(1, 49) = 0, \quad v(1, 50) = v(49, 50) = 1;$$

$$v(1, 49, 50) = 1.$$

In general, in an n-person game there are 2^n subsets of players ($2^3 = 8$ in the above example), and the characteristic function v assigns values—not necessarily different—to each of these subsets.

The characteristic function of a game must satisfy two conditions:

(1) $\qquad\qquad\qquad v(\phi) = 0,$

(2) $\qquad\qquad\qquad v(S \cup T) \geq v(S) + v(T).$

---

3. More specifically, this minimum amount is based on the assumption that a single opposition coalition forms that minimizes the maximum payoff to S in a two-person zero-sum game. See R. Duncan Luce and Howard Raiffa, *Games and Decisions: Introduction and Critical Survey* (New York: John Wiley & Sons, 1957), p. 191. Although this assumption leads to a rather pessimistic estimate of a coalition's value, it is not unreasonable in voting games, where a majority (not necessarily simple) wins the same amount irrespective of what other coalition(s) forms.

4. See L. S. Shapley, "Simple Games: An Outline of the Descriptive Theory," *Behavioral Science*, 7 (Jan. 1962), pp. 59–66.

Condition (1) is a technical requirement and specifies that a coalition consisting of no members has a value equal to 0. Condition (2), which is called *superadditivity*, specifies that two subsets of players S and T that form a coalition and act in concert, provided $S \cap T = \phi$ (S and T are disjoint—they have no members in common), can obtain at least as much and possibly more by combining rather than acting separately.[5] (In other words, no coalition can achieve a higher value by splitting into smaller coalitions.) This condition provides an incentive for the formation of coalitions since their members profit from joint action, at least if the inequality is strict (i.e., "$>$") in condition (2). It is easy to show that both conditions are satisfied in the game previously described.

To return to this game, we suggested earlier that with no prior information about the order in which voters would join to form coalitions, we could assume that each of the three players would approach one of the other two players with probability equal to $\frac{1}{2}$. With the addition of the third (and last) player, we can now identify all of the ways in which value can be added in the formation of the grand coalition:

(1a) 1-vote player joins 49-vote player, bringing value of 0; then followed by 50-vote player bringing value of 1.

(1b) 1-vote player joins 50-vote player, bringing value of 1; then followed by 49-vote player bringing value of 0.

(2a) 49-vote player joins 1-vote player, bringing value of 0; then followed by 50-vote player bringing value of 1.

(2b) 49-vote player joins 50-vote player, bringing value of 1; then followed by 1-vote player bringing value of 0.

(3a) 50-vote player joins 1-vote player, bringing value of 1; then followed by 49-vote player bringing value of 0.

(3b) 50-vote player joins 49-vote player, bringing value of 1; then followed by 1-vote player bringing value of 0.

The first concept of voting power that we shall define is based on the *incremental* (or marginal) value a player contributes to a coalition when he joins it, weighted by the probability that the player will join the coalition. For the 1-vote player, this expectation is

$$\tfrac{1}{6}(0 - 0) + \tfrac{1}{6}(1 - 0) + \tfrac{1}{6}(0 - 0) + \tfrac{1}{6}(0 - 0) + \tfrac{1}{6}(0 - 0) + \tfrac{1}{6}(0 - 0) = \tfrac{1}{6}$$

since only in case (1b), when he joins the 50-vote player, does the 1-vote player increase the value of a coalition from 0 to 1. Similarly, the expectation of the 49-vote player is

$$\tfrac{1}{6}(0 - 0) + \tfrac{1}{6}(0 - 0) + \tfrac{1}{6}(0 - 0) + \tfrac{1}{6}(1 - 0) + \tfrac{1}{6}(0 - 0) + \tfrac{1}{6}(0 - 0) = \tfrac{1}{6},$$

---

5. "$S \cup T$," called the *union* of sets S and T, is the set of all members belonging to S or T or both. "$S \cap T$," called the *intersection* of sets S and T, is the set of all members belonging to both S and T.

and the expectation of the 50-vote player is

$$\tfrac{1}{6}(1 - 0) + \tfrac{1}{6}(0 - 0) + \tfrac{1}{6}(1 - 0) + \tfrac{1}{6}(0 - 0) + \tfrac{1}{6}(1 - 0) + \tfrac{1}{6}(1 - 0) = \tfrac{4}{6} = \tfrac{2}{3}.$$

These expectations are called the *Shapley value* of the game,[6] which we denote by the "power" vector $(\tfrac{1}{6}, \tfrac{1}{6}, \tfrac{2}{3})$ for the players with "vote" vector (1, 49, 50). The fractional components of the power vector sum to one, since one and only one player is decisive in each of the six possible orderings in which the grand coalition can form. They provide us with a notion of how the payoff from winning is to be apportioned among players based on their "bargaining power" in all coalitions.

It is immediately apparent in this example that there is no linear relationship between the players' proportions of votes and their proportions of voting power as given by the Shapley value. The player with 1 percent of the votes has $16\tfrac{2}{3}$ percent of the voting power, which is the same amount of power as that held by the player with 49 percent of the votes. The 50-vote player, on the other hand, with only one more vote than the 49-vote player, has four times as much power as either of the other two players.

Although these deviations from voter weights are extreme, there is in general no analytic relationship between votes and voting power, except when (occasionally) some, or all, voters are weighted equally.[7] When all voters cast the same number of votes, the Shapley value assigns the same voting power to each voter.[8] When voters are unequally weighted, the exact calculation of Shapley values can be quite difficult, especially for large voting games, although at least one algorithm has been developed that facilitates the approximate calculation of such values.[9]

We earlier interpreted the Shapley value as an expected value that a player can get out of a game, based on his incremental contributions to the grand coalition as successively more members are added in all possible orders. The value was originally proposed, however, not as an indicator of the distribution of power in a committee—this was later suggested as an *application* of the value by Shapley and Shubik[10]—but as a measure

---

6. L. S. Shapley, "A Value for N-Person Games," in *Annals of Mathematics Studies (Contributions to the Theory of Games,* ed. H. W. Kuhn and A. W. Tucker), 28 (Princeton, N.J.: Princeton University Press. 1953), pp. 307–17.

7. William H. Riker and Lloyd S. Shapley, "Weighted Voting: A Mathematical Analysis for Instrumental Judgments," in *Representation: Nomos X,* ed. J. Roland Pennock and John W. Chapman (New York: Atherton Press, 1968), n. 9, p. 208.

8. Other isolated examples have been found in which the distribution of votes and voting power agree—e.g., a four-member committee with vote vector (5, 3, 3, 1), where the decision rule is simple majority—but no constructive method has been discovered that generates all such cases.

9. Guillermo Owen, "Multilinear Extensions of Games," *Management Science,* 18 (Jan., Part 2, 1972), pp. P-64–P-79; see also S. C. Littlechild and G. Owen, "A Simple Expression for the Shapley Value in a Simple Case," *Management Science,* 20 (Nov. 1973), pp. 370–72. A related concept of value is developed in G. Owen, "Values of Games without Side Payments," *International Journal of Game Theory,* 1 (1971/72), pp. 95–109.

10. L. S. Shapley and Martin Shubik, "A Method of Evaluating the Distribution of Power in a Committee System," *American Political Science Review,* 48 (Sept. 1954), pp. 787–92.

of the a priori value to players for participation in a game. This measure provides a way of distributing among all players coalition values given by the characteristic function, which, Shapley proved, satisfies three conditions. Very roughly, these are:

1. The value of a game to a player depends only on the characteristic function of the game—the structural relationship between the values of coalitions and their membership—and not on the manner in which the players are labeled.

2. The values for all players are additive and sum to one (in a simple game), which is the value of the game for all players in the grand coalition.

3. The value of a player in two separate games is equal to his value in the single composite game comprising the separate games.

It is remarkable that the Shapley value not only satisfies these three conditions but that it is also the *only* measure that does so. In the social sciences it is both rare and gratifying that an index, applicable to real-life situations (as we shall show presently), can be uniquely defined by an apparently reasonable set of conditions that it satisfies.

Although the first two conditions seem unexceptionable, Luce and Raiffa point out that the third condition reflects a property of games that may not jibe with commonsensical notions about composite games. In particular, they argue, there is no reason to expect that a composite game, which may have its own peculiar structure, will be played as if it were two separate games.[11] Indeed, we shall give a real-life example later in which a composite game cannot be treated as two separate games because of the interdependency of institutional structures in the different games that compose it.

It is useful to introduce the notion of "pivot" in calculating the Shapley-Shubik index of voting power based on the Shapley value. For all the different ways in which the buildup of the grand coalition can occur through a sequence of additions of one player, there will be one player in each sequence who will be decisive to a coalition's becoming winning. We call this player, who makes the difference between a coalition's being winning or losing, the *pivot*, and we designate the position he occupies in the sequential buildup of the coalition to be the *pivotal position*. The pivots

---

The Shapley–Shubik index of voting power is a limiting case of a more general concept of bargaining power proposed in John C. Harsanyi, "Measurement of Social Power in *n*-Person Reciprocal Power Situations," *Behavioral Science*, 7 (Jan. 1962), pp. 81–91. In Harsanyi's bargaining model, a player's bargaining power is based not only on his pivotal power but also on his ability to change the behavior of other players through incentives, threats, and compromise. Because of the computational difficulties involved in the calculation of Harsanyi's measure, which in general requires an iterative procedure for finding a numerical value, we shall not attempt to apply it.

11. Luce and Raiffa, *Games and Decisions*, p. 248.

in our earlier example are given below in parentheses for each of the different ways in which the grand coalition can form:

$$
\begin{array}{lll}
1 & 49 & (50), \\
1 & (50) & 49, \\
49 & 1 & (50), \\
49 & (50) & 1, \\
50 & (1) & 49, \\
50 & (49) & 1.
\end{array}
$$

In four of these arrangements, the pivotal player is the second voter to join the coalition, making the pivotal position second in these arrangements; in the other two arrangements, the pivotal player is the third voter to join the coalition, making the pivotal position the third position in these arrangements.

If, as before, we consider these different arrangements to be equally likely, the 50-vote player will be pivotal with probability equal to $\frac{4}{6} = \frac{2}{3}$, and each of the two other players will be pivotal with probability equal to $\frac{1}{6}$. Thus, the Shapley value can be interpreted as a measure of the *pivotalness* of each player—that is, the probability that he will be a pivot in a coalition—given that all different sequences in which players join the grand coalition are equiprobable.

In the foregoing treatment, we have implicitly assumed that coalitions are of two types—winning or losing. In simple majority games of n players, where all players have the same numbers of votes, a coalition is *winning* if it contains more than $n/2$ members and *losing* if it contains $n/2$ or fewer members. Of course, if n is an even number, *blocking* coalitions of exactly $n/2$ members, which neither win nor permit their complements to win, are also possible. This suggests that a measure of the "blocking power" of players, based on their pivotalness in *defeating* motions, would also be germane to the study of voting power.

To illustrate the calculation of this blocking measure in our previous example, assume that players cast "negative" votes against a motion in the six different arrangements specified earlier. In each of these arrangements, the player who casts the decisive negative vote is given in parentheses below:

$$
\begin{array}{lll}
1 & (49) & 50, \\
1 & (50) & 49, \\
49 & (1) & 50, \\
49 & (50) & 1, \\
(50) & 1 & 49, \\
(50) & 49 & 1.
\end{array}
$$

Note that while the pivotal positions of each of the three players are not necessarily the same as those based on their ability to cast a decisive "positive" vote that makes a coalition winning, the proportion of pivots of each player remains the same, which is in general the case whatever are the weights of the players or the decision rule of the voting body. For the purpose of measuring voting power, then, the winning and blocking measures, based on all possible orders in which members join the grand coalition, yield the same result.

In proposing their index of voting power, Shapley and Shubik remarked, "Our definition of the power of an individual member depends on the chance he has of being critical to the success of a winning coalition."[12] Yet, although the Shapley value uniquely satisfies the three conditions given earlier, it is not the only reasonable measure of voting power that can be anchored to the concept of a critical vote, as we shall show in section 5.3.

5.3. THE BANZHAF INDEX OF VOTING POWER

It is important to note that the Shapley-Shubik index is based on the assumption that the pivotalness of a player is proportional to the number of *different orders* (or arrangements) in which he is pivotal. This may not be the same as the number of *distinct* (*winning*) *coalitions* in which his presence is critical to the outcome—that is, which would be rendered nonwinning (i.e., blocking or losing) were he to resign or defect. For example, the 50-vote player in our example is pivotal in the two arrangements

$$1 \quad 49 \quad (50),$$
$$49 \quad 1 \quad (50),$$

but ignoring the order in which the first two players join, his defection is critical in only one coalition, (1, 49, 50), based on these different arrangements.

The defection of the 50-vote player is also critical in the two-member coalitions, (1, 50) and (49, 50). The first is based on the single arrangement,

$$1 \quad (50) \quad 49,$$

and the second on the single arrangement,

$$49 \quad (50) \quad 1.$$

Similarly, the defection of the 1-vote player is critical in the coalition (50, 1), based on the single arrangement,

$$50 \quad (1) \quad 49,$$

and the defection of the 49-vote player is critical in the coalition (50, 49), based on the single arrangement,

$$50 \quad (49) \quad 1.$$

12. Shapley and Shubik, "A Method for Evaluating the Distribution of Power in a Committee System," p. 787.

Altogether, there are three coalitions in which the defection of the 50-vote player is critical and one each in which the defections of the 1-vote and 49-vote players are critical, which makes for a total of five coalitions in which the defection of at least one player is critical. The subtraction of

the 50-vote player from coalitions (1, 49, 50), (1, 50), or (49, 50),

the 1-vote player from coalition (50,1),

the 49-vote player from coalition (50, 49),

would transform each of these winning coalitions into a nonwinning coalition.

We call such winning coalitions *minimal winning* with respect to the players whose defection can change their status to nonwinning. Note that a coalition [e.g., (1, 49, 50)] may be minimal winning with respect to the defection of some players (the 50-vote player) but not others (the 1-vote and 49-vote players).[13] We call a defection a *critical defection* if and only if it transforms a minimal winning coalition (with respect to at least one player) into a nonwinning coalition.

Of the five minimal winning coalitions we have listed, the two given for the 1-vote and 49-vote players, (50, 1) and (50, 49), duplicate two of the three, (1, 49, 50), (1, 50), and (49, 50), given for the 50-vote player, except for the order in which the members are listed. Thus, only three of the five minimal winning coalitions are distinct: (1, 49, 50), (1, 50), and (49, 50).

As a second measure of voting power, John F. Banzhaf III has defined the power of a player to be equal to the number of his critical defections in *distinct minimal winning coalitions* (DMWC's) as a proportion of the total number of critical defections of all players combined. Thus, in our three-member voting body, the defection of the 50-vote player is critical in all three DMWC's, and the defections of the 49-vote and 1-vote players are critical in one each, so there are a total of five defections which are critical in the three DMWC's. The voting power of the 50-vote player is therefore $\frac{3}{5}$, and the voting power of each of the other two players is $\frac{1}{5}$.

This index of voting power has been used by Banzhaf in studies of weighted voting and multimember electoral districts,[14] and, combined

---

13. The coalition (1, 49, 50) is minimal winning with respect to the defection of the smaller (nonwinning) coalition (1, 49), but not to the individual players in this smaller coalition. In our subsequent treatment, we shall restrict the analysis to coalitions that are minimal winning with respect to the defection of *single players*, but the definition of the power of players in minimal winning coalitions could obviously be extended to *nonwinning coalitions* of players.

14. Banzhaf, "Weighted Voting Doesn't Work: A Mathematical Analysis"; John F. Banzhaf III, "Multimember Electoral Districts—Do They Violate the 'One Man, One Vote' Principle?" *Yale Law Journal*, 75 (July 1966), pp. 1309–88.

with a measure of citizen voting power, the Electoral College.[15] By Banzhaf's definition, a player's voting power is proportional to the number of DMWC's in which his presence is *necessary* for the coalition to remain minimal winning. (The 50-vote player's presence in a coalition is *sufficient* for him to be blocking but not to be winning, where his presence is always necessary.)[16] Rooted in a necessary-and-sufficient-condition notion of causality, the Banzhaf definition of voting power avoids the recipe-type manipulative element that Riker finds in the Shapley-Shubik index.[17] For in the case of the Shapley-Shubik index, there are usually other players besides the pivot whose presence is necessary for a coalition to remain winning in each arrangement, but they are counted in the index only in the arrangements where they are in the pivotal position. In the Banzhaf index, on the other hand, all the minimal winning coalitions in which a player's defection is critical are counted—once and only once—however the members may be arranged before and after the pivot.

By comparison with the Shapley-Shubik power values of $(\frac{1}{6}, \frac{1}{6}, \frac{2}{3})$ for the voting body whose member weights are (1, 49, 50), the Banzhaf index, which gives power values of $(\frac{1}{5}, \frac{1}{5}, \frac{3}{5})$, does not change the (equal) voting power of the two least weighty members relative to each other, but it does increase somewhat their voting power relative to the weightiest member. This change is not dramatic, however, and most analysts would probably be hard pressed to say, in the absence of other information, which index better reflects the realities of power relationships in the three-member voting body. If an analyst did have a preference for one index, it seems reasonable to demand that this measure also better capture the power realities in bodies whose members are differently weighted—for example, in bodies with member weights of (1, 1, 2) or (1, 2, 3).

In fact, the "reality" of voting in all three bodies is the same under simple majority rule: Whichever of the indices of voting power one chooses, the power distributions are the same for all three voting bodies—$(\frac{1}{6}, \frac{1}{6}, \frac{2}{3})$ using the Shapley-Shubik index, and $(\frac{1}{5}, \frac{1}{5}, \frac{3}{5})$ using the Banzhaf index. It is hard to think of an argument that could be used to justify the choice of one index as uniformly superior to the other when applied to these very differently weighted bodies.

---

15. John F. Banzhaf III, "One Man, 3.312 Votes: A Mathematical Analysis of the Electoral College," *Villanova Law Review*, 14 (Winter 1968), pp. 304–32.

16. In a weighted majority game, a single player's presence in a coalition is sufficient for him to be winning if the player alone constitutes a winning coalition. Such a player's presence is also necessary, since no winning coalition can form without him.

17. To expunge this element from the index, Riker suggests a modification that takes into account the proportion of arrangements (not DMWC's) in which a player's presence is necessary, but this modification is subject to the same criticism that we offer in section 5.4 of the model on which the original Shapley-Shubik index is based. See William H. Riker, "Some Ambiguities in the Notion of Power," *American Political Science Review*, 58 (June 1964), pp. 341–49. For a variation on Banzhaf's index based on player preferences as well as weights, see Peter C. Fishburn, *The Theory of Social Choice* (Princeton, N.J.: Princeton University Press, 1972), pp. 53–55.

We call voting games *power equivalent* in which the distribution of power among the players is the same.[18] Besides the three games listed above, there is an infinite number of other three-person voting games consisting of players of different weights and using different decision rules that are power equivalent. In general, if two or more games have the same characteristic function, they are power equivalent.

5.4. COALITION MODELS OF THE TWO POWER INDICES

Both indices of voting power that we have defined depend only on the characteristic function of a game. They can be defined independently of voting procedures, types of voting, and so on. Although these considerations were relevant to our analyses in the previous chapters, they have no bearing on the measurement of voting power as it has been defined in this chapter. Despite their dependence on the same information, however, the two indices use it in different ways. If we cannot make an unequivocal choice between these indices on the basis of the examples so far presented, some consideration of the voting models on which they are based may prove helpful.

Both models focus on players who are crucial to the outcome of the voting process. If a player is never crucial to an outcome, he is by definition a dummy since he cannot contribute anything to a coalition or be the pivot in any arrangement of players. This role is played, for example, by the 1-vote player in the weighted voting body (1, 3, 3, 4) under a decision rule of simple majority (6 out of 11). Both indices of voting power give the same power vector, $(0, \frac{1}{3}, \frac{1}{3}, \frac{1}{3})$, for members of this body, which also shows that the 4-vote member in this body is not advantaged in relation to the two 3-vote members.

Both the Shapley-Shubik index and the Banzhaf index postulate the existence of minimal winning coalitions. The *processes* by which such coalitions form and break up, however, are based on rather different conceptions of coalition formation. Consistent with the Shapley-Shubik index is a model that assumes members of a voting body can be ordered according to their likelihood of supporting a bill. The manager of the bill who seeks to enlist the support of a minimal majority of members will typically bargain with undecided members (e.g., by modifying provisions in the bill or promising support on other bills) in order to build up support for it:

Presumably the least hostile will become a supporter for the lowest price (i.e., the smallest promise), the next least hostile for the next

---

18. Power equivalence should not be confused with the concept of "strategic equivalence" in game theory, which is a relation between two games which means roughly that they contain the same strategic possibilities (e.g., inducements to form coalitions). For a formal definition of strategic equivalence, and its relationship to the Shapley value, see Ewald Burger, *Introduction to the Theory of Games* (Englewood Cliffs, N.J.: Prentice-Hall, 1963), Definition 26, pp. 165–66, and Theorem 54, pp. 189–90.

higher price, and so forth. Since we can assume the manager of the bill has limited resources, we can assume also that he will buy the least hostile undecided first, the next least hostile second, and so forth. Thus, clearly the highest price he must pay goes to the legislator in the pivot position.[19]

This seems to be a plausible description of coalition-formation processes in many committees, legislatures, and even collectivities in which a formal vote may not be taken when one coalition appears to be dominant. What is a good deal less plausible, however, is the assumption that all orderings of members not only occur but are equally likely.

It is seldom recognized that in order for all orderings to occur, there must exist a number of attitudinal dimensions (e.g., liberalism/conservatism) along which members can be arrayed (e.g., from very liberal to very conservative) that is equal to the number of different orderings. For some three-member committees, perhaps, it may not be implausible that members (a, b, c) take positions on a series of issues such that

1. a is most supportive, b next, and c least;

2. a is most supportive, c next, and b least;

3. b is most supportive, a next, and c least;

4. b is most supportive, c next, and a least;

5. c is most supportive, a next, and b least;

6. c is most supportive, b next, and a least.

If the six dimensions underlying these preference orders occur on issues that come before the committee with more or less equal frequency over time, then the previously described model of the Shapley-Shubik index seems quite satisfactory.

Yet, with the expansion of a three-member committee to four members, the number of possible arrangements of these members jumps to twenty-four, and with the expansion to five members the number of possible arrangements soars to 120. In general, if there are n members, there are $n!$ (read "n factorial") different arrangements or orders.[20]

In calculating the Shapley-Shubik index for weighted voting bodies with as few as ten members—not to mention bodies like the U.S. Electoral College with fifty-one bloc members—the number of different arrangements of the members is truly enormous. It is inconceivable that there are

---

19. Riker and Shapley, "Weighted Voting: A Mathematical Analysis for Instrumental Judgments," p. 203.

20. $n! = (n)(n - 1) \ldots (2)(1)$. Its interpretation in the preceding example is that in an n-member voting body, there are n ways of selecting the most supportive member; once he is chosen, there are $(n - 1)$ ways of selecting the next most supportive member, $\ldots$, and finally only one way of selecting the (last remaining) least supportive member.

$n!$ attitudinal dimensions along which members of such bodies can be arrayed that take the form of issues equally likely to come before the voting body.

Empirically, the concrete issues on which members of voting bodies vote usually tap only a very small number of different attitudinal dimensions.[21] This fact tends to enhance the pivotal power of those voters who are least committed to positions on the relevant dimensions, as Glendon A. Schubert showed in an application of the Shapley-Shubik index to the bloc voting of justices in the 1936 term of the U.S. Supreme Court.[22]

Riker and Shapley argue that if there is no dimension or scale along which voters can be ordered and the pivotal voter identified,

> this is not a serious drawback, since we are ultimately concerned not with votes on individual bills, but with the distribution of power inherent in the voting system itself. Those legislators who are assumed to have just missed the pivot position, in a doubtful case, will nevertheless be recognized in the power index, to the extent that other, equally likely arrangements (in time or in viewpoint) exist in which they would be pivotal.[23]

It is precisely the existence of "equally likely arrangements" that we have questioned.

This equiprobability assumption is dropped in the Banzhaf index, where defections from distinct minimal winning coalitions (DMWC's) that render them nonwinning replace pivots in different arrangements as the new unit of analysis. The order in which members of a voting body

---

21. Statistical methods (e.g., Guttman scaling and factor analysis) for deriving attitudinal dimensions of voting, and their application to voting in actual legislative and judicial bodies, are described in Lee F. Anderson, Meredith W. Watts, Jr., and Allen R. Wilcox, *Legislative Roll-Call Analysis* (Evanston, Ill.: Northwestern University Press, 1966); and Duncan MacRae, Jr., *Issues and Parties in Legislative Voting: Methods of Statistical Analysis* (New York: Harper & Row, 1970).

22. After demonstrating that the middle bloc of two justices had as much voting power as the three-justice liberal bloc and the four-justice conservative bloc when these blocs are treated as single players (i.e., not as the nine justices that compose them) in the calculation of the Shapley-Shubik index, Schubert confirmed that these theoretical values for the blocs closely approximate the empirical values that justices derive from being members of winning coalitions when the payoff from winning is divided equally among the members of the winning coalition. See Glendon A. Schubert, "The Study of Judicial Decision-Making as an Aspect of Political Behavior," *American Political Science Review*, 52 (Dec. 1958), pp. 1022–24. For a similar attempt to calculate the empirical power of blocs in the U.S. House of Representatives in the Eighty-sixth Congress, see William H. Riker and Donald Niemi, "The Stability of Coalitions on Roll Calls in the House of Representatives," *American Political Science Review*, 56 (March 1962), pp. 58–65; in this study, Riker and Niemi found that blocs are stable on some issues but not on others, as measured by changes in their pivotal power over time. In another empirical application of the Shapley-Shubik index, power differences between parties at the state and federal level were used to measure partisan "disharmony" in government. See William H. Riker and Ronald Schaps, "Disharmony in Government," *Behavioral Science*, 2 (Oct. 1957), pp. 276–90.

23. Riker and Shapley, "Weighted Voting: A Mathematical Analysis for Instrumental Judgments," p. 203.

join a coalition is irrelevant to this definition of voting power. In constructing a model of this index, therefore, we can ignore the history that led up to the formation of a minimal winning coalition and instead start by immediately postulating the existence of minimal winning coalitions.

We then conceive a process in which it is observed over time that the (potential) defection of some members is critical more frequently than others, given that all DMWC's form over time. The Banzhaf index of voting power simply formalizes this observation by defining a player's voting power to be equal to his proportion of critical defections. The concept of "critical defection" in this model conveys better than the concept of "pivot" in the model of the Shapley-Shubik index how a member of DMWC can use the threat of breaking up an already formed minimal winning coalition as bargaining power to extract concessions or payments from a coalition leader.

In most real voting bodies, of course, party, ideology, and other constraints render improbable, if not impossible, the formation of certain minimal winning coalitions. In Chapter 6 we shall review evidence pertinent to this question as well as consider the conditions under which minimal winning coalitions are likely to form and break up. For now it is important to note that the major parameter that differentiates the two models on which the two different indices of voting power are based is the *time* that minimal winning coalitions are observed.

The Shapley-Shubik index envisions a world in which the attitudinal dimensions that underlie players' preferences are sufficiently heterogeneous that over time the order in which players join the (putative) grand coalition will reveal no particular pattern but instead will be essentially random; voting power accrues to a player on the basis of his pivotalness (his probability of being pivotal) when all orders in which players join the grand coalition are considered equally likely. The Banzhaf index, on the other hand, envisions a world in which minimal winning coalitions have already formed and power accrues to a player on the basis of the credibility of his threat to defect, which is measured by his proportion of critical defections in all minimal winning coalitions. Thus, the determination of a player's power by each index may be conceptualized as occurring at historically different points in time—in one case (Shapley-Shubik) during the buildup of the grand coalition, in the other case (Banzhaf) after the formation and during the disintegration of minimal winning coalitions.

To be sure, the breakup and formation of coalitions are two sides of the same coin: After the breakup of a winning coalition, new coalitions form, and typically one becomes winning. However, what is absent from the idea of a pivot in the Shapley-Shubik index is the notion that a player's presence is *necessary* to a winning coalition's continued existence, which is basic to the idea of a critical defection—and the possible breakup of minimal winning coalitions—in the Banzhaf index.

In summary, the Shapley-Shubik index counts the different ways that the grand coalition can form and keeps track of which player assumes the pivotal position in each, whereas the Banzhaf index counts the different ways in which minimal winning coalitions can break up—however they formed—and keeps track of which players are critical to each. Unfortunately, although the formation/disintegration coalition models on which the two indices are based offer some insight into the conceptualization of power implicit in each index, they offer no insight into *quantitative* differences in the values generated by the two indices for different voting bodies. Since no work appears to have been done that elucidates general mathematical relationships between the indices, it is useful to consider what differences arise in their application to real voting bodies. First, however, we shall consider some (mostly) hypothetical voting bodies to illustrate the calculation of the indices.

5.5. CALCULATION OF THE POWER INDICES

To facilitate the calculation of the indices and highlight numerical relationships between them, it is convenient to introduce some notation from combinatorial mathematics. The standard notation for *combinations*,

$$\binom{m}{n},$$

will be used to denote the number of combinations that can be formed from m objects taken n at a time, or $m!/[n!\,(m-n)!]$. Recall from note 20 that the exclamation point (!) indicates a factorial and means that the number it follows is to be multiplied by every positive integer smaller than itself [e.g., $4! = (4)(3)(2)(1) = 24$].

To illustrate the meaning of combinations with a simple example, suppose we wanted to know the number of ways of choosing a subset of three voters from a set of four, which we will designate by the set $\{a, b, c, d\}$. Clearly, one of the four voters will be excluded from each subset of three voters that is chosen, yielding four different subsets of three voters: $\{a, b, c\}$, $\{a, b, d\}$, $\{a, c, d\}$, and $\{b, c, d\}$. This number, found by complete enumeration of all the subsets, can be calculated directly as the number of combinations of four objects taken three at a time:

$$\binom{4}{3} = \frac{4!}{3!\,1!} = \frac{[(4)(3)(2)(1)]}{[(3)(2)(1)][(1)]} = 4.$$

Now consider a voting body of four members whose five votes are distributed $(1, 1, 1, 2)$. If the decision rule is simple majority (3 out of 5), each of the 1-vote members will be the pivotal member when preceded by the two other 1-vote members,

$$1 \quad 1 \quad (1) \quad 2,$$

or when preceded by the 2-vote member,

$$2 \quad (1) \quad 1 \quad 1.$$

In the former sequence, there is a total of $2! = 2$ ways of ordering the two 1-vote members prior to the pivot, and $1! = 1$ way of ordering the one 2-vote member subsequent to the pivot, or $2!1! = (2)(1) = 2$ ways of ordering all nonpivotal members. Similarly, in the latter sequence there are $1!2! = 2$ ways of ordering the nonpivotal members, making for a total of $2 + 2 = 4$ different arrangements, or *permutations*, in which each 1-vote member is pivotal in both sequences.[24] By contrast, there are only two DMWC's—(1, 1, 1) and (2, 1)—in which the defection of a single 1-vote member is critical.

The 2-vote member will be pivotal in two sequences,

$$1 \quad (2) \quad 1 \quad 1,$$
$$1 \quad 1 \quad (2) \quad 1,$$

where he casts a decisive (third, as well as other) vote. In the former sequence, any one of the three 1-vote members may precede him in $\binom{3}{1} = 3$ possible ways; for each of these ways there are $1! \, 2! = 2$ ways of ordering the nonpivotal members that precede and follow him, making him pivotal in $\binom{3}{1} 1! \, 2! = 6$ permutations. Similarly, in the latter sequence there are $\binom{3}{2} 2! \, 1! = 6$ permutations in which he is pivotal, making for a total of $6 + 6 = 12$ permutations in which the 2-vote member is pivotal. By contrast, there are only six DMWC's—$\binom{3}{1} = 3$ of the form (1, 2), where "1" represents any of the three 1-vote members; $\binom{3}{2} = 3$ of the form (1, 1, 2), where "1, 1" represents any of the three distinct pairs of 1-vote members—in which the defection of the 2-vote member is critical. To summarize, for each 1-vote member there are

4 permutations (i.e., arrangements) in which he is pivotal;

2 combinations (i.e., DMWC's) in which his defection is critical.

---

24. Recall that the Shapley-Shubik index is based on the assumption that if there are n members, there are $n!$ scales along which all members can be ordered from the most supportive to the least supportive. Under this interpretation, if the pivot votes favorably,

(1) all members who precede him will also vote favorably;

(2) members who follow him may or may not vote favorably;

(3) whether they do or not, however, the pivot casts the decisive vote.

Since each permutation of members represents a different scale, it is necessary to count only the number of permutations in which each voter is in the pivotal position; because all members who precede him are assumed to vote favorably, this is the only position in which he casts the decisive vote.

For the 2-vote member, there are

12 permutations (i.e., arrangements) in which he is pivotal;

6 combinations (i.e., DMWC's) in which his defection is critical.

Since there are three 1-vote members, there are 3(4) + 12 = 24 permutations in which the three 1-vote and one 2-vote members are pivotal, which agrees (as it should) with the number of ways one can permute four objects (4!). Similarly, there are 3(2) + 6 = 12 combinations in which the defection of at least one member is critical.

The Shapley-Shubik index and the Banzhaf index give the same power values for the voters in this example. Each of the three 1-vote voters has $\frac{4}{24}$ proportion of pivots and $\frac{2}{12}$ proportion of critical defections, or $\frac{1}{6}$ of the voting power, whereas the 2-vote member has $\frac{12}{24}$ proportion of pivots and $\frac{6}{12}$ proportion of critical defections, or $\frac{1}{2}$ of the voting power.

Now suppose that the 2-vote member in the four-member voting body actually represents two 1-vote members in a five-member voting body. Since the individual power of each member in a five-member voting body is $\frac{1}{5}$ = 0.20, the power of two members is 0.40. If the two members decide to form a coalition and vote as a bloc, however, their power as a (non-winning) coalition would be 0.50—the same as the 2-vote member in the example above—which is a 25 percent increase over their combined individual power. If a third member joins their coalition, the coalition would be minimal winning and therefore have voting power equal to 1.00, which is a 67 percent increase over the three members' combined individual power of 0.60.

This example illustrates that with the serial addition of each member, the proportion of pivots or critical defections held by the coalition increases from 0.20 to 0.50 to 1.00. From these proportions, we can calculate the "added value" that each voter brings to the coalition when he joins it, which we call his *incremental contribution*. Thus, the incremental contribution of the second voter who joins the coalition is

$$0.50 - 0.20 = 0.30,$$

and the incremental contribution of the third and decisive voter is

$$1.00 - 0.50 = 0.50.$$

Unlike the Shapley-Shubik index, we do not assume in this calculation that the pivotal player gets all the marginal gain when the coalition becomes minimal winning, but instead we assume that it is apportioned among players according to their incremental contributions at each stage. Presumably, the greater a player's contribution, the greater will be his payoff should the coalition become minimal winning.

Assuming an opposition coalition forms that also tries to become minimal winning (which we did not previously assume), the greatest payoff typically goes to a player who joins *before* the pivotal player (unlike the above example, where the third—and pivotal—player to join the coalition makes the largest incremental contribution).[25] This is true whether the calculation of voting power is based on the Shapley-Shubik index or the Banzhaf index.[26]

The calculation of serial incremental contributions shifts attention from the point at which a coalition becomes minimal winning backward to the stages which preceded it. It suggests a dynamic model of coalition-formation processes that may be used to explain *why* voters join a coalition in a particular order—in terms of the payoffs they receive—rather than one that assumes all arrangements occur and are equally likely. We reserve for the next chapter, however, further discussion of coalition models.

The correspondence between permutations and the arrangements in which a player is pivotal, and combinations and the DMWC's in which he is critical, was illustrated by the preceding example. On certain occasions, power indices can be determined without appealing to these concepts and the detailed calculations they entail, as the following example illustrates.

Consider a voting body of one hundred members (e.g., the U.S. Senate), wherein each member has one vote and there is a chairman (in the Senate, the presiding officer is the vice-president) who can break ties. When all regular members are present and voting, what is the power of the chairman?

Clearly, if one or more members is absent and there is an odd number of members present and voting, the members can never split evenly and there will never be a tie. In such situations, a chairman who can only break ties would never vote and hence would be powerless.

If there is an even number of members present and voting, Shapley and Shubik argue that the power of the chairman is the same as that of any regular member. The reason is that he, like all other members, is pivotal

---

25. See Steven J. Brams and William H. Riker, "Models of Coalition Formation in Voting Bodies," in *Mathematical Applications in Political Science, VI,* ed. James F. Herndon and Joseph L. Bernd (Charlottesville, Va.: University Press of Virginia, 1972), pp. 79–124; and Steven J. Brams, "A Cost/Benefit Analysis of Coalition Formation in Voting Bodies," in *Probability Models of Collective Decision-Making,* ed. Richard G. Niemi and Herbert F. Weisberg (Columbus, Ohio: Charles E. Merrill Publishing Co., 1972), pp. 101–24.

26. Empirical support for this finding comes from United States national party conventions that operated under majority rule, where the decisive stage seems to be set when a presidential candidate captures 42 percent or more of the votes on a ballot. Since no candidate has ever lost the nomination once he gained more than 41 percent of the votes on a ballot, support after the 42 percent threshold appears superfluous. Ulysses S. Grant set this threshold when he polled 41 percent of the votes on the thirty-fifth ballot of the 1880 Republican convention and went on to lose to James A. Garfield on the thirty-sixth ballot. In Democractic conventions prior to 1936, in which the decision rule was two-thirds, the threshold was 56 percent; Martin Van Buren polled 55 percent of the votes on the first ballot of the 1884 convention, with James K. Polk eventually emerging as the winner on the ninth ballot. See Richard C. Bain, *Convention Decisions and Voting Records* (Washington, D.C.: Brookings Institution, 1960).

only in arrangements in which he casts the decisive vote. Although there will be many arrangements in which the chairman will not be able to vote, these will be cases in which his vote would not have mattered, anyway. Consequently, even though the chairman casts his vote less often than other members, he will be pivotal in the same number of arrangements as every other member. As suggestive evidence of this power, Shapley and Shubik observed that "in the present Senate (1953–1954) the tie-breaking power of the Vice President [Richard M. Nixon] ... has been a very significant factor."[27]

The above calculation may underestimate a chairman's power, given his (nonvoting) power to rule motions and discussion "out of order." Although such a procedural ruling can normally be challenged by a member and appealed to a vote of the entire membership, in many voting bodies it is extremely rare that a majority will overrule a chairman. To the extent that a chairman can effectively veto motions by allowing only those to which he is already favorably disposed to come to a vote, his power will be greatly enhanced.

More precisely, a chairman with a veto in a 100-member voting body will be pivotal in the 51st through 101st positions, for he will always be decisive given the support of at least 50 regular members. (With only the power to break ties, he would be pivotal only in the 51st position, given a fifty–fifty split of regular members.) Hence, his power according to the Shapley-Shubik index will be $\frac{51}{101} = 0.51$, which is more than that of all of the regular members combined.

The Banzhaf index gives the same power value as the Shapley-Shubik index to the tie-breaking chairman, but not to the chairman with the veto. To illustrate first the calculation for the tie-breaking chairman with a simple example, consider a four-member voting body (a, b, c, d) with a tie-breaking chairman, e. Now the defection of member d (who could be any other regular member) will be critical in the following six DMWC's:

$$(a, b, d), (a, c, d), (a, e, d), (b, c, d), (b, e, d), (c, e, d).$$

Likewise, the defection of chairman e will be critical in six DMWC'S,

$$(a, b, e), (a, c, e), (a, d, e), (b, c, e), (b, d, e), (c, d, e),$$

so his power is identical to that of a regular member (0.20). In general, every member (regular or chairman) will be critical in every DMWC containing himself and *exactly* two of the four remaining other members, or in a total of $\binom{4}{2} = 6$ DMWC's.

Now if a chairman is endowed with veto power, his presence will be necessary in all DMWC's. The number of coalitions that are minimal

27. Shapley and Shubik, "A Method for Evaluating the Distribution of Power in a Committee, System," p. 788.

winning with respect to his defection will be equal to the number of coalitions containing himself and *at least* two regular members:

$$\binom{4}{2} + \binom{4}{3} + \binom{4}{4} = 6 + 4 + 1 = 11.$$

By comparison, the number of coalitions that are minimal winning with respect to each regular member will be equal to the number of coalitions containing himself, the chairman, and *exactly* one of the other three regular members:

$$\binom{1}{1}\binom{1}{1}\binom{3}{1} = (1)(1)(3) = 3.$$

Thus, the power of the chairman in this example will be

$$\frac{11}{4(3) + 11} = \frac{11}{23} = 0.48,$$

whereas his power according to the Shapley-Shubik index would be $\frac{3}{5} = 0.60$, since he would be pivotal in the third, fourth, and fifth positions. Hence, a chairman with a veto has more power than all other members combined by one index (Shapley-Shubik), and less by the other index (Banzhaf). Further illustrations of the effects of vetoes on voting power will be given in section 5.7.

In this section we have presented several examples that illustrate the calculation of the Shapley-Shubik and Banzhaf indices of voting power. We have shown how the former index treats all permutations of members as equally likely and the latter index counts only combinations in which the defection of one or more voters is critical. In the case of the Shapley-Shubik index, we have also given examples where the pivotalness of a "distinctive" member (a chairman who can break ties or veto measures) can be determined immediately, obviating the necessity of making complex calculations in large voting bodies. Calculations need not be very complex, however, to reveal some surprising consequences of the power indices, as we show in the next section.

5.6. THREE PARADOXES OF VOTING POWER

The example given above of a five-member voting body where first two, and then three, members formed a coalition that voted as a bloc illustrated the advantages that a coalition may have over single members in enhancing the power of its members. It has not been generally recognized by theorists, however, that under certain circumstances coalitions may actually diminish the share of voting power that their members have.[28]

---

28. An exception is Lloyd Shapley, but he gave only a trivial example of this phenomenon— a three-player voting body with a decision rule of unanimity. Clearly, if two players form a

As an example, consider the voting body of player weights (2, 2, 3), where the decision rule is simple majority (4 out of 7). By both the Banzhaf and Shapley-Shubik indices, the voting power of each player is $\frac{1}{3} = 0.33$, since the 3-vote player is critical (and pivotal) in the same number of two-player coalitions (i.e., two) in which each 2-vote player is also critical. Thus, the 2-vote and 3-vote players have the same voting power.

Now assume that the 3-vote player breaks up into constituent 1-vote members. In the restructured voting body of player weights (2, 2, 1, 1, 1), there are four combinations in which each of the three 1-vote players is critical,

$$2\ 1\ (1)\ 1\ 2 \qquad \binom{2}{1}\binom{2}{1} = 4,$$

and a total of eight combinations in which each of the two 2-vote players is critical,

$$2\ (2)\ 1\ 1\ 1 \qquad \binom{1}{1}\binom{3}{0} = 1,$$

$$2\ 1\ (2)\ 1\ 1 \qquad \binom{1}{1}\binom{3}{1} = 3,$$

$$1\ 1\ (2)\ 2\ 1 \qquad \binom{1}{0}\binom{3}{2} = 3,$$

$$1\ 1\ 1\ (2)\ 2 \qquad \binom{1}{0}\binom{3}{3} = 1.$$

This gives to each of the 2-vote players a Banzhaf power value of

$$\frac{8}{2(8) + 3(4)} = \frac{8}{28} = \frac{2}{7} = 0.29,$$

and to each of the 1-vote players a Banzhaf power value of $\frac{4}{28} = \frac{1}{7} = 0.14$. What is surprising is that the combined individual voting power of the three 1-vote members, $\frac{12}{28} = \frac{3}{7} = 0.43$, is greater than that which all would receive (0.33) if they formed a coalition that voted as a bloc.

The 3-member bloc also suffers according to the Shapley-Shubik index, which attributes to each of the three 1-vote players voting power of

$$\frac{16}{2(36) + 3(16)} = \frac{16}{120} = \frac{2}{15} = 0.13.$$

---

coalition, they decrease their share of power from $\frac{2}{3}$ to $\frac{1}{2}$. Shapley suggests that "as a general rule of thumb, in games like this where there is a good deal of blocking, and relatively few winning coalitions, it pays not to contract players." See *Notes of Lectures on Mathematics in the Behavioral Sciences*, ed. Harvey A. Selby (Providence, R.I.: Mathematical Association of America, 1973), p. 57. This rule, however, does not apply to games like that illustrated in the text.

This sums to $\frac{2}{5}$ = 0.40 for all three 1-vote players, which is again greater than their coalition voting power of 0.33.

The conventional wisdom that the whole is greater than—or at least equal to—the sum of its parts is clearly violated by both indices of voting power in the above example. The general conditions under which this phenomenon occurs, and its expected frequency, have not been investigated. The fact that power considerations may underlie not only the formation but also the breakup of coalitions suggests that theories that account for coalition changes solely on the basis of the ideological incompatibility of their members (which we shall examine in detail in Chapter 6) may be in need of revision: Gains in power may also motivate the defection of coalition members.

Although the breakup of the 3-vote coalition into three 1-vote members is advantageous to all three members, two of the 1-vote members can do even better by simply expelling the third member. Since the 1-vote player is a dummy in a voting body with player weights (2, 2, 2, 1), each 2-vote coalition shares one-third of the power. Thus, each of the constituent 1-vote members of each 2-vote coalition has approximately 0.17 proportion of the voting power, which is better than each of the three 1-vote members can realize in the voting body (2, 2, 1, 1, 1), in which (as we showed above) each 1-vote member's power is 0.14 by the Banzhaf index, 0.13 by the Shapley-Shubik index.

Even in a simple voting body like the above comprising only seven voters, there are many combinations of members that may form and break up. (To be more precise, there are fifteen ways of partitioning the members into between two and seven nonempty coalitions, which is also the number of ways of writing the integer 7 as the sum of positive integers.[29]) We shall not compare the distributions of voting power across other possible partitions, for the above examples are sufficient to demonstrate our main point: There is not necessarily strength in numbers. Since greater size does not necessarily imply greater voting power—at least when power is apportioned equally among individual members[30]—there may exist incentives that lead to the breakup of coalitions. As a consequence, coalition processes may be quite unstable as coalitions form and break up; indeed, much theorizing about coalition behavior involves an attempt to isolate conditions that lead to stable configurations, as we shall show in Chapter 6.

Perhaps stranger than the *paradox of (large) size* illustrated above is what might be called the *paradox of new members*. This occurs when one or more players is added to a voting body and the voting power of at least

---

29. See Ivan Niven, *Mathematics of Choice, or How to Count without Counting* (New York: Random House, 1965), chap. 6.

30. To be sure, if voting power is considered a collective or public good, then the power of each member of a coalition is the power of the coalition itself. It seems that in most political coalitions, however, power is usually viewed as a private good to be divided among its members.

one of the original players increases—rather than decreases or stays the same—in the new and larger game.

To illustrate this phenomenon, assume a 1-vote player is added to the voting body whose player weights are (3, 2, 2). [As was shown above, the power values of the three players in this body are $(\frac{1}{3}, \frac{1}{3}, \frac{1}{3})$, given a simple majority decision rule of 4 out of 7.] Assuming a simple majority decision rule of 5 out of 8 in the enlarged voting body (3, 2, 2, 1), the 1-vote player will be critical in one combination,

$$2\ 2\ (1)\ 3 \qquad \binom{1}{0}\binom{2}{2} = 1,$$

each of the 2-vote players will be critical in a total of three combinations,

$$3\ (2)\ 2\ 1 \qquad \binom{1}{1}\binom{1}{0}\binom{1}{0} = 1,$$

$$3\ 1\ (2)\ 2 \qquad \binom{1}{1}\binom{1}{0}\binom{1}{1} = 1,$$

$$2\ 1\ (2)\ 3 \qquad \binom{1}{0}\binom{1}{1}\binom{1}{1} = 1,$$

and the 3-vote player will be critical in a total of five combinations,

$$2\ (3)\ 2\ 1 \qquad \binom{2}{1}\binom{1}{0} = 2,$$

$$2\ 1\ (3)\ 2 \qquad \binom{2}{1}\binom{1}{1} = 2,$$

$$2\ 2\ (3)\ 1 \qquad \binom{2}{2}\binom{1}{0} = 1.$$

By the Banzhaf index, the voting power of the 1-vote player is

$$\frac{1}{1 + 2(3) + 5} = \frac{1}{12} = 0.08,$$

the power of each of the 2-vote players is $\frac{3}{12} = \frac{1}{4} = 0.25$, and the power of the 3-vote player is $\frac{5}{12} = 0.42$. Compared with the equal power of 0.33 of each of the three players in the original game, the power of the 2-vote players decreases, but the power of the 3-vote player actually increases. On the basis of this redistribution of power caused by the addition of the 1-vote player, it seems reasonable to suppose that the 3-vote player would favor an expansion in the size of the voting body by one member. This conclusion is also supported by the Shapley-Shubik power values, which are the same as the Banzhaf values (.42, .25, .25, .08) for the players (3, 2, 2, 1) in the enlarged game.

Lest one should think that a player's greater power in an enlarged game is an artifact of a change in the decision rule [from 4 (out of 7) in the original game to 5 (out of 8) in the enlarged game], consider what the power of the three original players would be if they had operated under a decision rule of 5 (out of 7), the same as that assumed in the enlarged game. Then each of the 2-vote players would be critical in one combination,

$$3\ (2)\ 2 \qquad \binom{1}{1}\binom{1}{0} = 1,$$

and the 3-vote player would be critical in a total of three combinations,

$$2\ (3)\ 2 \qquad \binom{2}{1} = 2,$$

$$2\ 2\ (3) \qquad \binom{2}{2} = 1.$$

Thus, by the Banzhaf index the power of each of the 2-vote players is

$$\frac{1}{2(1) + 3} = \frac{1}{5} = 0.20,$$

and the power of the 3-vote player is $\frac{3}{5} = 0.60$. By the Shapley-Shubik index, each of the 2-vote players has power equal to

$$\frac{1}{2(1) + 4} = \frac{1}{6} = 0.17;$$

the 3-vote player has power equal to $\frac{4}{6} = \frac{2}{3} = 0.67$. Comparing the power values of each of the 2-vote players in the original (0.20 and 0.17) and enlarged (0.25) games, each clearly benefits from the addition of a 1-vote player.

The lesson we derive from the paradox of new members is that not all players suffer when one or more players is added to a voting body.[31] An old player may actually gain power, as the above examples demonstrated, though the more typical pattern seems to be that the voting power of all the original players is diluted when new player(s) are added. As with the paradox of size, however, there have not been any systematic studies of the conditions that lead to this phenomenon or its expected frequency of occurrence.

The third paradox involves placing restrictions on winning coalitions that are allowed to form. We may suppose, for example, that two players are involved in a quarrel and refuse to join together to help form a winning

---

31. As an illustration of this statement when *two* players are added to a voting body, consider a voting body of player weights (4, 4, 3) enlarged to a voting body of player weights (4, 4, 3, 1, 1). Assuming a simple majority decision rule in both cases (6 out of 11 in the original body, 7 out of 13 in the enlarged body), the (equal) voting power of the three original players remains exactly the same in the enlarged body since the two new players are both dummies.

coalition.[32] Although one might suspect that they could only succeed in hurting each other, it is a curious fact that the quarrel between two players may actually redound to *their* benefit by increasing both their individual and combined voting power. We call this phenomenon the *paradox of quarreling members.*

As an example, consider again the voting body consisting of weights (3, 2, 2), where we showed previously that for a decision rule of 5 out of 7 the Banzhaf power values are $(\frac{3}{5}, \frac{1}{5}, \frac{1}{5})$ and the Shapley-Shubik power values are $(\frac{2}{3}, \frac{1}{6}, \frac{1}{6})$. Assuming that the quarreling players in this voting body are the two 2-vote players, there will (as before) be one combination in which each 2-vote player is critical,

$$3 \ (2) \ 2 \qquad \binom{1}{1}\binom{1}{0} = 1.$$

Now, however, the 3-vote player will be critical in only two combinations,

$$2 \ (3) \ 2 \qquad \binom{2}{1} = 2,$$

since the third combination,

$$2 \ 2 \ (3) \qquad \binom{2}{2} = 1,$$

in which he was critical previously, is precluded by the quarreling restriction that prevents the grand coalition containing both 2-vote players (as well as the 3-vote player) from forming. Thus, by the Banzhaf index the power of the 3-vote player will be twice as great as the power of each 2-vote player, yielding the power values $(\frac{1}{2}, \frac{1}{4}, \frac{1}{4})$ for the players (3, 2, 2). (The Shapley-Shubik index gives the same power values in this case.) Since the values for the two 2-vote players are greater than their power values (according to both indices) if they do not quarrel and coalition formation is unrestricted, there is an incentive for them to quarrel to increase their share of the voting power. Thus, just as power considerations may lead to the breakup of coalitions, these considerations may also inspire conflicts among the members of a voting body independently of ideological considerations. The expected frequency of occurrence of such conflicts, and the kinds of instabilities they give rise to, have not been studied.

Although we call the phenomena described in this section "paradoxes," this term is not meant to imply that they in any way invalidate the power indices. Quite the contrary: They illustrate their usefulness in showing up aspects of voting power whose existence would have been difficult to

---

32. For a formalization of this idea, based on the Shapley value, that suggested the name for this paradox, see D. Marc Kilgour, "A Shapley Value for Cooperative Games with Quarrelling," in *Game Theory as a Theory of Conflict Resolution,* ed. Anatol Rapoport (Dordrecht-Holland: D. Reidel Publishing Co., 1974), pp. 193–206. Kilgour establishes the conditions under which the paradox described in the text occurs for the Shapley-Shubik but not the Banzhaf index.

demonstrate convincingly in the absence of precise quantitative concepts. Taken together, the paradoxes of size, new members, and quarreling members suggest that there may be instabilities in power relationships and structures that have heretofore not been evident. Related to these instabilities are opportunities for effecting changes in the status quo to the benefit of certain players—which will, of course, be pleasing to some but disturbing to others.

If our illustrative examples are descriptive of real phenomena, the identification of these phenomena may have to await theoretical work that clarifies and generalizes the conditions under which the paradoxes occur and their expected frequency. In the absence of general theoretical results, we now turn to some real-life applications of the power indices and offer a preliminary assessment of the intuitive reasonableness of the values obtained.

5.7. EMPIRICAL APPLICATIONS OF THE POWER INDICES

When the framers of the United Nations met in San Francisco in 1945, they provided for two separate voting bodies: a General Assembly, in which all member states would be equally represented; and a Security Council with permanent and nonpermanent members. It was thought that representation in the Security Council, which was given "primary responsibility for the maintenance of peace and international security" (Article 24 of the UN Charter), should reflect the differences in national power that separate large and small states. Since there was no obvious way in which the power of states could be weighted that would reflect differences in population, wealth, natural resources, military strength, and so on, five states (China, France, Soviet Union, United Kingdom, and United States) were singled out from all others and made permanent members of the Security Council, each being given a veto on all substantive resolutions. In its original form, the Security Council also included six nonpermanent members, without a veto, and the decision rule was a majority of seven without a veto.

It is not difficult to count the number of permutations in which a nonpermanent member is pivotal and the number of combinations in which his defection is critical. If we let P_i stand for permanent member i, and N_j stand for nonpermanent member j, then for N_1 to be pivotal (he could be any other nonpermanent member), he must be preceded by all five P_i's and exactly one N_j (say, N_2):

$$P_1 \ P_2 \ P_3 \ P_4 \ P_5 \ N_2 \ (N_1) \ N_3 \ N_4 \ N_5 \ N_6$$

Since there are $\binom{5}{5}\binom{5}{1} = 5$ ways of choosing all the P_i's and any one of the five nonpivotal N_j's to precede N_1, there are 5 combinations in which the defection of any of the nonpivotal N_j's is critical to the minimal winning coalition containing one nonpermanent member and all five permanent members. Because the six members that precede N_1, and the four members that follow N_1, can be ordered in 6! 4! ways, there are $\binom{5}{5}\binom{5}{1}$ 6! 4! = 86,400 permutations in which N_1 is pivotal.

The combinations in which the defection of permanent member P_1 is critical, and permutations in which he is pivotal, include all ways in which P_1, preceded by four other P_i's, is decisive in the seventh, eighth, ninth, tenth, and eleventh positions. We have illustrated one possible arrangement for each position below and given the number of combinations and permutations for each:

	Combinations	Permutations
$P_2\ P_3\ P_4\ P_5\ N_1\ N_2\ (P_1)\ N_3\ N_4\ N_5\ N_6$	$\binom{4}{4}\binom{6}{2}$	$\binom{4}{4}\binom{6}{2}$ 6! 4!
$P_2\ P_3\ P_4\ P_5\ N_1\ N_2\ N_3\ (P_1)\ N_4\ N_5\ N_6$	$\binom{4}{4}\binom{6}{3}$	$\binom{4}{4}\binom{6}{3}$ 7! 3!
$P_2\ P_3\ P_4\ P_5\ N_1\ N_2\ N_3\ N_4\ (P_1)\ N_5\ N_6$	$\binom{4}{4}\binom{6}{4}$	$\binom{4}{4}\binom{6}{4}$ 8! 2!
$P_2\ P_3\ P_4\ P_5\ N_1\ N_2\ N_3\ N_4\ N_5\ (P_1)\ N_6$	$\binom{4}{4}\binom{6}{5}$	$\binom{4}{4}\binom{6}{5}$ 9! 1!
$P_2\ P_3\ P_4\ P_5\ N_1\ N_2\ N_3\ N_4\ N_5\ N_6\ (P_1)$	$\binom{4}{4}\binom{6}{6}$	$\binom{4}{4}\binom{6}{6}$ 10! 0![33]

The total number of combinations in which P_1's defection is critical is 57, and the total number of permutations in which P_1 is pivotal is 7,939,680.

Summing the number of combinations in which the five permanent and six nonpermanent members are critical gives the total number of critical defections in all DMWC's:

$$6(5) + 5(57) = 315.$$

The total number of arrangements in which all permanent and nonpermanent members are pivotal is

$$6(86,400) + 5(7,939,680) = 40,216,800.$$

This latter figure is equal to 11!, or the number of different ways of permuting all eleven members.

The value of the Shapley-Shubik index for each of the permanent members is $\dfrac{7,939,680}{40,216,800} = \dfrac{76}{385} = 0.197$, and for each of the nonpermanent members is $\dfrac{86,400}{40,216,800} = \dfrac{1}{462} = 0.002$. The combined power of the five permanent members is 0.987, and of the six nonpermanent members 0.013.

---

33. By definition, 0! = 1.

Although this disparity between the permanent and nonpermanent members is reduced by the Banzhaf index, which gives values of $\frac{57}{315} = 0.181$ for each of the permanent members and $\frac{5}{315} = \frac{1}{63} = 0.016$ for each of the nonpermanent members, the combined power of the permanent members is still very substantial (0.905) as compared with that of the nonpermanent members (0.095). Comparing the Banzhaf values with the Shapley-Shubik values, the power of the permanent members is decreased by 8 percent, but the power of the nonpermanent members is increased by 633 percent.

It is not hard to understand why the nonpermanent members benefit so much when the order in which members join a coalition is eliminated from the calculation of voting power. In a single coalition in which N_1 is pivotal, there are 6! 4! = 17,280 arrangements in which the other members can be differently ordered prior to and subsequent to the pivot. In a single coalition in which P_1 is pivotal, on the other hand, there may be as many as 10! 0! = 3,628,800 such arrangements. (This occurs when P_1 occupies the eleventh position in the grand coalition; see the permutation calculation in the last arrangement given above for P_1.) Since $\frac{3,628,800}{17,280} = 210$, a single coalition (i.e., the grand coalition) that is minimal winning with respect P_1 (as well as the other permanent members) counts 210 times as much in the Shapley-Shubik index as does a single coalition that is minimal winning with respect to N_1 (as well as all other permanent and nonpermanent members).[34]

This seems a sensible bias only if one believes that a power index should be weighted in favor of the more powerful members *to a degree greater than the proportion of DMWC's in which they are critical*. The degree of disproportional weighting in the Shapley-Shubik index is based on the order in which votes are cast or in which members join a coalition. If this order is inconsequential to the outcome, the justification of the index as a valid measure of voting power is untenable. As Banzhaf put it,

> It seems unreasonable to credit a legislator with different amounts of voting power depending on when or for what reasons he joins a particular voting coalition. His joining is a use of his voting power—not a measure of it—for it is reasonable to assume that each legislator will make the most effective use of his voting power under the circumstances.[35]

James S. Coleman has suggested three alternative measures of power, applicable to voting bodies and other collectives, based on the propor-

---

34. James S. Coleman, "Control of Collectivities and the Power of a Collectivity to Act," in *Social Choice*, ed. Bernhardt Lieberman (New York: Gordon and Breach, 1971), p. 276.

35. Banzhaf, "Weighted Voting Doesn't Work: A Mathematical Analysis," n. 32, p. 331.

tion of coalitions in a voting body that are winning (though not necessarily minimal winning with respect to the defection of any members).[36] If there are n members of a voting body, and each member can either favor or oppose a motion, then the total number of coalitions favoring (or complementary coalitions opposing) the motion (including the coalition with no members and the grand coalition with all members) is 2^n, since each voter may or may not be a member of this coalition (or the complementary coalition). (This number is the same as the number of ways of partitioning the voting body into two subsets.) Coleman defines the power of a body to act as the number of coalitions in it that can take action (i.e., are winning) divided by the total number of coalitions.

In the case of the UN Security Council, we showed that there are 57 coalitions (i.e., combinations) that are minimal winning with respect to the defection of P_1 (as well as all other permanent members). These include the 5 coalitions that are minimal winning with respect to N_1 (as well as all other nonpermanent members), because a coalition cannot be minimal winning with respect to the defection of a nonpermanent member and not be minimal winning with respect to the defection of a permanent member. Since there are no other winning coalitions (i.e., every winning coalition is also minimal winning with respect to each permanent member), and there are eleven permanent and nonpermanent members of the Security Council (SC), its *power to act* (A) is

$$A(\text{SC}) = \frac{57}{2^{11}} = \frac{57}{2,048} = 0.028.$$

That is, less than 3 percent of all possible coalitions have the power to pass resolutions.

Obviously, if every member of the Security Council had a veto, or equivalently, if unanimity were required on all resolutions, there would be only one coalition (the grand coalition) that could pass resolutions, giving the Security Council a power to act of less than 0.05 percent. Trial juries, which under Anglo-American law usually—but not always—consist of twelve jurors and must reach a unanimous verdict, have even less power to act, which is commonly regarded as a safeguard that favors the defendant since the evidence for conviction must be "beyond a reasonable doubt."

In those cases in which a collectively has the power to act, one or more of its members may have the power to prevent action. Coleman defines the preventive power of a member as the proportion of winning coalitions in which his presence is necessary (i.e., that are minimal winning with respect to the member's defection). In the case of the Security Council, the defection of a permanent member P_i is always critical, so its *power to prevent action* (P) in the 57 cases in which the Council can act is

$$P(P_i) = \frac{57}{57} = 1.0.$$

36. Coleman, "Control of Collectivities and the Power of a Collectivity to Act," pp. 277–87.

There are only 5 coalitions that are minimal winning with respect to the defection of a nonpermanent member N_j, so its power to prevent action is

$$P(N_j) = \frac{5}{57} = 0.088.$$

The power of jurors on a trial jury to prevent action is obviously one, because in the single case in which a guilty verdict is reached—the formation of the grand coalition—each juror has a veto.

Finally, Coleman defines the power of an individual to initiate action as the number of ways in which he can, by changing his vote, change a negative collective action into a positive one. In the case of the UN Security Council, there are $2,048 - 57 = 1,991$ ways in which no action is taken. Since a permanent member is decisive in all cases in which the four other permanent members, and at least two nonpermanent members, favor an action $\left[\binom{4}{4} \left\{ \binom{6}{2} + \binom{6}{3} + \binom{6}{4} + \binom{6}{5} + \binom{6}{6} \right\} = 57 \right]$, his *power to initiate action (I)* is

$$I(P_i) = \frac{57}{1,991} = 0.029.$$

Similarly, a nonpermanent member will be decisive in all cases in which the permanent members, and exactly one of the other five nonpermanent members, favor an action $\left[\binom{5}{5}\binom{5}{1} = 5 \right]$, which gives

$$I(N_j) = \frac{5}{1,991} = 0.003.$$

From these figures and those given previously, Coleman concludes that

> the power to initiate action is small for the permanent as well as the temporary members. The permanent members of the Security Council, in devising this constitution, in effect sacrificed the power of the collectivity to act . . . , as well as their own power to initiate action, and in return kept complete power to prevent action by the collectivity.[37]

This power is to be contrasted with the power of a single individual to initiate actions in certain situations—a United States president to commit armed forces to a conflict, a district attorney (in some states) to charge a person with a crime (though in other states it is necessary for a grand jury to issue an indictment formally accusing a person of a crime), and even an ordinary citizen to file a civil suit, gather signatures to initiate a referendum, and so on.

---

37. Coleman, "Control of Collectivities and the Power of a Collectivity to Act," p. 283.

TABLE 5.1 VALUES OF POWER INDICES, AND
PERCENT CHANGES IN VALUES, FOR MEMBERS OF
UN SECURITY COUNCIL BEFORE AND AFTER 1965

Power indices	Before 1965	After 1965	Percent change
Shapley-Shubik index			
All P_i	0.9870	0.9814	−0.6
All N_j	0.0130	0.0189	+45.4
Banzhaf index			
All P_i	0.9048	0.8346	−7.8
All N_j	0.0952	0.1654	+73.7
Power to act			
$A(SC)$	0.0278	0.0259	−6.8
Power to prevent action			
$P(P_i)$	1.0000	1.0000	0.0
$P(N_j)$	0.0877	0.0990	+12.9
Power to initiate action			
$I(P_i)$	0.0286	0.0266	−7.0
$I(N_j)$	0.0025	0.0026	+4.0

In 1965, apparently to increase the voting power of the nonpermanent members of the UN Security Council relative to the permanent members, four nonpermanent members were added to the Security Council, increasing its size from eleven to fifteen. The decision rule was changed from seven to nine, still requiring the assent of the five permanent members but now requiring at least four (out of ten) instead of two (out of six) nonpermanent members.

The effects of these changes on the power indices are shown in Table 5.1. The percent changes in the values of these indices indicate that the power of the permanent members fell slightly, as measured by both the Shapley-Shubik and Banzhaf indices, while the power of the nonpermanent members increased substantially. Yet, because the combined power of the nonpermanent members was low initially—less than 2 percent according to the Shapley-Shubik index and less than 10 percent according to the Banzhaf index before 1965—the increase in absolute terms was hardly earthshaking. In fact, because the 66.7 percent increase in nonpermanent members (from six to ten) was greater than their 45.4 percent increase in power by the Shapley-Shubik index, the power of *individual* nonpermanent members actually dropped according to this index, though the 73.7 percent increase in the Banzhaf index gave them a slight net gain (see Table 5.1).

The power of the Security Council to act decreased marginally after 1965 from 2.78 to 2.59 percent. Since the permanent members retained their veto power, their absolute power to prevent action remained unchanged, though their power to initiate action fell slightly (by 6.8 percent). The power of the nonpermanent members to prevent action increased somewhat (by 12.9 percent), but their miniscule power to initiate action (0.0025) stayed practically the same. At best, it seems, the effects of the rule change were cosmetic, which is to say in more blunt language that "this reform accomplished nothing."[38]

Under its present rules, the UN Security Council can also be represented as the weighted majority game,

$$(7, 7, 7, 7, 7, 1, 1, 1, 1, 1, 1, 1, 1, 1, 1)$$

where each of the permanent members has seven votes, each of the nonpermanent members one vote, and the decision rule is 39 (out of 45) votes.[39] In this representation, the permanent members, who together control 78 percent of the votes, have 98 percent of the voting power according to the Shapley-Shubik index and 83 percent according to the Banzhaf index. In general, both indices, but especially the Shapley-Shubik index, usually—though not always—give a power advantage to the largest members that is disproportional to their voting weights.[40]

We have seen how the voting rules of the UN Security Council vest absolute "preventive" power in the hands of each of the permanent members and thereby afford the Council relatively little power to act. Provision for this power was originally justified on the grounds that without the assent, or at least acquiescence (through abstention), of all permanent members, the efficacy of collective security measures, especially in conflict areas, would be undermined. In practice, these rules have offered the permanent members a way of blocking resolutions that they consider inimical to their interests.

Nevertheless, the exercise of veto power by permanent members has not completely stultified action on the part of the UN. Article 12 of the UN Charter provides that if (and only if) the Security Council is unable to reach agreement on peace-keeping measures ("ceases to deal with matters" that are "relative to the maintenance of international peace and security"), the General Assembly may take action with the approval of at least a two-thirds majority of its members (Article 18).

The precedent for such a transfer of decisional authority was established in 1950, when the Security Council was blocked by vetoes cast by the Soviet Union against various resolutions to resist aggression in Korea.

---

38. William H. Riker and Peter C. Ordeshook, *An Introduction to Positive Political Theory* (Englewood Cliffs, N.J.: Prentice-Hall, 1973), p. 172.

39. Anatol Rapoport, *N-Person Game Theory: Concepts and Applications* (Ann Arbor, Mich.: University of Michigan Press, 1970), pp. 218–19.

40. The magnitude of this bias for the Shapley-Shubik index is discussed in Riker and Shapley, "Weighted Voting: A Mathematical Analysis for Instrumental Judgments." Generally speaking, smaller members are advantaged only when the larger members are evenly matched and they (the smaller members) can tip the "balance of power."

TABLE 5.2 SHAPLEY-SHUBIK INDICES, AND
PERCENT CHANGES IN VALUES, FOR MEMBERS OF
UN SECURITY COUNCIL (SC) AND GENERAL
ASSEMBLY (GA) BEFORE AND AFTER 1965[a]

Players	Before 1965	After 1965	Percent change
Soviet Union	0.0450	0.0444	−1.3
Each of other P_i	0.0205	0.0199	−2.9
Each N_j member of SC	0.0147	0.0164	+11.6
Each other member of GA	0.0070	0.0066	+5.7

[a] From Schwödiauer, "Calculation of A Priori Distributions for the United Nations."

The General Assembly then responded by passing the so-called Uniting for Peace resolution, which provided that the Assembly could consider recommendations for "collective measures . . . to maintain or restore international peace and security" if the Security Council, due to a lack of unanimity of permanent members, failed to exercise its "primary responsibility."[41]

Basing his study on this interrelationship between the Security Council and the General Assembly, Gerhard Schwödiauer demonstrated how one can define a composite game that comprises voting in both voting bodies.[42] Unfortunately, because of the structural interaction between the Security Council and the General Assembly, voting cannot be considered as a separate game in each body for the purpose of computing the Shapley value for the players who are members of both bodies in the composite game. There is, moreover, no weighted majority representation of this composite game.

The calculation of Shapley-Shubik indices for players in this game involves some rather complex combinatorial computations which we shall not elaborate here. The values of the indices for the players in this game are given in Table 5.2 before and after the change in the voting rules of the Security Council in 1965. (Note that the Soviet Union is distinguished from the other permanent members of the Security Council because it in effect casts three votes in the General Assembly—its own vote plus the

41. Stephen S. Goodspeed, *The Nature and Function of International Organization*, 2d ed. (New York: Oxford University Press, 1967), p. 227.

42. Gerhard Schwödiauer, "Calculation of A Priori Power Distributions for the United Nations," Research Memorandum No. 24 (Vienna: Institut für Höhere Studien, July 1968). For the purposes of his calculations, Schwödiauer assumed the size of the General Assembly to be 122, which was its membership in 1968. Since then the number of members has grown to 138 (1974).

votes of the Ukraine and White Russia.) The percent changes given in Table 5.2 afford few surprises and are in line with our previous results for members of the Security Council alone, which indicated a slight shift of power from the permanent to the nonpermanent members.

The inclusion of ordinary members of the General Assembly in the composite game allows one to compute the shift in the proportion of power held by the Security Council and General Assembly after the Charter revision. With the increase in size of the Security Council, and the concomitant increase in opportunities for the formation of majorities among its members, it is hardly surprising that the proportion of power held by the Security Council as a body after the Charter revision increased by 9.3 percent.

What is surprising, however, is that the power of a permanent member of the Security Council (excluding the Soviet Union), both before and after 1965, is not much greater than the power of a nonpermanent member in this larger game. Whereas the power ratio by the Shapley-Shubik index of a permanent (P_i) to nonpermanent (N_j) member since 1965 has been over 50:1 in the Security Council alone (see Table 5.1), the ratio is only about 6:5 when the General Assembly is also included (see Table 5.2). Before 1965 these ratios were about 90:1 and 4:3, respectively.

The relatively small differences in power values that separate permanent from nonpermanent members of the Security Council in the enlarged game would seem to indicate that the permanent members would not have much to lose if they relinquished their vetoes, thereby conferring equal status on all members of the Security Council. In fact, Schwödiauer shows such a gesture would involve no magnanimity on their part: It would actually raise the power of the Soviet Union by 38 percent and the power of the other permanent members by a whopping 185 percent! But it is the nonpermanent members of the Security Council who would most benefit from abolition of the veto with a spectacular 244 percent jump in power to bring them to the level of the (indistinguishable) permanent members. All these astounding bonuses to the permanent and nonpermanent members of the Security Council would, of course, be at the expense of the ordinary members of the General Assembly, who would suffer an 80 percent devaluation of power.

These startlingly nonintuitive results have a rather simple explanation: With the permanent members deprived of their vetoes, the number of blocking coalitions in the Security Council would be cut drastically, which would in turn sharply reduce the opportunities that ordinary members of the General Assembly would have to act and thus be pivotal members in a minimal winning coalition. But whereas it might be to the advantage of the permanent members to relinquish their vetoes so as to increase their opportunities to be pivotal members of minimal winning coalitions, we should not forget that this also means they would relinquish most of their opportunities to prevent action in the composite game. It is this privilege which the permanent members would probably be unwilling to sacrifice in the interest of enhancing the power of the UN to act—perhaps against their own interests.

The Shapley-Shubik index has been used to calculate the distribution of power in several voting bodies, including the U.S. Supreme Court,[43] the New York City Board of Estimate,[44] and a proposed body of representatives from Canadian provinces.[45] Banzhaf has applied his index to the calculation of the power of members of the Board of Supervisors of Nassau County, New York, among other bodies, and demonstrated that absolutely no voting power was accorded to the three members (out of a total of six) with the fewest votes (22 percent of the total).[46] Incredible as it may seem, these dummy members—half the representatives on the Board of Supervisors—could under no circumstances make any difference in the outcome and might as well have never voted.[47]

The largest weighted voting body to which the Shapley-Shubik index has been applied is the U.S. Electoral College, where the electors from each state almost always cast their votes as blocs in the election of a president. For United States presidential elections in 1964 and 1968, in which the number of electoral votes cast by the fifty states and the District of Columbia ranged from three to forty-three, Irwin Mann and L. S. Shapley calculated (with the aid of a computer) Shapley-Shubik values for all states. Their calculations revealed a slight bias in favor of the largest states, with the proportion of power held by the largest state (New York), with forty-three electoral votes, about 5 percent greater than its proportion of electoral votes; and the proportion held by the smallest states (Alaska, Delaware, Nevada, Vermont, and Wyoming) and the District of Columbia, with three electoral votes each, about 3 percent less than their proportion of electoral votes.[48] Defining the *power discrepancy* as the difference

---

43. Glendon A. Schubert, *Quantitative Analysis of Judicial Behavior* (Glencoe, Ill.: Free Press, 1959), chap. 4.

44. Samuel Krislov, "The Power Index, Reapportionment, and the Principle of One Man, One Vote," *Modern Uses of Logic in Law* (now *Jurimetrics Journal*), June 1965, pp. 37–44.

45. D. R. Miller, "A Shapley Value Analysis of the Proposed Canadian Constitutional Amendment Scheme," *Canadian Journal of Political Science*, 6 (March 1973), pp. 140–43.

46. Banzhaf, "Weighted Voting Doesn't Work: A Mathematical Analysis," pp. 338–40. These dummy members would also be dummies were the Shapley-Shubik index used.

47. On the basis of these findings, this voting system was overturned by the courts and a new, more complex system that gives all representatives some voting power was instituted in its place. After it received judicial sanction from the U.S. Supreme Court, it was overturned in a referendum. See *New York Times*, Nov. 17, 1974, p. 43.

48. Irwin Mann and L. S. Shapley, "Values of Large Games, VI: Evaluating the Electoral College Exactly," Memorandum RM-3158-PR (Santa Monica, Calif.: RAND Corporation, May 1962), p. 13. For excerpts from this memorandum and a previous one on the Electoral College, see Irwin Mann and L. S. Shapley, "The A Priori Voting Strength of the Electoral College," in *Game Theory and Related Approaches to Social Behavior*, ed. Martin Shubik (New York: John Wiley & Sons, 1964), pp. 151–64. See also Paul T. David, Ralph M. Goldman, and Richard C. Bain, *The Politics of National Party Conventions* (Washington, D.C.: Brookings Institution, 1960), p. 174, for some early results of Mann and Shapley's calculations.

between these figures, we see that there was an 8 percent power discrepancy between the largest and smallest states.

No values of the Banzhaf index could be obtained for the Electoral College, although this index has been used in a summary index that also measures the effect of a citizen's vote on the outcome in each state.[49] We shall discuss this and other effects of the Electoral College in Chapter 7, where we argue that the Shapley-Shubik values indicating a slight—and seemingly innocuous—large-state bias greatly understate the distortions produced by the Electoral College.

One of the most interesting applications of the Shapley-Shubik index is not to a weighted voting body but to the political subsystem comprising the United States president, members of the Senate, and members of the House of Representatives.[50] These actors are interconnected by rules that allow for the enactment of bills if supported by at least a simple majority of senators, a simple majority of representatives, and the president; or at least two-thirds majorities in both the Senate and House without the support of the president (i.e., in order to override his veto).

To be sure, this conceptualization ignores the fact that a majority of the Supreme Court can in effect veto a law by declaring it unconstitutional; and it ignores the countervailing power that Congress (and the states) has to amend the Constitution and thereby nullify Supreme Court rulings. Nevertheless, although other actors may affect the outcome, it seems useful to abstract those relationships among that set of actors who have the most immediate impact on the enactment of bills into laws.

If one considers all orders in which the 533 individual players in this game (as of 1954, when Shapley and Shubik calculated their index) are pivotal, the Shapley-Shubik values for a single representative, senator, and the president are in the proportions 2:9:350. Thus, a president has about 175 times as much power as a representative and nearly 40 times as much power as a senator. Collectively, the power vested in the House, Senate, and presidency are in the approximate proportions 5:5:2, giving the president about one-sixth of all power. If Congress did not have the power to override a presidential veto, the proportions would be approximately 1:1:2, with the House having slightly less power than the Senate and the Senate having about half the power of the president. Assuming that the president in his legislative role represents in effect one house with a single

---

49. From a calculation suggested by Lawrence D. Longley (personal communication, Nov. 29, 1973) and figures given in Table 3 of Lawrence D. Longley and John H. Yunker, "The Changing Biases of the Electoral College" (Paper delivered at the 1973 Annual Meeting of the American Political Science Association, Sept. 4–8), a 15 percent power discrepancy was found between the largest and smallest states using the Banzhaf index, compared to the 8 percent discrepancy using the Shapley-Shubik index. It could not be determined, however, whether this discrepancy helped the largest states more than it hurt the smallest states (as was the case with the Shapley-Shubik index) or vice versa.

50. Shapley and Shubik, "A Method for Evaluating the Distribution of Power in a Committee System," p. 790.

TABLE 5.3   VALUES OF POWER INDICES FOR TRICAMERAL LEGISLATURE

House	Shapley-Shubik index	Banzhaf index
1-member house	0.38	0.23
3-member house	0.32	0.34
5-member house	0.30	0.43

member, it is certainly not obvious in a tricameral legislature that the power possessed by the three houses is in inverse relation to their size.

To illustrate this calculation, Shapley and Shubik give a worked-out example (in the appendix to their article) of a tricameral legislature with one house containing five members, a second house containing three members, and a third house containing one member (e.g., a president), where the approval of three members in the first house, two members in the second house, and one (i.e., the only) member in the third house is required for the passage of bills. Whereas the Shapley-Shubik index reveals that the power of the houses—assumed to be equal to the sum of the power of their members—is inversely related to their size, the Banzhaf index shows, surprisingly, the power of the houses to be directly related to their size (see Table 5.3).

This example perversely illustrates how the different indices can be used to support diametrically opposed conclusions about where the most power resides in complex, interconnected institutions. As an "explanation" of this discrepancy, one could say that the smallest house (e.g., the president) is advantaged by the Shapley-Shubik index because it can be pivotal in more permutations of the members of the larger houses than vice versa, whereas the largest house is advantaged by the Banzhaf index because its more numerous members are together critical in more DMWC's. This particularistic explanation is not very revealing, however, for it abstracts only from the calculations but does not illuminate any general rationale underlying them. Given the many instances of real-life multicameral institutions and the speculative and atheoretical nature of the literature that portrays power relationships among them, it seems worthwhile to try to develop a deeper and more general understanding of the manner in which they share power that clarifies "discrepancies" such as those considered above. This example underscores the need for more rigorous models than the verbal coalition models given earlier to show up the consequences entailed by using each index.[51]

---

51. In any weighted voting body—including bodies in which some members have veto power—both indices of voting power will always generate the same ranking of *individual*

As a further practical justification for studying relationships between the power indices, the Banzhaf index has been accepted by the New York State Court of Appeals as a basis for assigning weights to representatives on New York County Boards of Supervisors.[52] It seems desirable, therefore, to analyze its formal properties and to compare these properties with those previously described for the Shapley value in order to determine what similar and different aspects of a situation these two indices tap.

So far we have given examples of how the Shapley-Shubik and Banzhaf indices can be used to compare the power values of players in a voting game. It will be recalled that at the beginning of this chapter we suggested that these power indices could also be used as a guide for determining whether a game is even worth playing. As an application related to the latter usage, one may want to determine which side to join in a game if one already is a player. This consideration introduces a dynamic element into the analysis of voting games by raising the question of what changes a player might make in his position (and the structure of a game) that would enhance his power.

Looking at changes in party affiliation in the French National Assembly in 1953–54, William H. Riker attempted to determine whether the members who switched parties increased their opportunities to pivot, as measured by the Shapley-Shubik index.[53] There were 34 such migrations from one party to another in the two-year period studied which involved 61 changes of party affiliation by 46 (out of 627) members. For the purpose of computing power indices, Riker assumed that the parties in the National Assembly, which numbered between fifteen and twenty in the period studied, acted as blocs. Because party discipline was quite tight, especially as exercised by the parties on the left and right (which included a majority of members of the National Assembly), this was not an unreasonable assumption.

In the weighted majority game defined by the parties as bloc players, Riker assumed that the power of an individual member of a party was

---

members from most to least powerful, given the possibility of ties in the ranks. It is not difficult to demonstrate that a discrepancy in power rankings by the two indices can occur only for *collective* actors—connected by decision rules—whose power is assumed to be equal to the sum of the power of the actors they include. For further applications of the power indices to the president, the two houses of Congress, and its committees in their various roles (proposal of constitutional amendments, treaty ratification, impeachment, and so on), see Steven J. Brams and Lee Papayanopoulos, "Legislative Rules and Legislative Power" (Paper delivered at the Seminar on Mathematical Models of Congress, Aspen, Col., June 16–23, 1974).

52. Ronald E. Johnson, "An Analysis of Weighted Voting as Used in Reapportionment of County Governments in New York State," *Albany Law Review*, 34 (Fall 1969), pp. 317–43; Robert W. Imrie, "The Impact of Weighted Vote on Representation in Municipal Governing Bodies of New York State," *Annals of the New York Academy of Sciences* (*Democratic Representation and Apportionment: Quantitative Methods, Measures, and Criteria*, ed. L. Papayanopoulos), 219 (New York: New York Academy of Sciences, 1973), pp. 192–99.

53. William H. Riker, "A Test of the Adequacy of the Power Index," *Behavioral Science*, 4 (April 1959), pp. 276–90.

equal to the power value of his party divided by the number of its members. For each migration, Riker calculated the incremental change in power of the migrating member(s) from the old game before his migration to the new game after his migration. If this increment were positive, the member gained from the migration.

In only ten of the thirty-four migrations was this increment positive, and overall the algebraic sum of all positive and negative increments was negative. This would seem to indicate that the migrating members were not paying much attention to changes in their relative power positions. By a different criterion, however, the evidence was less clear-cut. In forty-five of the sixty-one individual changes of party affiliation, a migrating member's power was less than the power of an average member of the National Assembly. In fact, many of the members who migrated were quite severely disadvantaged before migrating, suggesting the conclusion that members were motivated to migrate when they perceived that large power discrepancies separated the members of different parties—even if they were not able to discern those parties that offered the greatest power advantages.

At best the evidence seems inconclusive that power considerations motivated party migrations. Undoubtedly, ideological and other considerations induced some members to quit one party and join another. It seems useful, nevertheless, to try to establish the baseline effect of possible power calculations before trying to evaluate the effects of non-power-related substantive factors.

5.8. SUMMARY AND CONCLUSION

In this chapter we have looked at power in terms of the effects that actors have on outcomes. We assumed that the manner in which actors influence the choice of outcomes depends solely on constitutional and other rules and not on the content of their actions, their ideological views, and so forth. These rules, together with the weights of actors, are sufficient to define the characteristic function of a game that assigns values to all subsets of players in the game.

Restricting our models to weighted majority games, which assign values of 1 to a coalition if it is winning and 0 if it is losing, we offered two definitions of voting power based on two different bargaining models. The first assumed a coalition-formation process wherein the order in which members join the grand coalition is random—each sequence is as likely as every other. A player's voting power was defined to be the proportion of different voting orders or arrangements in which he casts the pivotal or decisive vote. This measure of power, called the Shapley-Shubik index, had the advantage of being the only measure that satisfied three conditions for distributing the a priori value of a game among players, but it had the disadvantage of requiring that there be as many attitudinal dimensions as voting orders along which players can be scaled in terms of their preferences for different issues.

The second measure of voting power, called the Banzhaf index, made no assumptions about the process by which minimal winning coalitions first come into being. Instead, it was based on a bargaining model that presupposed the existence of minimal winning coalitions (with respect to at least one member). A player's voting power was measured by his ability to threaten the disruption of a minimal winning coalition, which was assumed to be proportional to the number of winning coalitions that would become losing were he to defect. We suggested that this index was more related to the breakup of coalitions, the Shapley-Shubik index to the formation of coalitions. In addition, we showed how the concept of a "critical defection" in the Banzhaf index incorporated a necessary-and-sufficient notion of causality in its definition of power.

Several examples were used to illustrate the calculation of the Shapley-Shubik and Banzhaf indices and elucidate the relationship between pivots and permutations in the former, and critical defections and combinations in the latter. For both indices, examples were given that illustrated three anomalous aspects of voting power: (1) the paradox of size, wherein coalition members can increase their power by splitting up; (2) the paradox of new members, wherein the addition of new members to a voting body can increase the power of some of the original members; and (3) the paradox of quarreling members, wherein a quarrel between two (or more) members that prevents them from joining together to help form a winning coalition can increase their individual and combined power. All three paradoxes, we argued, are suggestive of instabilities in power relationships and structures, but estimates of the pervasiveness of such instabilities will require further research. As real-life examples of power relationships in voting bodies, empirical applications of the power indices to voting in the United Nations, the U.S. Electoral College, the French National Assembly, and the political subsystem comprising the U.S. Congress and president were among those discussed.

Some limitations of the power indices were highlighted by Coleman's concepts relating power and action. In the voting game whose players are members of the UN Security Council and/or General Assembly, for example, the values of the Shapley-Shubik index indicated that it would be very much to the advantage of the permanent members of the Security Council to relinquish their vetoes. This index, however, tended to cover up the absolute power of the permanent members to prevent action by their vetoes, which is a feature whose consequences were nicely illuminated by the application of one of Coleman's indices to the UN Security Council.

There is probably no single index or model of voting power that will be completely satisfactory for all purposes. On balance it seems that the best-known and most widely applied index—the Shapley-Shubik index—has some major defects as a causally based explanatory concept, not to mention some of the unrealistic assumptions of the model on which it is based. Although the Banzhaf index seems grounded in a more plausible model of bargaining power, it, too, has some serious deficiencies arising from its suppression of information about the initiation and prevention of action. Finally, a drawback of Coleman's indices is that they tend to obscure how power is apportioned in a collectivity, which makes summary power comparisons of its members difficult.

Although the Shapley-Shubik and Banzhaf indices extract certain information from different models of coalition formation and disintegration, they provide almost no insight into coalition processes and outcomes. To develop a better understanding of coalition behavior, we need models that utilize more information than is embodied in the characteristic function of a game. We turn to these in the next chapter.

6

COALITION GAMES

6.1. INTRODUCTION

In Chapter 5, we abstracted from the strategies available to players and considered the payoffs that all possible coalitions could ensure for themselves, as defined by the characteristic function of a game. We showed that the characteristic-function form of a game captures the strategic possibilities implicit in coalition alignments, but it does not reveal what payments will be made to individual players in the game. By making certain assumptions about the ways coalitions might form and break up, however, we were able to define the power of players as it might be revealed in different bargaining processes. Since these power values do not depend on players' strategies, they are best viewed as strictly a priori assessments of how the total payoff from winning might be divided among all players in an n-person game.

In the mathematical literature of game theory, there are many other "solution concepts" that prescribe how the winnings will be distributed among the players in an n-person game.[1] Minimally, these concepts satisfy two conditions. If x_i, $i = 1, 2, \ldots, n$, is the payoff to player i, and N is the grand coalition, then

$$x_i \geq v(\{i\}),$$
$$\sum_{i \text{ in } N} x_i = v(N).$$

The former condition is called *individual rationality* and states that the payoff to each player i will be at least as much as he would obtain were he in a coalition by himself (as defined by the characteristic function v). The latter condition is called *group rationality* and states that the sum of the payoffs to all players in the game will be equal to the value of the grand coalition.[2] A vector $X = (x_1, x_2, \ldots, x_n)$ of payoffs to players that

---

1. It has been estimated that there exist twenty to thirty such concepts in the literature. See Martin Shubik, "Game Theory: Economic Applications," *International Encyclopedia of the Social Sciences* (New York: Macmillan and Free Press, 1968), vol. 6, p. 69.

2. Payoffs that satisfy this condition are also considered Pareto optimal, for there are no other sets of payoffs in which one player is better off and no other player worse off. This follows from the fact that all players by definition are in the grand coalition, so there are no

satisfies the conditions of individual and group rationality is called an *imputation*. The principal problem in finding a solution to a game is to determine the imputations that are likely to be the outcomes of the game.

Unlike the power indices discussed in the last chapter, most solutions contain a large (usually infinite) number of different imputations. Consequently, their applicability to different empirical situations—except in some controlled experimental settings—has been difficult to ascertain.[3] Furthermore, because the characteristic function from which most of these solutions are derived affords no information on sociological or political constraints that may operate on coalition formation, the solutions tend not to be compelling, except, perhaps, in an abstract mathematical sense. With one exception—besides the power indices discussed in Chapter 4— we shall not discuss such solutions, whose relevance to real-world politics seems limited at best.

To infuse the mathematical theory with greater political relevance requires that we move from a more abstract to a more concrete level of analysis. Toward this end, we shall introduce into the analysis institutions that restrict the manner in which coalitions can form. We begin the task of incorporating institutional details by making certain assumptions about the "stability" of coalition alignments. The import of these assumptions is to restrict the composition of coalitions, but not their size directly. We shall then show how the question of size can be attacked by placing restrictions on the values that the characteristic function itself may assume. Empirical cases and data that bear on the different coalition models will also be examined. But first, in order to place the analysis of this chapter within a broader perspective, we shall assay some of the relationships between power and coalitions—uncovered in the previous chapters—that heretofore have not been evident.

6.2. POWER AND COALITIONS

A *coalition* is a subset of players; its incentive to form may be to coordinate the selection of joint strategies (see section 2.9), to increase the voting power of its members (see section 5.5), or to achieve some other end. The usual end sought in voting games is to win, for by definition only a winning coalition can enforce its selection of outcome(s) on other players. Each player, then, faces the problem of what other players to join with in a coalition that can win.[4]

---

other players from whom to win anything. Since the payoffs to all players exhaust the total payment available, no redistribution of payoffs can benefit one player without hurting another.

3. For a review and comparison of some of the more prominent solution concepts, see Anatol Rapoport, *N-Person Game Theory: Concepts and Applications* (Ann Arbor, Mich.: University of Michigan Press, 1970).

4. In situations involving only three actors, Theodore Caplow has postulated control over other actors as a goal, from which he has developed a theory about what coalitions will form

In two-person zero-sum games, like the quantitative voting games discussed in Chapter 3, this problem did not arise, because under no circumstances was there anything to gain by cooperating with one's opponent. This may not be true in two-person nonzero-sum games, however, as we showed in the case of some international relations games in Chapter 1.

It is in *n*-person games, where there are at least three players, that questions about coalition formation assume central importance. To be sure, if one assumes that there is no cooperation—that is, there is no communication—among the players, as we did in defining the concept of sophisticated voting in Chapter 2, then the central question (as in two-person games) concerns the choice of individually optimal strategies. But when voters are allowed to communicate and coordinate their strategies, outcomes that result from the selection of previously optimal strategies may be vulnerable, as was illustrated by some of the three-person cooperative voting and vote-trading games discussed in Chapters 2 and 4. In these games, the primary consideration of players was not the selection of strategies but the selection of coalition partners. This selection, it will be recalled, was not made *in vacuo*; rather, it depended on who could agree on a "best" outcome and not be tempted to defect to another coalition (e.g., because of the existence of a paradox of voting).

A coalition's ability to enforce an agreement seems clearly related to its power, and how it might be valued as defined by the characteristic function. By comparison, the power of an individual player seems most closely related to how crucial he is in the formation and breakup of coalitions (as we suggested in Chapter 5). This raises the interesting question: Is a player powerful because of his position in a coalition, or does a coalition form because players are powerful in it (i.e., can enforce an agreement as well as obtain something of value)? In politics, it seems, the power of players in a game is inextricably linked with the coalitions of which they are members, and any scheme that attempts to disentangle the players from the coalitions they constitute necessarily simplifies reality.

If one is to try to understand the interrelated features of a complex system, rather than mask them as an undifferentiated whole, such simplification seems unavoidable. For analytic purposes, it appears that one may either start with the players and their preferences and ask what coalitions will form if communication is allowed (Chapters 2 and 4), or start with the coalitions and ask how their value will be apportioned among the players as a function of their bargaining power (Chapter 5). Having illustrated both approaches in previous chapters, we shall now attempt to explicate

---

in triads for different distributions of strength (power) among the three actors. See Theodore Caplow, *Two against One: Coalitions in Triads* (Englewood Cliffs, N.J.: Prentice-Hall, 1968). For a review of empirical research related to this theory, see Morton D. Davis, *Game Theory: A Nontechnical Introduction* (New York: Basic Books, 1970), pp. 179–82; and David T. Burhans, Jr., "Coalition Game Research: A Reexamination," *American Journal of Sociology*, 79 (Sept. 1973), pp. 389–408.

the linkage between power and coalitions further by imposing certain restrictions on the formation of coalitions in cooperative games, where, generally speaking, players in a coalition can obtain more than they can obtain by acting alone. This focus on problems of organization and control in n-person simple games, in which winning and losing are the only outcomes, differs sharply from our focus on two-person quantitative international relations and voting games, discussed in Chapters 1 and 3.

6.3. RESTRICTIONS ON COALITION ALIGNMENTS

The power of players in voting games typically depends on many other factors besides the relative frequency with which they are pivotal or critical in coalitions. Yet this is all the information about the bargaining power of players that we were able to glean from a game in characteristic-function form. To supplement this rather spare representation of a game requires that we make additional assumptions about environmental constraints that may prevent possible coalitions from forming.

One such constraint is the organizational ties of players, which may limit their freedom to select other players as coalition partners. In many legislatures, for example, the structure of the party system is all-important in determining what coalitions form. When strict party discipline prevails, a legislator always votes with his party and has no opportunity to seek out potential coalition partners among nonparty members (see section 4.10).

This is not the case in the U.S. Congress, where most bills are passed by coalitions containing members of both parties. Nevertheless, there are limitations on the formation of coalitions in Congress, which in part determine how power is distributed among its members. To try to take account of these limitations, R. Duncan Luce and Arnold A. Rogow made the following assumptions in their model of Congress: [5]

1. There are two parties in Congress, one majority (party M) and one minority (party N), denoted by the sets M and N.

2. M and N are the majority and minority parties in *both* houses of Congress (i.e., one party is not the majority party in one house and the minority party in the other).[6]

---

5. R. Duncan Luce and Arnold A. Rogow, "A Game-Theoretic Analysis of Congressional Power Distributions for a Stable Two-Party System," *Behavioral Science*, 1 (April 1956), pp. 83–95; for a somewhat modified version of the Luce-Rogow model, see Reinhard Selten, "Anwendungen der Spieltheorie auf die politische Wissenschaft," *Politik und Wissenschaft*, ed. Hans Maier, Klaus Ritter, and Ulrich Matz (Munich: C. H. Beck, 1970), pp. 287–320, esp. pp. 310–14. A geometric approach for incorporating coalition restrictions in the calculation of the Shapley value, with an application to the Israeli Knesset, is given in Guillermo Owen, "Political Games," *Naval Research Logistics Quarterly*, 18 (Sept. 1971), pp. 345–55.

6. Since the ratification of the Seventeenth Amendment in 1913 that provided for popular election of United States senators, there has been only one Congress (65th, 1917–18) in

3. The members of each party are divided into two disjoint subsets that consist of *loyalists* (M_l and N_l), whose loyalty to their parties is absolute, and *potential defectors* (M_d and N_d), who may or may not vote with their party. By assumption,

$$M = M_l \cup M_d; \qquad N = N_l \cup N_d.$$

4. The president (P) may or may not be a potential defector (i.e., be willing to desert loyalists of his own party).

The president is included in this legislative scheme because his support is necessary to enact bills which are not supported by two-thirds majorities in both houses of Congress.

It is assumed in the model that the defectors who leave one party do so to join the other party and do not join together to form a separate coalition. This assumption is a reflection of the fact that the two-party system has persisted over a long period of time. Although "liberals" and "conservatives" of the two parties have often joined together to pass or defeat a particular bill, this has so far not resulted in an irreparable breakdown of the two-party system. If there were indications that the system were not stable—that it might split into more than two parties or reduce to one—this assumption of the model could be altered.

The assumptions of the model, coupled with the decision rules of Congress, limit the number of coalitions that can pass a bill. These limitations depend in part on the number of potential defectors. Luce and Rogow assume that the defectors are added to a coalition in order of decreasing willingness to defect until they just create, with the members of the other party, exactly a simple or two-thirds majority (if possible). In other words, the majorites that include both defectors and regular party members are (unique) minimal winning majorities.

The possible situations that may occur in a two-party system, as defined here, include one or the other of the following dichotomous events:

(a) Either party M does or does not have a two-thirds majority in both houses.

(b) Either the president is a member of party M or party N.

(c) Party M plus the defectors from party N form either a simple majority or a two-thirds majority in both houses.

One of the following trichotomous events may also occur:

(d) Party N plus the defectors from party M do not form a simple majority in at least one house; or they form a simple, but not a two-thirds, majority in both houses; or they form a two-thirds majority in both houses.

---

which one party was not the majority (or plurality) party in both houses of Congress. See Joseph Nathan Kane, *Facts about the Presidents: A Compilation of Biographical and Historical Data*, 2d ed. (New York: H. W. Wilson Co., 1968), pp. 362–63.

The situations associated with events (c) and (d) indicate the ways in which winning coalitions can form across party lines. If defections swell the ranks of one party to a two-thirds majority, then it can override a presidential veto and hence alone constitutes a winning coalition. On the other hand, if defections give a party only a simple majority in both houses, then this majority must receive the support of the president to form a winning coalition.

The three dichotomous events and one trichotomous event given above yield (2)(2)(2)(3) = 24 theoretical possibilities, which are doubled to 48 by the assumption 4 that either permits or prevents the president's defection. Twelve of these theoretical possibilities, however, are incompatible with a partition of the legislative actors into two coalitions. For example, it is not possible for party M to have a two-thirds majority in both houses and for the addition of defectors from party N to reduce this to a simple majority. The 36 situations that can actually occur are shown in Table 6.1.[7]

6.4. THE STABILITY OF ALIGNMENTS

Associated with each of these 36 situations is a distribution of power among the actors. To illustrate how this distribution is determined, consider situation 1.a.ii in Table 6.1, where (1) the president is a potential defector; (2) the number of members in party M plus the defectors from party N form only a simple majority; and (3) the number of members in party N plus the defectors from party M form only a simple majority. Obviously, given only simple majorities in situation 1.a.ii, any winning coalition must include the president. The three that can form are:

$$M \cup P; M \cup N_d \cup P; N \cup M_d \cup P.$$

Instead of trying to determine which of these coalitions will form, which is a question we consider later in this chapter, Luce and Rogow ask where the "location of power" is. Their answer is that only the set of actors in *all three* coalitions in the above case has power. This is the set $M_d \cup P$—the potential defectors from the majority party and the president —which is the intersection of the three coalitions.

The reasoning that Luce and Rogow offer in support of their answer is very similar to that presented in support of the Banzhaf index (and model) in Chapter 5 (see section 5.4). If the potential defectors from party M and/or the president should defect from any of the three winning coalitions, each would become a losing coalition. No other legislative actor has power because his presence is not necessary in one or more of the three admissible coalitions.

---

7. Actually, the "locations of power" associated with only 27 situations are listed in Table 6.1. The reason is that the 9 "locations" where the party of the president is "either" apply to 18 different situations, depending on whether the president is a member of party M or party N. This makes for a total of 36 situations, rather than the 27 shown.

TABLE 6.1   POWER DISTRIBUTIONS IN A TWO-PARTY SYSTEM[a]

Situation	Presidential defection	Party of president	Size of party M majority	Size of Party M plus party N defectors (N_d)	Size of party N plus party M defectors (M_d)	Locations of power
1.a. i	Possible	Either	Simple	Simple	Not majority	M, P
. ii					Simple	M_d, P
.iii					Two-thirds	M_d
1.b. i				Two-thirds	Not majority	M
. ii					Simple	M_d
.iii					Two-thirds	M_d
2.a			Two-thirds	Two-thirds	Not majority	M
b					Simple	M_d
c					Two-thirds	M_d
3.a. i		M	Simple	Simple	Not majority	M, P
. ii					Simple	M, P
.iii					Two-thirds	M_d
3.b. i				Two-thirds	Not majority	M
. ii					Simple	M
.iii					Two-thirds	M_d
4.a. i	Not possible	N	Simple	Simple	Not majority	Deadlock[b]
. ii					Simple	N, M_d, P
.iii					Two-thirds	N, M_d
4.b. i				Two-thirds	Not majority	M, N_d
. ii					Simple	M_d, N_d
.iii					Two-thirds	M_d, N_d
5.a		M	Two-thirds	Two-thirds	Not majority	M
b					Simple	M
c					Two-thirds	M_d
6.a		N	Two-thirds	Two-thirds	Not majority	M
b					Simple	M_d
c					Two-thirds	M_d

[a] Adapted from Luce and Rogow, "A Game Theoretic Analysis of Congressional Power Distributions for a Stable Two-Party System." Reprinted from *Behavioral Science*, Volume 1, No. 2, 1956, by permission of James G. Miller, M.D., Ph.D., Editor.

[b] This is the only situation in which no winning coalition can form.

We can formalize this discussion in terms of the concept of "ψ-stability" (ψ is the Greek letter "psi"). Let τ (the Greek letter "tau") denote a partition of a legislature into coalitions (always two in the above case). For example, if the president is a member of the majority party and forms a coalition with his party in Congress, the partition that divides the legislative system is

$$\tau = (M \cup P, N).$$

We call a partition that divides the legislative system into two disjoint coalitions a *coalition structure*.

Let $\psi(\tau)$ denote the set of coalitions that can form not only along party lines but also across party lines through defections from one party to the other, as specified by a rule of admissible coalition changes, ψ, applied to τ. In our example, potential defectors from party N (N_d) can desert their party and join the president (P) and members of party M (M) in the coalition,

$$P \cup M \cup N_d,$$

or potential defectors from party $M(M_d)$ and the president (P) can join members of party N in the coalition,

$$N \cup M_d \cup P.$$

There are other changes in τ permitted by ψ (e.g., $M_l \cup N_d \cup P$ or $N_l \cup M_d \cup P$), but the two coalitions given above that include all members of either M or N, along with $M \cup P$ in τ itself, are the only ones in situation 1.a.ii in which it is possible to ascertain that one coalition is winning.[8]

Let X denote the power distribution (or imputation) x_i, $i = 1, 2, \ldots, n$, where i stands for legislator i and x_i the power that accrues to him. The total power of all legislators is assumed to be equal to one,

$$\sum_{i=1}^{n} x_i = 1,$$

which is the assumption of group rationality, and no legislator can have power less than zero,

$$x_i \geq 0,$$

which is the assumption of individual rationality. Given a coalition structure τ, the question we seek to answer is what (if any) imputation X forms a "stable pair" with τ such that the legislators have no inducement to realign themselves into another coalition that is in $\psi(\tau)$. In other words, we wish to find equilibrium outcomes that specify not only the legislators who are members of admissible coalitions but also payoffs to these members as well.

---

8. It seems that Luce and Rogow meant to exclude situations in which the loyalists of one party combine with the potential defectors of the other party (see examples in parenthetic expression in above sentence) since no assumptions are made about the size of such groupings in Table 6.1. But they are not precluded by their assumption that the defectors cannot combine to form a third coalition, nor even by their assumption that "the only defections are from one party to the other" (Luce and Rogow, "A Game-Theoretic Analysis of Congressional Power Distributions for a Stable Two-Party System," p. 88), since parties are not considered indecomposable units. Although it seems reasonable to exclude such coalitions from $\psi(\tau)$ on the grounds that such "double defections" make a mockery of the party system, in reality "liberal" and "conservative" coalitions that comprise members of both parties are quite common in Congress.

Consider coalition $M \cup P$ in the coalition structure τ, which itself is included in the set of admissible coalitions that are in $\psi(\tau)$. If any legislator *i not in* this coalition has $x_i > 0$, then the sum of x_j for legislators j *in* $M \cup P$ must be less than 1. But since $M \cup P$ is winning, it has value (or power) equal to 1. If it forms, each of its members can receive more than they did for some power distribution where $x_i > 0$ and i is not in this coalition.

Given that each legislator wants to increase his power, he cannot do so if he is not a member of $M \cup P$, or, by a similar argument, a member of the other admissible winning coalitions, $M \cup N_d \cup P$ and $N \cup M_d \cup P$. In other words, he would have to be a member of all three winning coalitions to share in the distribution of power x_i, $i = 1, 2, \ldots, n$. Thus, this distribution is "in equilibrium" for a coalition structure τ only if $x_i = 0$ for any legislator i not in all three of the coalitions.

Formally, the *pair* (X, τ), representing an imputation X and a coalition structure τ, is ψ-*stable* if for each coalition S in $\psi(\tau)$,

$$v(S) \le \sum_{i \text{ in } S} x_i.$$

That is, the value of any coalition S that *could* form (according to the rule of admissible coalition changes, ψ) cannot exceed that which all of its members presently receive.[9] Thus, in the above example, the three admissible winning coalitions each have a value equal to 1, which does not exceed the value of 1 distributed over members of the set $M_d \cup P$, who are in all three winning coalitions and together share all the value (i.e., have all the power). Other coalitions that could form would be losing and have value equal to 0, which cannot exceed the sum of $x_i (= 0)$ for all of their members i. Thus, the members of no winning or losing coalitions in $\psi(\tau)$ can disrupt the pair (X, τ) since there are no coalitions that can form in which they all could do better.[10]

It is important to point out what this model, and its solution consisting of two parts—an imputation and a coalition structure—does *not* tell us.

---

9. The assumptions of individual and group rationality are special cases of this condition, called *coalition rationality*, for coalitions S in $\psi(\tau)$. See R. Duncan Luce and Howard Raiffa, *Games and Decisions: Introduction and Critical Survey* (New York: John Wiley & Sons, 1957), pp. 194–95. The set of imputations in a game satisfying the inequality specified in the text constitutes the *core* of the game; imputations in the core are not dominated by any other imputations.

10. In this application, Luce and Rogow drop the technical restriction of ψ-stability that player i must be in a single-member coalition in τ if $x_i = 0$; otherwise, he must be in a coalition in τ with at least one other player where he receives more than he can get by himself. (This restriction provides an incentive for a player to join a multimember coalition by postulating membership in it to be profitable.) Arguing that this restriction is not compelling in a legislature that considers many bills in an ongoing process, Luce and Rogow contend that a legislator's motivation to join a coalition on a particular bill will be sustained by long-term considerations and not the immediate gains that he receives from cooperating with other legislators on that bill. Luce and Rogow, "A Game-Theoretic Analysis of Congressional Power Distributions for a Stable Two-Party System," p. 90.

First, the fact that a coalition structure forms that is in some sense stable with respect to a distribution of power among its players says nothing about how the partition τ from which all other structures are generated comes into existence. Second, the rule ψ permitting certain changes in the coalition structure depends entirely on the empirical game being modeled and is therefore unspecified as a theoretical concept. Third, the bargaining process by which a stable pair (X, τ) is reached through the repeated application of the rule ψ is not given by the model, which in certain situations may perversely result in short-run gains at the expense of long-run losses for some players.[11] Fourth, ψ-stable pairs are not in general unique, which leaves open the question of identifying the conditions under which one of several possible stable pairs will be chosen. Fifth, while the model tells us in the particular situation we have discussed (1.a.ii) that power is concentrated entirely in the hands of the president, and defectors from party M, it provides no information—unlike the Shapley-Shubik or Banzhaf index—of exactly what that power distribution is.

The concept of ψ-stability, nonetheless, offers advantages that these a priori power indices do not. It provides a notion of power that depends not only on what players in a game can get for themselves but also on their ability to form coalitions whose members cannot be tempted by offers that could disrupt the coalition. Furthermore, the concept captures the reality of possible ideological and institutional restrictions on coalition alignments (e.g., parties, pressure groups, and public opinion), as well, conceivably, as the less tangible forces of tradition, culture, and history. Together, these constraints on social change constitute what might be called "social friction," which smoothes some associations but mucks up others. Probably its main effect in real-life situations is to impart social stability to situations by inhibiting (if not preventing) radical changes that could upset the status quo. Its origins are irrelevant in applications of the ψ-stability model; it is simply postulated as a limitation on coalition realignments.

Social friction may not be easy to identify empirically, especially in situations wherein the rules of association are either implicit or vague. But what may be even more difficult to determine empirically are the payoffs to players. Even if loyalists and defectors in the previous example could be identified operationally from their voting records, how would one measure the payoffs to legislators from being in "locations of power"? When several individuals occupy positions of power, and the model does not specify how power is apportioned among them, testing the consequences of the model becomes fraught with difficulties. With these difficulties in mind, the relevance of the Luce-Rogow model for understanding the distribution of congressional–presidential power will now be considered.

---

11. For an example, see Luce and Raiffa, *Games and Decisions*, pp. 231–33.

6.5. EMPIRICAL CONCLUSIONS OF THE LUCE-ROGOW MODEL

The analysis of the preceding section for situation 1.a.ii can be repeated in each of the thirty-five other situations given in Table 6.1 to determine the players in the legislative game who occupy positions of power in every situation. Note that this game is characterized by the same decision rules that were used in the calculation of the power indices for the United States president, members of the Senate, and members of the House (see section 5.6). What has changed are the players in the game (except for the president), who are no longer distinguishable by their membership in the Senate or House, but instead by their party membership and party loyalty. When these "political" factors are incorporated in a model of this legislative system, what power distributions render the two-party structure stable in the face of possible party defections?

When we examine those in positions of power in the thirty-six possible situations, several conclusions emerge:

(1) Deadlock is very unlikely—there is only one situation (4.a.i) in which the constraints are so stringent that no winning coalition can form. This occurs when
   (a) the president is a member of the minority party and is loyal to it;
   (b) the majority party, even with defectors from the minority party, has only a simple majority; and
   (c) the minority party, even with defectors from the majority party, does not have a simple majority.

(2) Coalitions across party lines are frequent—in 63 percent of the non-deadlock situations, the presence of defectors from one or the other parties is necessary in a winning coalition.

(3) The president is weak (i.e., his presence in a winning coalition is not necessary) whenever the majority party has a two-thirds majority—whether he is a member of it or not—or when either party, together with defectors from the other party, has a two-thirds majority.

(4) The president is strong (i.e., his presence in a winning coalition is necessary) only when neither party can assemble more than a simple majority, even with the aid of defectors from the other party.

(5) The minority party is strong only when the president is a member of it, loyal to it, and the majority party has only a simple majority, even with the aid of defectors from the other party.

(6) Loyalists are not necessary in more than half of the nondeadlock situations, although loyalists of the majority party are much more powerful comparatively than loyalists of the minority party—the presence of the former is necessary in 37 percent of the nondeadlock situations, the presence of the latter in only 6 percent.

Most of these conclusions, upon reflection, do not depart a great deal from commonsensical generalizations that have often been made about power relationships between the executive and legislative branches and within the legislative branch. Paraphrased, agreement is usually reached (1), even if it means crossing party lines (2).[12] The president is weak if he does not hold the balance of power (3), strong if he does (4). A minority party needs its president's support (5), and loyalists (especially those of the minority party) suffer because of their intransigence (6). In connection with the last conclusion, we might add that potential defectors from at least one party, whether they remain loyal to their party or not, are *always* necessary in winning coalitions in every nondeadlock situation.

To be sure, the implicit assumption underlying these conclusions is that the thirty-five nondeadlock situations are equivalent and can be considered equiprobable for the purpose of uttering probabilistic statements about the relative frequency of particular sets of situations. In fact, however, the majority party has rarely had a two-thirds majority in both the Senate and House, especially in recent times.[13] Thus, twelve of the thirty-five nondeadlock situations postulated in Table 6.1 occur a good deal less frequently in reality than is suggested by their approximately one-third representation in this table.

Although this fact would appear to undermine seriously the realism of the Luce-Rogow model, it does not nullify the model's usefulness as a heuristic device for showing up the manner in which hypotheses may be generated from a rather simple set of assumptions. In principle, the application of the model could be extended to perhaps more "realistic" situations (e.g., the cases of "double defection" mentioned in note 8 above) through a reinterpretation of the rule of admissible changes, ψ. Moreover, probabilistic statements about the likelihood of the players' being in positions of power could reflect the relative frequencies of occurrence of different empirical situations. As a related refinement in the Luce-Rogow model, Luce and Raiffa suggest that a probability might be assigned to

each pair consisting of a coalition S and a coalition structure τ . . . which is interpreted as the probability that a change to S will be considered when the players are arranged in coalitions according to τ. We do not intend these to be interpreted as subjective probabilities existing in the player's [*sic*] minds; rather they are objective descriptions of the probability that a certain event will occur. . . . These

12. Note that the filibuster in the Senate may be viewed as a last-resort extra-constitutional response to this fact for a minority that has no other means to prevent action.

13. Since the administration of Ulysses S. Grant (1869–76), there have been only three Congresses out of forty-nine (through the 93rd, 1973–74), or 6 percent, that have had party majorities of two-thirds or greater in both houses of Congress. These are the 74th and 75th (1935–38) during the administration of Franklin D. Roosevelt and the 89th (1965–66) during the administration of Lyndon B. Johnson. See Kane. *Facts about the Presidents*, pp. 361–63.

probabilities would . . . subsume a great number of facts about ease of communication, social limitations, intellectual limitations of the players, and so on. It is perfectly clear that it would be extremely difficult to get objective estimates of them, except, possibly, in situations which have recurred so often that frequencies can be observed.[14]

Whatever the merit and feasibility of these refinements in the model, most of the conclusions Luce and Rogow reach in their very simple representation of the legislative game do not conflict in any significant way with the observations of students of the congressional process. Nevertheless, it is reasonable to ask that a formal model do more than reaffirm the findings of descriptive studies. Even if the model also enables one to codify and order these findings in a parsimonious theoretical structure, a cogent argument can be made that it is intellectual "overkill" to construct a rather complicated mathematical apparatus that succeeds only in generating obvious—or, by the lights of its critics, mundane and trivial—findings.

At least one consequence of the Luce and Rogow model does not seem totally obvious, even if it is based on situations that have occurred only infrequently. These are the three situations 5.a/b/c in Table 6.1, in which the party of the president commands a two-thirds majority in Congress. In these situations, the president is *never* in a position of power. We thus may infer from the model that a president has as much to fear from a two-thirds majority of his own party, which can override his veto, as from the opposition party. This seems contrary to common sense and the rhetoric of presidential candidates who stress in their campaigns the importance of electing members of their own party to Congress.

In support of this proposition derived from their model, Luce and Rogow point out that 69 percent of the vetoes that Franklin Roosevelt cast were during the three (out of six) full-term Congresses in his administration (1933 to 1945)[15] in which one or both houses was controlled by a two-thirds or better Democratic majority.[16] Actually, however, this statistic is based on a misreading of their own table, for there were in fact four Congresses in which Roosevelt enjoyed at least two-thirds Democratic majorities in the Senate. Of these, only in the Seventy-fourth and Seventy-fifth Congresses (1935–1938) did he also enjoy at least two-thirds Democratic majorities in the House as well.

Comparing the average number of vetoes cast by Roosevelt in his two-thirds majority Congresses with the other four Congresses in which there

---

14. Luce and Raiffa, *Games and Decisions*, pp. 226–27.

15. The 79th Congress (1945–46) was not counted since Roosevelt died early in the first session.

16. Luce and Rogow, "A Game Theoretic Analysis of Congressional Power Distributions for a Stable Two-Party System," n. 11, p. 93.

TABLE 6.2   COMPARISON OF VETOES CAST IN SIMPLE AND
TWO-THIRDS MAJORITY CONGRESSES[a]

SIZE OF CONGRESSIONAL MAJORITY	(AVERAGE) NUMBER OF VETOES CAST PER CONGRESS	
	Franklin Roosevelt	*Lyndon Johnson*
Simple	91	8
Two-thirds	133	14
Two-thirds as % of simple	146	175

[a] Computed from information given in U.S. Senate, *Presidential Vetoes.*

were at least simple, but not two-thirds, Democratic majorities in both houses, we see that Roosevelt cast an average of 46 percent more vetoes in his two-thirds majority Congresses than in his simple majority Congresses (see Table 6.2). Lyndon Johnson, the only other president to have had two-thirds majorities of his own party in both houses of Congress since the popular election of senators was constitutionally mandated in 1913, displayed similar behavior. Although he did not cast nearly so many vetoes as did Roosevelt (see Table 6.2), Johnson cast 75 percent more vetoes in the two-thirds majority Eighty-ninth Congress (1965 to 1966) than the simple majority Ninetieth Congress (1967 to 1968).[17]

The president who holds the record for vetoes overridden by Congress in both absolute (15) and percentage (52 percent) terms is Andrew Johnson (1865 to 1869), who is the only president ever to be impeached. He is also one of only two presidents—the other being James Monroe (1817 to 1825), who cast only one veto during his two terms—to have had two-thirds majorities of his own party in Congress during his entire administration.[18]

From the figures we have given, it would appear that Presidents Roosevelt and Johnson were at their most effective when overwhelming majorities from their own party did *not* control Congress. To be sure, these figures do not take into account their "positive" power in getting bills they favored enacted. They further ignore the fact that Roosevelt constantly held out the threat of veto on legislation pending before Congress,[19]

17. Computed from information given in U.S. Senate, *Presidential Vetoes* (Washington, D.C.: U.S. Government Printing Office, 1969). It is also interesting to note that Roosevelt's nine vetoes that were overridden by Congress, an average of 2.00 vetoes per Congress were overrridden in his two-thirds majority Congresses, an average of only 1.25 vetoes per Congress in his simple-majority Congresses. In the case of Johnson, none of his vetoes was overridden in either his simple or two-thirds majority Congresses.

18. U.S. Senate, *Presidential Vetoes*; Kane, *Facts about the Presidents*, pp. 360–61.

19. Carlton Jackson, *Presidential Vetoes: 1792–1945* (Athens, Ga.: University of Georgia Press, 1967), p. 205.

which is obviously a form of power relevant to his control over outcomes. Moreover, these figures in no way reflect the possibly crucial effect that individual and collective actors not included in the model—party leaders, congressional committees, and even chairman of these committees, among others—may have had on legislative outcomes. Yet, although the veto evidence is pertinent to only one aspect of presidential power, it does suggest that there may be an unwitting diminution of this power when a president's party has two-thirds or better majorities in Congress. More systematic evidence is required, however, to establish the truth (or falsity) of this proposition conclusively.[20]

We have shown in this section how Luce and Rogow attempted to go beyond an a priori notion of power based only on the characteristic function.[21] By building into their model coalition possibilities that depend on party loyalty in a two-party system, they were able to determine those players in positions of power by virtue of their presence being necessary in all winning coalitions in specified situations. Thus, in a logical sense, the solution concept of ψ-stability as applied to this system coincides with Banzhaf's notion of power. Indeed, it seems that a theoretical framework based on the logical concept of necessity could unify the study of power in a wide variety of settings, including those in which institutions like political parties have an evident effect. By introducing such institutions and other relevant empirical details into models, genuinely comparative studies of the distribution of power in political systems would appear to be feasible within a unified framework.

In the model developed in subsequent sections, no attempt is made to assess how power must be distributed among players to guarantee that the coalitions of which they are members be stable. Instead, an attempt is made to determine how stability is related to the size of coalitions independent of institutional constraints peculiar to particular situations.

6.6. THE CONCEPT OF WINNING

Having dealt in Chapter 5 with the question of how winnings might be divided among players according to their bargaining power in minimal winning coalitions, we have shifted the focus in this chapter to a discussion of coalition partners—specifically, those to whom the benefits from

---

20. Other information relevant to the measurement of power over outcomes would be the acceptance rate by the U.S. Senate of a president's appointment nominations. See Joseph P. Harris, *The Advice and Consent of the Senate: A Study of the Confirmation of Appointments by the United States Senate* (Berkeley, Calif.: University of California Press, 1953).

21. For another attempt to define a ψ-function of admissible coalition changes that formalizes a theory of coalition formation first developed by William A. Gamson, see Walter Isard, *General Theory: Social, Political, Economic, and Regional with Particular Reference to Decision-Making Analysis* (Cambridge, Mass.: MIT Press, 1969), pp. 402–406, which is based on William A. Gamson, "A Theory of Coalition Formation," *American Sociological Review*, 26 (June 1961), pp. 372–82.

winning could be expected to accrue (i.e., those in positions of power). This shift was accomplished by embroidering the bare mathematical theory related to the characteristic function with sociological assumptions that prohibited the formation of certain coalitions. Continuing in this vein, we shall now consider a more general theory—not tied to particular institutions—that attempts to answer what winning coalitions will form and persist, rather than what payoffs members of winning coalitions can expect to receive.

We focus our attention on winning coalitions for the simple reason that there is not much else in politics to which one can attach value. True, many political actors claim to be interested in maximizing their "power," but aside from the measures of voting power discussed in the last chapter and the ψ-stable solution that identifies actors in positions of power discussed in this chapter, there have been no generally accepted operationalizations of this concept, not to mention models that justify these operationalizations. Some effort has been expended in developing models for measuring how power is distributed in a political system,[22] but the meaning of power outside the domain of voting situations has been far from clear. And even within this somewhat restricted domain, we have seen that there are alternative models from which to choose.

All the models we have so far discussed treat power as control over outcomes. Indeed, insofar as outcomes can be valued, the characteristic function may be interpreted as a function that assigns values not to coalitions, but rather to the outcomes they can secure. In other words, coalitions *qua* coalitions have no value themselves; it is only the outcomes they can bring about which have empirical significance.

Luce and Raiffa term the problem of assigning values to these outcomes "profound":

It seems to us that the most important development for empirical verification will be a practical method to calculate the characteristic function of actual situations. Probably the most significant contribution social scientists can make in this area is a feasible method for the approximate determination of characteristic functions.[23]

Almost two decades since this was written, it is sad to report that not much progress has been made. This is really not surprising, because a determination of the characteristic function depends ultimately on a determination of the power of coalitions (including those with one member). For if the payoffs to coalitions are related to the outcomes they

---

22. See, for example, Steven J. Brams, "Measuring the Concentration of Power in Political Systems," *American Political Science Review*, 62 (June 1968), pp. 461–475, and references cited therein; and Steven J. Brams, "The Structure of Influence Relationships in the International System," in *International Politics and Foreign Policy: A Reader in Research and Theory*, 2d ed., ed. James N. Rosenau (New York: Free Press, 1969), pp. 583–99.

23. Luce and Raiffa, *Games and Decisions*, p. 259.

can secure, and power means control over outcomes, then the determination of the characteristic function obviously depends on determining the power of coalitions. Once again (see section 6.2) we observe that the study of power and the study of coalitions are inseparable from one another.

Inseparable as the concepts of "power" and "coalition" are, they are amenable to simplification. Declaring power to be an "imprecise notion," Riker boldly replaces it with the notion of winning:

> What the rational political man wants, I believe, is to win, a much more specific and specificable motive than the desire for power. Furthermore, the desire to win differentiates some men from others. Unquestionably there are guilt-ridden and shame-conscious men who do not desire to win, who in fact desire to lose. These are the irrational ones of politics. With these in mind, therefore, it is possible to define rationality in a meaningful way without reference to the notion of power. Politically rational man is the man who would rather win than lose, regardless of the stakes. This definition . . . is consonant with . . . definitions of power. The man who wants to win also wants to make other people do things they would not otherwise do, he wants to exploit each situation to his advantage, and he wants to succeed in a given situation.[24]

In one stroke, then, Riker simplifies the goal of political man to one that involves achievement of a specifiable outcome—winning—where all power accrues to the victors.

Of course, there is a price to be paid for such a simplification—namely, the exclusion of empirical situations to which the concept of winning is inapplicable (e.g., a nuclear exchange that devastates all participants and leaves no winners). After reviewing some situations to which Riker's model does seem applicable, we shall suggest how the goal of winning might be modified to account for situations wherein the winner-loser mentality may subsume other ends as well.

We use the word "mentality" to underscore the fact that any statement we make about a postulated goal is necessarily based on the *perceptions* of actors. Ordinarily, this causes no problem in political situations like elections, where the method of counting votes and the decision rule for selecting a winner are known and accepted by the contestants—at least in situations where information is assumed to be complete. But it should be noted that although elections usually distinguish unambiguously the winner from the losers, the winner may not be perceived as the true victor. The Democratic presidential primary in New Hampshire in 1968 is a case in point: The incumbent president, Lyndon Johnson, captured 50 percent of the vote to Eugene McCarthy's 42 percent, with the remaining 8 percent split among minor candidates. Despite the fact that Johnson's candidacy

---

24. William H. Riker, *The Theory of Political Coalitions* (New Haven, Conn.: Yale University Press, 1962), p. 22.

was unannounced (his name did not even appear on the ballot), and he never set foot in New Hampshire, a majority of voters took the trouble to write in his name. Nonetheless, McCarthy was hailed as the victor by political pundits and the press since he surpassed his "expected" vote.[25] In such situations, expectations become the benchmark against which reality is tested.

6.7. THE SIZE PRINCIPLE

Besides positing the goal of winning, Riker makes several other assumptions in his game-theoretic model:

1. *Rationality.* Players are rational, which means that they will choose the alternative that leads to their most-preferred outcome—namely, winning.[26] Riker does not argue, however, that all actors are rational with respect to this goal but rather that the winner-loser mentality pervades, and conditions the behavior of, participants in such situations as elections and total wars. (Note that the notion of selecting one's most-preferred alternative is a more straightforward formulation of rationality than we were able to postulate in Chapter 2, where sophisticated voting and the possible formation of coalitions made it rational for voters sometimes not to vote for their most-preferred outcome in order to prevent adoption of their worst. But since in Riker's model winning and losing are assumed to be the only possible outcomes, this problem does not arise.)

2. *Zero-sum.* Decisions have a winner-take-all character—what one coalition wins the other coalition(s) loses. In other words, the model embraces only situations of pure and unrelieved conflict where all value accrues to the winner; cooperation among participants that redounds to the mutual benefit of all is excluded.

3. *Perfect information.* As defined in section 1.3, this means that all players are fully informed about the state of affairs or moves of other players at all times. (Riker also assumes information is complete, but to simplify the subsequent discussion we shall speak only of perfect—or imperfect—information, with the understanding that this also refers to complete—or incomplete—information.)

4. *Allowance for side payments.* Players can communicate with each other and bargain about the distribution of payoffs in a winning coalition, whose value is divided among its members. (We implicitly assumed exchanges among members of a coalition involving some common medium like money were possible in our earlier discussion of ψ-stability,

---

25. Richard M. Scammon and Ben J. Wattenberg, *The Real Majority* (New York: Berkley Publishing Corp., 1972), pp. 27–28, 87–96.

26. Strictly speaking, it is not necessary to postulate winning as the most-preferred outcome. Rather, "rationality" may be defined in terms of the choice of the *most-valued* outcome, where the value associated with winning coalitions and the payoffs to their members are stipulated by assumptions 5 and 6 in the text.

where an imputation and a coalition structure were stable precisely because the value of a coalition could not exceed the sum of payments to its members given by the imputation. This condition made it impossible for *every* member of such a coalition to receive more than he was assigned by the imputation, however the value of a coalition was distributed through bargaining among its members.)

These four assumptions are standard assumptions in *n*-person game theory. The fundamental problem for players in *n*-person games is to select coalition partners who, through the collective choice of strategies, maximize the achievement of particular goals. To specify more precisely the goal of winning, Riker makes the following additional assumptions:

5. *Positive value.* Only winning coalitions have positive value. (The grand coalition, however, has zero value since there are no losers from whom to extract value, and so do blocking coalitions, since the complement of a blocking coalition is a blocking coalition.)

6. *Positive payoffs.* Imputations associated with winning coalitions are such that all members receive positive payoffs. This assumption, of course, provides an incentive for players to join a winning coalition.

7. *Control over membership.* Members of a winning coalition have the ability to admit or eject members from it.

These last three assumptions, besides the assumption that the goal of players is to form winning coalitions, are the "sociological assumptions" —as distinguished from the four previously postulated "mathematical assumptions"—which Riker uses to restrict the range of the characteristic function. The sociological assumptions in effect constitute political constraints that enable him to specify the size of winning coalitions that is optimal, and therefore can be expected to form.[27]

To derive the optimal size of winning coalitions from these assumptions, Riker argues as follows.[28] At the point at which a coalition is minimal winning (i.e., where the subtraction of any member would render it losing), the characteristic function may be decreasing, constant, or increasing as the coalition grows in size past this point. That is, the value of a winning coalition may be maximal, minimal, or neither when it is "just winning."

If the characteristic function is decreasing, it is always advantageous for a coalition to eject superfluous members who are not necessary to its remaining winning (allowed by assumption 7), because they decrease the total amount that can be paid to all members. Even if the characteristic function does not decrease, but remains constant past the minimal

---

27. William H. Riker and Peter C. Ordeshook, *An Introduction to Positive Political Theory* (Englewood Cliffs, N.J.: Prentice-Hall, 1973), pp. 179–80.

28. The discussion that follows closely parallels that in Riker and Ordeshook, *Introduction to Positive Political Theory,* pp. 181–87. For further details, see Riker, *Theory of Political Coalitions,* pp. 40–46, 247–78; and William H. Riker, "A New Proof of the Size Principle," in *Mathematical Applications in Political Science, II,* ed. Joseph L. Bernd (Dallas: Southern Methodist University Press, 1967), pp. 167–74.

winning point—that is, winning by whatever amount is the sole determinant of value—the ejection of superfluous members means that the same total amount can be divided among fewer members, to the advantage of at least one of these (remaining) members.[29] The model itself does not specify how the ejected members are singled out by the other members for removal, but we shall suggest possible mechanisms used for this purpose in our later discussion of empirical examples.

If the characteristic function is increasing, then players will be motivated to expand the coalition past the minimal winning point to gain some—but not all—of the additional value associated with increasing size (allowed by assumption 4), for some of this value must also go to the new members (by assumption 6). But to say that players are motivated to gain more than what comes from winning itself is to say that winning is not the only operative goal, or that it may take different forms.[30] Since no other goal is postulated in the model, this cannot be the justification for greater than minimal winning coalitions.

Alternatively, members of a minimal winning coalition may wish to acquire additional members if they are not certain their coalition is sufficiently large to be winning. But they have perfect information (by assumption 3), so this cannot be the case. Thus, we must conclude that the assumptions of the model preclude characteristic functions that are increasing in the range of winning coalitions. Since this argument pertains to coalitions of *any* size past the minimal winning point, it also precludes characteristic functions that are both increasing and decreasing past this point.

Given the assumptions of the model, then, there are no circumstances wherein an incentive exists for coalitions of greater than minimal winning size to form. On the other hand, the fact that there is a positive value associated with winning coalitions (assumption 5), each of whose members receive positive payoffs from winning (assumption 6), is a sufficient incentive for such coalitions to form. The incentive for winning coalitions to form, but not to be of greater than minimal winning size, means that the realization of the goal of winning takes form in the creation of only minimal winning coalitions, which Riker calls the *size principle*.

Several things should be noted about Riker's derivation of this principle. First, the goal of winning is implicit in the assumptions of the model,

---

29. But when the spoils dispensed are some collective good, which is not perfectly divisible but instead is equally available to all members, there will be no incentive for a coalition to eject members since the payoff to each member will by definition remain the same whatever the size of the winning coalition. See Thomas S. McCaleb, "The Size Principle and Collective Consumption Payoffs to Political Coalitions," *Public Choice*, 17 (Spring 1974), pp. 107–109.

30. For example, winning quickly and decisively in most wars may be worth more than winning in a prolonged and exhausting struggle, but the model does not allow for "degrees" of winning—at least Riker doubts that zero-sum situations in which the characteristic function increases past the minimal winning point ever exist in the real world.

particularly the assumption of rationality that motivates players to obtain the benefits of being in a winning coalition, as stipulated in assumptions 5 and 6. Second, the size principle is a statement about an outcome, the size of winning coalitions, and not about the process of coalition formation. And finally, the role that the zero-sum assumption (assumption 2) plays in the proof of the size principle, which was not evident in the argument presented above, deserves mention. Given that only winning coalitions have positive value (assumption 5), the zero-sum assumption implies that losing coalitions must have complementary negative value. Since such a coalition has no things of value to distribute among its members, it would form only as a pretender to eventual winning status. Indeed, the possibility that a losing coalition could eventually become winning provides a strong incentive for a winning coalition to pare off superfluous members, who would be vulnerable to offers from a losing coalition that could promise them greater rewards in a (prospective) minimal winning coalition.

Although not employing the formal apparatus of game theory used in Riker's proof of the size principle, David H. Koehler has argued that in an environment in which resources are scarce, but the zero-sum condition is not necessarily operative (e.g., in legislatures), coalitions will still tend toward minimal winning size since their leaders must limit the number of members to whom they can make offers in order to maximize payments to individual members who are being simultaneously wooed by other coalitions.[31] There is, in other words, an intimate connection between the size of a coalition and the resources its leaders have available to pay members to join it.

It is worth reiterating (see section 6.1) that Riker's focus on the size of winning coalitions differs significantly from most solution concepts that have been developed in the mathematical theory of games. The focus of this latter research has been on finding "stable" ways in which payoffs will be divided among a set of players by imposing constraints on the payments players can receive. Sometimes these constraints have been coupled to a coalition structure, as in the case of ψ-stability. Although a number of "reasonable" sets of constraints has been proposed, however, efforts at restricting the admissible payoffs to players (i.e., finding preferred *individual* outcomes), rather than the admissible values of different-sized coalitions (i.e., finding preferred *coalition* outcomes à la Riker), have not led to solutions that seem particularly useful in the study of the behavior

---

31. David H. Koehler, "The Legislative Process and Minimal Winning Coalition," in *Probability Models of Collective Decision Making*, ed. Richard G. Niemi and Herbert F. Weisberg (Columbis, Ohio: Charles E. Merrill Publishing Co., 1972), pp. 149–64. The zero-sum assumption is retained, however, in another legislative bargaining model; see David B. Meltz, "Legislative Party Cohesion: A Model of the Bargaining Process in State Legislatures," *Journal of Politics*, 35 (Aug. 1973), pp. 649–81.

of real coalitions.[32] Only recently have efforts been made to tie together theoretically these two different kinds of outcomes and to develop a richer and more general theory of coalition behavior.[33]

6.8.   THE INFORMATION EFFECT

To return to our discussion of the size principle, one might argue that one does not require the formalism of a mathematical model to grasp the idea that the smaller the size of a winning coalition, the more each of its members individually profits and would therefore be expected to work to reduce its size; hence, the expected result would be minimal winning coalitions. Yet this commonsensical "explanation" of the size principle suffers from a difficulty endemic to all proverbial wisdom: It does not offer limiting conditions on the veracity of the size principle that a mechanism suggested by the assumptions of a model does.

We have already reviewed how the assumptions of Riker's model lead to a situation in which the realization of the goal of winning takes form in the creation of only minimal winning coalitions. As an explanation of why the size principle should hold, however, this argument has logical force only. To connect Riker's model with reality, we must now consider how its assumptions can be interpreted as conditions that limit the operation of the size principle when its theoretical concepts are operationally defined and it is posited as an empirical law. In translating the size prin-

---

32. Riker, *Theory of Political Coalitions*, p. 38. Yet Riker, in an article with William Zavonia comparing a game-theoretic solution with other kinds of explanations of bargaining behavior in experimental games, concludes that the game-theoretic solution is the best predictor of the distribution of payoffs among the subjects. See William H. Riker and William James Zavonia, "Rational Behavior in Politics: Evidence from a Three Person Game," *American Political Science Review*, 64 (March 1970), pp. 48–60. For a nontechnical summary of several game-theoretic solution concepts that highlights their relevance to coalition behavior, see Michael Leiserson, "Game Theory and the Study of Coalition Behavior," in *The Study of Coalition Behavior: Theoretical Perspectives and Cases from Four Continents*, ed. Sven Groennings, E. W. Kelley, and Michael Leiserson (New York: Holt, Rinehart and Winston, 1970), pp. 255–72. For a recent empirical study of the distribution of coalition payoffs in war, see Harvey Starr, *War Coalitions: The Distributions of Payoffs and Losses* (Lexington, Mass.: D. C. Heath and Co., 1972).

33. See Kenneth A. Shepsle, "On the Size of Winning Coalitions," *American Political Science Review*, 68 (June 1974), pp. 509–18. Shepsle analyzes issues posed in an exchange between Riker and Robert Butterworth based on the latter's misunderstanding of the definition of the characteristic function of a game. See Robert Lyle Butterworth, "A Research Note on the Size of Winning Coalitions"; William H. Riker, "Comment on Butterworth, 'A Research Note on the Size of Winning Coalitions'"; and Butterworth, "Rejoinder to Riker's 'Comment,'" *American Political Science Review*, 65 (Sept. 1971), pp. 741–48. Nevertheless, although Shepsle finds support for the size principle in different solution concepts of n-person games, he also shows that minimal winning coalitions are characterized by an instability somewhat akin to the cooperative solution to Prisoner's Dilemma-type situations (discussed in sections 1.8 and 4.7) and that of the "3/2's rule" (discussed in section 7.7). For further debate on this and related points, see Robert Lyle Butterworth, "Comment on Shepsle's 'On the Size of Winning Coalitions,'" and Kenneth A. Shepsle, "Minimum Winning Coalitions Reconsidered: A Rejoinder to Butterworth's 'Comment,'" *American Political Science Review*, 68 (June 1974), pp. 519–24.

ciple from a theoretical statement derived from the assumptions of a game-theoretic model into a descriptive statement about the real world that is capable of empirical confirmation or disconfirmation, Riker interprets it as follows: "In social situations similar to n-person games with side-payments, participants create coalitions just as large as they believe will ensure winning and no larger."[34]

The introduction of the beliefs or perceptions of individuals about a subjectively estimated minimum simply acknowledges the real-world fact that players do not have perfect information about the environments in which they act and the actions of other players. Consequently, players are inclined to form coalitions that are larger than minimal winning size as a cushion against uncertainty, and oversized winning coalitions thus become a quite rational response in an uncertain world. In this manner, perfect information serves as a limiting condition on the truth of the size principle: In its absence, coalitions will *not* tend toward minimal winning size.[35] The other assumptions of the model also restrict the applicability of the size principle in the real world, but Riker singles out the effect of information on the operation of the size principle for special attention probably because it is the concept most easily interpreted, if not operationalized, of the concepts embodied in the assumptions of his model.

The enlargement of coalitions above minimal winning size due to the effects of imperfect information is what Riker calls the *information effect*. Since this effect is endemic in an uncertain world, there naturally exist many examples of nonminimal winning coalitions. For Riker these examples represent situations in which coalition leaders miscalculated the capabilities, or misread the intentions, of opponents—given that the other assumptions of his model were reasonably well met—not because they were irrational but rather because they lacked accurate and reliable information on which to act.

In the absence of such information, leaders may end up paying more for winning than winning is "objectively" worth, according to Riker. Characterizing the present-day international strategic situation as an Age of Maneuver, he argues that the search for coalition partners who may tip the strategic balance becomes more and more desperate as the superpowers approach a situation of parity.[36] Since the point at which this situation occurs—and that where one coalition enjoys just a slight edge over the other—is not entirely clear, the superpowers may make overpayments to smaller nations as a hedge against uncertainty. Although this

---

34. Riker, *Theory of Political Coalitions*, p. 47.

35. Another related reason why they may not form is that when there is imperfect information, maximizing the probability of minimum winning coalitions may not be consistent with maximizing the probability of winning, especially in small groups. See Richard G. Niemi and Herbert F. Weisberg, "The Effects of Group Size on Collective Decision Making," in *Probability Models of Collective Decision Making*, pp. 140–43.

36. This conclusion is reinforced to the degree that there is parity in nuclear weapons, for stalemate at this level fosters competition at other levels.

may offer short-run advantages to secure their allegiance—or at least to prevent their drift to the other side—it may in the long-run prove ruinous.

As a case in point, the material and human costs to the United States of the Vietnam war proved to be very great indeed, which its critics maintained far exceeded the value of preserving a divided country and its corrupt government from a Communist take-over. On the other hand, supporters of the Vietnam war maintained that its value cannot be reckoned in terms of preventing the take-over of a relatively small country but rather in halting a Communist advance through all of Southeast Asia.

Whatever the merits of these different arguments, there is no doubt that early projections of the costs of United States involvement greatly underestimated the eventual costs of the war. Whether one believes that the price paid in the end was worth the peace agreement that was reached depends, of course, on the value one attaches to this outcome.

The fact that United States policy makers in the beginning erred drastically in their projections of the future costs of the war is precisely the kind of miscalculation that Riker argues may lead to incomplete, if not Pyrrhic, victories. In fact, he suggests, the decline of empires historically may be viewed as a product of leaders' paying too much for coalition partners who were thought to be necessary to winning. This practice ultimately rendered them vulnerable to the challenges of new leaders who could fashion minimal winning coalitions that could offer their members more than could an oversized coalition that divided the spoils of victory among more players.

In an uncertain world, it would seem, there is no way for empires to survive indefinitely. For no matter how brilliant their leaders are—specifically, how well they understand the implications of the size principle—miscalculations are inevitable without perfect information. To be sure, brilliance coupled with good intelligence-gathering capabilities may retard the day of reckoning. But eventually the survival of a winning coalition in an essentially zero-sum environment is doomed by the mistakes of leaders who fail to trim an oversized coalition. If they cannot be sure of the point at which their coalition is minimal winning, this point—an unstable equilibrium at best—becomes even more difficult to achieve and maintain. Hence, any balance of power subject to the exigencies of uncertainty must, in the long run, collapse.

6.9. EMPIRICAL EVIDENCE FOR THE SIZE PRINCIPLE

However oversized coalitions come into existence, the prediction of the size principle is that they will be relatively short-lived. In seeking out evidence that supports the principle, Riker examines all instances in the modern (mostly) European state system of "overwhelming majorities." He finds three such cases, all being the products of total war in which one coalition of states became dominant in the system upon defeating an opposing coalition. These examples of overwhelming majorities in world politics—the Concert of Europe allies (England, Austria, Prussia, and Russia) after the Napoleonic wars, and the Allied powers after World Wars I and II—arose in situations that all seem to approximate the assumptions of the model (except for perfect information). During the wars, the governments on each side were intent on destroying the governments on the other side, suggesting that the wars were basically zero-sum in character. Nations for the most part sought allies who would enhance their capabilities of winning, which seems indicative of the rational pursuit of the postulated goal of winning. Finally, side payments reckoned in promises and threats, as well as material goods, were exchanged.

The conclusion of the total wars in all of these instances resulted in a grand coalition of winners, which is by definition worthless in a zero-sum game. Accordingly, each of these coalitions was almost immediately plagued by internal strife, with the Congress of Vienna splitting into two camps (Austria and England versus Prussia and Russia) after the Napoleonic wars; England, France, and the United States dividing over the future role of Germany after World War I; and the Soviet Union and the United States fighting for allies and hegemony after World War II. The hopes for permanent peace enshrined first in the League of Nations and now in the United Nations founder on the size principle precisely because of the undiscriminating inclusiveness of these international organizations.

Riker's second source of evidence on the dissolution of overwhelming majorities is from American presidential politics. Looking at all instances in which one party effectively demolished another party in a presidential election—and the losing party virtually disappeared from the national scene—he finds that the period of one-party dominance following such elections is soon undercut by party leaders who force out certain elements in policy and personality disputes. This leads to the dominant party's shrinkage and eventual displacement by another party. As in total wars, victory in elections in indivisible; hence, Riker argues, elections can properly be modeled as zero-sum games, where the players are individuals and groups who unite behind the banners of political parties.

The three instances of overwhelming majorities in American politics analyzed in *The Theory of Political Coalitions* are the Republican Party after the 1816 election that destroyed the Federalist Party, the Democratic Party after the 1852 election that destroyed the Whig Party, and the Republican Party after the 1868 election that signaled the temporary demise of the Democratic Party, at least outside the South. Because Riker's book was published in 1962, there is of course no mention of the 1964 election, in which Lyndon Johnson crushed Barry Goldwater with a 16-million-vote popular majority, but Riker and Ordeshook have since analyzed the aftermath of this election in terms of the size principle. Arguing that Johnson dissipated his overwhelming majority by progressively alienating (1) many Southerners with his strong pro-civil rights stance, and then (2) liberals with his escalation of the Vietnam war, Riker and Ordeshook contend that by 1968 Johnson probably did not have the support of even a minimal majority of the electorate and hence chose not to run for a second term.[37]

A similar fate befell Richard Nixon after his landslide victory in 1972. Unable to extricate himself from the crisis of Watergate—not to mention the energy crisis and a faltering economy—he was forced to resign a year-and-one-half into his second term. It seems no accident that whatever the time and circumstance, every United States president or political party that has emerged with an overwhelming majority has always faced a mounting tide of opposition, to which the president or party has succumbed in the end. In zero-sum politics, it seems, an instability occasioned by the miscalculations of leaders is unavoidable in a world enshrouded by uncertainty.

The evidence offered above is historical in nature and not particularly susceptible to quantification and rigorous empirical testing. Nonetheless, it has the virtue of being systematic: Riker did not ransack history for isolated examples that support the size principle but rather considered all instances of relevant cases within the spatially and temporally defined limits he set. To the extent that these cases—and the classes of events they are drawn from—are representative of zero-sum situations generally, they allow him to draw more general conclusions than could be adduced from anecdotal evidence alone. Yet, one is still perhaps left with a sense of uneasiness about what exactly is being tested if none of the key concepts has been operationally defined.

The appeal of the size principle may be better appreciated from a quantitative example that Riker offers, which also illustrates the "dynamics" of coalition-formation processes.[38] We enclose the word "dynamics" in quotation marks, however, because—for reasons discussed later in this chapter—the goal of winning does not seem to provide an adequate foundation on which to build a dynamic theory of coalition-formation processes. Nevertheless, it is instructive to consider a coalition situation unfolding over time wherein the participants acted as if they repeatedly invoked the size principle, even if repeated application of the size principle does not constitute a dynamic theory.

Following the demise and eventual disappearance of the Federalist Party, James Monroe was reelected president almost unanimously in 1820—only one vote was cast against him in the Electoral College. But the grand coalition of Republicans that he headed soon fell into disarray, consistent with the size principle, and as the presidential election of 1824

---

37. Riker and Ordeshook, *Introduction to Positive Political Theory*, pp. 194–96. Whether or not Johnson was conscious beforehand of the effects his actions would have is irrelevant to the test of the size principle, which asserts that he could not have prevented the breakup of his grand coalition no matter how hard he tried. That is, the game-theoretic logic of the model says that some of his "excess" supporters would invariably have been disaffected and attracted to the opposition coalition, whomever he tried to please with his policies.

38. Riker, *Theory of Political Coalitions*, pp. 149–58.

approached, factions formed around five candidates, whom Riker characterized as follows:

John Quincy Adams, of Massachusetts, Secretary of State, favorite of the former Federalists, although, as a moderate, he had been driven out of the Federalist Party in 1808.

William Crawford, of Georgia, Secretary of the Treasury, was the candidate of that alliance of Jeffersonians which had produced a series of presidents from Virginia and vice-presidents from New York.

John C. Calhoun, of South Carolina, Secretary of War, was the candidate of South Carolina and himself.

Andrew Jackson of Tennessee, the hero of New Orleans, governor of the Florida territory, and later senator from Tennessee, turned out to be the most popular candidate.

Henry Clay, of Kentucky, representative and leader of the opposition in the House, ultimately became the founder of the Whig Party.

Calhoun settled for the vice-presidency, narrowing the list to four candidates, none of whom won a majority of electoral votes in the Electoral College:

Jackson: Carried 11 states with 99 electoral votes;

Adams: Carried 7 states with 84 electoral votes;

Crawford: Carried 3 states with 41 electoral votes;

Clay: Carried 3 states with 37 electoral votes.

Consequently, the election was thrown into the House of Representatives, where each state had a single vote that went to the candidate favored by a plurality of its representatives. Following considerable bargaining in the House, Adams replaced Jackson as the leader in the standings:

Adams: 10 votes (states);

Jackson: 7 votes (states);

Crawford: 4 votes (states);

Clay: 3 votes (states).

Riker suggests that the desertion of representatives from states that Jackson carried was not surprising because the coalition of Adams, Crawford, and Clay initially constituted (see first distribution above) a unique minimal winning coalition with 13 votes (there were 24 states), making Jackson "strategically weak."

Because the Twelfth Amendment limits the number of candidates in the House to the three with the highest electoral-vote totals, Clay was eliminated. To whom would his three votes go? If Clay threw his support to Jackson, he would create a deadlock between Adams and Jackson, with Crawford then controlling the critical votes. By supporting Adams instead, he could himself be critical and elect Adams president with a minimal winning coalition.

This is exactly what he did in exchange for Adam's promise to appoint him Secretary of State, at that time traditionally a stepping-stone to the presidency. Thus was struck the so-called corrupt bargain of 1825, which also happens to be the outcome predicted by the size principle. In sum, both the desertions from Jackson and the Adams-Clay alliance, each of which involved the selection of coalition partners according to the size principle, triumphed over considerations of friendship, personal loyalty, and ideology (which are detailed in Riker's more extended historical treatment).

There is an important difference between the two kinds of historical evidence for the size principle that we have summarized, and this difference bears on the form that side payments take. The evidence from total wars and overwhelming majorities in American politics discussed earlier relates to the *disintegration* of grand coalitions, whereas the evidence from the presidential election of 1824–25 concerns the *formation* of minimal winning coalitions. (Recall that a similar distinction was made in the development of the bargaining models for measuring voting power in section 5.4.) Although the size principle is stated in coalition-formation terms, the "disintegration" evidence would appear more germane when side payments are mostly promised (or threatened, which is a negative kind of a promise). The reason for this is that players can renege on promises once they have won (e.g., as did Lyndon Johnson on the military policy he said he would pursue in Southeast Asia, creating the much-ballyhooed "credibility gap"), which may lose them support but not take away their victory. However, if side payments cannot be extravagantly promised, but must instead be delivered before the outcome is decided—or irrevocably committed, which amounts to the same thing (e.g., Adam's commitment of the Secretary of State post to Clay for his votes)—then coalition leaders must be selective in their commitments from the start and "formation" evidence would appear to be more germane. In any event, whether side payments take the form of "soft" or "hard" commitments, or some mixture in between, the crucial evidence that bears on the truth of the size principle is the existence of forces that move coalitions toward minimal winning (equilibrium) size, either from "above" or "below."

6.10. CRITICISMS OF THE SIZE PRINCIPLE[39]

Since the publication of *The Theory of Political Coalitions*, the size principle has stimulated many efforts aimed at testing its truth in a variety

---

39. This section is based largely on Steven J. Brams, "Positive Coalition Theory: The Relationship between Postulated Goals and Derived Behavior," pp. 3–40, esp. pp. 20–25,

of settings that go well beyond the arenas of American and international politics to which Riker mainly confined himself in mustering empirical support for the principle in his book. Evidence bearing on the principle has been adduced in African kingdoms and chiefdoms,[40] in local elections in Brazil,[41] in parliamentary elections in the French Fourth Republic[42] and other European democracies,[43] in cabinet formation in Denmark,[44] Israel,[45] Italy,[46] the Netherlands,[47] and West Germany,[48] among factions in the Japanese Liberal-Democratic party,[49] in the American Constitutional Convention of 1787,[50] and even on contested roll calls in the admittedly nonzero-sum legislative setting of the U.S. House of Representatives.[51]

---

from *Political Science Annual: An International Review*, edited by Cornelius P. Cotter, copyright © 1973, by The Bobbs-Merrill Company, Inc., reprinted by permission of the publisher.

40. Martin Southwold, "A Games Model of African Tribal Politics," in *Game Theory in the Behavioral Sciences*, ed. Ira R. Buchler and Hugo G. Nutini (Pittsburgh: University of Pittsburgh Press, 1969), pp. 23–43; and Martin Southwold, "Riker's Theory and the Analysis of Coalitions in Precolonial Africa," in *Study of Coalition Behavior*, pp. 336–50.

41. Phyllis Peterson, "Coalition-Formation in Local Elections in the State of São Paulo, Brazil," in *Study of Coalition Behavior*, pp. 141–59.

42. Howard Rosenthal, "Political Coalition: Elements of a Model and the Study of French Legislative Elections," *Calcul et Formalisation dous les Sciences de l'Homme* (Paris: Editions du Centre National de la Recherche Scientifique, 1968), pp. 237–85; Howard Rosenthal, "Voting and Coalition Models in Election Simulations," in *Simulation in the Study of Politics*, ed. William D. Coplin (Chicago: Markham Publishing Co., 1968), pp. 237–85; and Howard Rosenthal, "Size of Coalition and Electoral Outcomes in the Fourth French Republic," in *Study of Coalition Behavior*, pp. 43–59.

43. Eric C. Browne, "Testing Theories of Coalition Formation in the European Context," *Comparative Political Studies*, 3 (Jan. 1971), pp. 391–412; and Eric C. Browne and Mark N. Franklin, "Aspects of Coalition Payoffs in European Parliamentary Democracies," *American Political Science Review*, 67 (June 1973), pp. 453–69.

44. Erik Damgaard, "The Parliamentary Basis of Danish Governments: The Patterns of Coalition Formation," *Yearbook of the Political Science Association in Denmark, Finland, Norway, and Sweden*, 4 (1969), pp. 30–57.

45. David Nachmias, "A Note on Coalition Payoffs in a Dominant-Party System: Israel," *Political Studies*, 21 (Sept. 1973), pp. 301–305.

46. Robert Axelrod, *Conflict of Interest: A Theory of Divergent Goals with Applications to Politics* (Chicago: Markham Publishing Co., 1970), pp. 165–87.

47. Abraham DeSwaan, "An Empirical Model of Coalition Formation as an *N*-Person Game of Policy Distance Minimization," in *Study of Coalition Behavior*, pp. 424–44.

48. Peter H. Merkl, "Coalition Politics in West Germany," in *Study of Coalition Behavior*, pp. 13–42.

49. Michael Leiserson, "Factions and Coalitions in One-Party Japan: An Interpretation Based on the Theory of Games," *American Political Science Review*, 62 (Sept. 1968), pp. 770–87; and Michael Leiserson, "Coalition Government in Japan," in *Study of Coalition Behavior*, pp. 80–102.

50. Gerald M. Pomper, "Conflict and Coalitions at the Constitutional Convention," in *Study of Coalition Behavior*, pp. 209–25.

51. Koehler, "Legislative Process and Minimal Winning Coalition." But note Koehler's justification (see section 6.7) of the applicability of the size principle to nonzero-sum situations in which resources are scarce. Some preliminary results of Koehler's research are described in Steven J. Brams and Michael K. O'Leary, "An Axiomatic Model of Voting

To be sure, the authors of these studies have offered various qualifications, showing in some cases how other factors, like the desire to minimize ideological distance or the number of members in a coalition, can improve the predictive power of Riker's model and may also lead to other predictions, such as what the composition of coalitions and the distribution of payoffs (e.g., cabinet posts) among coalition members (e.g., political parties) will be. Unfortunately, most of these other factors adduced to explain coalition outcomes have largely an ad hoc flavor since they are not derived from the assumptions of a model. While admitting the importance of such a factor as ideology in vitiating the size principle through its preclusion of actors of dissimilar beliefs from joining the same coalition, Riker and Ordeshook also note that its inclusion as an explanatory variable may lead to circular reasoning: Ideological compatibility is determined from the history of previous party alignments, which is information then used to predict (or retrodict, i.e., predict a past state of affairs) what coalitions will form that are ideologically compatible.[52] An assumption should not be synonymous with the outcome it predicts but be directed at the determinants of the outcome.

On the other hand, it might be argued that ideological cleavages increase the perfectness of information, and thereby the applicability of the size principle, by making choices more predictable. Clearly, such plausible yet divergent claims made for the effects produced by ideology on the size principle leave its status as an explanatory variable unsettled, which vividly illustrates the kinds of logical problems that can be created by the introduction of new "explanatory" factors in a model. Only the logical development of the implications of a model from carefully articulated assumptions can make one aware of such problems. Free-floating hypotheses that incorporate factors like ideology may lead to good predictions but, at best, confused explanations.

Because ideology has played such a prominent role as a factor adduced to explain coalition behavior, it deserves special attention for the logical traps it has created. It is certainly understandable why this factor has been so extensively used in empirically oriented studies of coalition behavior since it seems such a reasonable, if not important, variable for leaders to take account of in drawing up lists of potential coalition partners. In fact, it has been shown, the deeper the ideological cleavages, the greater departures tend to be from the size principle.[53]

In principle, there is no reason why ideological considerations cannot be incorporated in rationalistic calculations related to the goal of winning.

Bodies," *American Political Science Review,* 64 (June 1970), pp. 468–69, where suggestions are made on the operationalization of the size principle in voting bodies. See also Barbara Hinckley, "Coalitions in Congress: Size and Ideological Distance," *Midwest Journal of Political Science,* 16 (May 1972), pp. 197–207; and Barbara Hinckley, "Coalitions in Congress: Size in a Series of Games," *American Politics Quarterly,* 1 (July 1973), pp. 339–60.

52. Riker and Ordeshook, *Introduction to Positive Political Theory,* p. 193.

53. Riker and Ordeshook, *Introduction to Positive Political Theory,* pp. 191–94.

As we have indicated, however, ideological considerations play havoc with the size principle when introduced as an empirical face-saving device, *deus ex machina*. This usually occurs when ideology is viewed as imposing a constraint on permissible coalitions, and that subset of all combinatorial possibilities *not* disallowed by ideological mismatches is compared with coalitions that actually formed. (No formalization of constraints based on ψ-stability has been attempted.) Michael Leiserson, for example, has constructed a model in which he assumes that political parties search for ideologically close partners and from which he predicts that coalitions will form that consist of the smallest number of neighboring parties (on a left-right continuum) sufficient to win.[54]

As Abraham DeSwaan has pointed out, however, this "minimal range" model, although successful in predicting actual cabinet coalitions in some countries, has led to less successful predictions in others where coalitions have included more members than necessary to win. As an alternative, he proposes a model based on the minimization of policy differences wherein actors "strive to become members of a winning coalition which they expect to adopt policies that are as close as possible to their most-preferred policies."[55] In addition to the empirical problem of ranking actors on specific policy issues, and statistically combining these rankings into scales on a limited number of factors, DeSwaan's theoretical model is based on some rather strong monotonicity and symmetry assumptions about the preference distributions of members. Nonetheless, he is able to generate some consequences that accord reasonably well with actual coalitions that formed in the Netherlands, although he confesses that "it is apparently possible to account for nearly every actual coalition on the basis of a commonsensical ideological ranking."[56]

This is a serious problem, for the justification of a formal model, as has been repeatedly stressed, depends to a considerable degree on the nonobviousness of the consequences that follow from its assumptions. Models like DeSwaan's and Lieserson's, although they assume, as do the previously discussed models, rational choice on the part of the actors involved, do not yield predictions very different from the simple combination rule that coalitions will form that consist of ideologically proximate actors.

Robert Axelrod has developed a model of coalition formation in which the general concept of "conflict of interest" is measured in terms of the ideological diversity of a set of actors. From the assumption that the less

---

54. Michael Leiserson, "Coalitions in Politics: A Theoretical and Empirical Study" (Ph.D. dissertation, Yale University, 1966). For a similar model, see Rosenthal, "Size of Coalition and Electoral Outcomes in the Fourth French Republic."

55. DeSwaan, "An Empirical Model of Coalition Formation as an *N*-Person Game of Policy Distance Minimization," p. 444. More generally, see Abraham DeSwaan, *Coalition Theories and Cabinet Formation* (San Francisco: Jorsey-Bass, 1973).

56. DeSwaan, "An Empirical Model of Coalition Formation as an *N*-Person Game of Policy Distance Minimization," p. 444.

conflict of interest among actors (as measured by the dispersion of their positions on policy issues), the more likely a coalition will form that includes them as members, Axelrod predicts the formation of minimal *connected* winning coalitions (i.e., those minimal winning coalitions consisting only of ideologically adjacent members ordered along an ordinal policy dimension). This prediction, of course, is very similar to that of Leiserson's model, but Axelrod claims that he eliminates the ad hoc quality of Leiseron's prediction by deriving the desire of actors to minimize conflictual behavior—and hence form minimal connected winning coalitions—from "a theoretical analysis of conflict of interest in various contexts."[57] Others have found his analysis persuasive,[58] but what relationship Axelrod's quantitative measure of conflict of interest has to minimal connected winning coalitions is not at all clear in his analysis.

Although conflict of interest may be conceptualized differently from conflictual behavior, in fact the genesis of the left-right scale for Italian political parties used by Axelrod to predict the formation and duration of cabinet coalitions in Italy would appear to be based in part on this very information, i.e., what coalitions actually did form and persist. This indeed seems an intellectual conflict of interest between predictor and predicted variables. It reduces Axelrod's test to one that measures how well a single left-right scale can predict the observed coalitions—that is, how consistent Italian political parties have been with respect to this scale.

Operationally, this kind of circularity, which we alluded to earlier, might be eliminated by defining ideological scales independently of coalitions that actually formed. James P. Zais and John H. Kessel observed this methodological canon in using separately developed attitudinal proximity scores to retrodict the likelihood of actual coalition alignments and vote totals in the 1968 Republican presidential nomination, but their failure to develop a model from which their likelihood measure can be rigorously derived does not make their box-score predictions, even though generated by a computer, particularly compelling.[59] The program that produces them, and that could produce other plausible predictions with an alteration in the measure used and values assigned, is not a substitute for deductions from the assumptions of a model that specify both why a particular set of

---

57. Axelrod, *Conflict of Interest,* p. 174.

58. David W. Rhode, "Policy Goals and Opinion Coalitions in the Supreme Court," *Midwest Journal of Political Science,* 16 (May 1972), pp. 208–24.

59. James P. Zais and John H. Kessel, "A Theory of Presidential Nominations, with a 1968 Illustration," in *Perspectives on Presidential Selection,* ed. Donald R. Matthews (Washington, D.C.: Brookings Institution, 1973), pp. 120–42; see also John H. Kessel, *The Goldwater Coalition: Republican Strategies in 1964* (Indianapolis: Bobbs-Merrill Co., 1968). For other computer models, see James P. Kahan and Richard A. Helwig, "'Coalitions': A System of Programs for Computer-Controlled Bargaining Games," *General Systems: Yearbook of the Society for General Systems Research,* 16 (1971), pp. 31–41; and for a review of the literature on computer models, see James D. Laing, Alexander Lebanon, Richard J. Morrison, and Howard Rosenthal, "Computer Models of Political Coalitions," in *Political Science Annual: An International Review, Volume Four—1973,* pp. 41–74.

relationships should obtain and the limiting conditions that circumscribe the validity of these relationships.

Our brief survey of some of the pitfalls of more empirically oriented "ideological" theories makes clear that the soundness of empirical research very much depends on the logic that supports it.[60] Yet this logic may not always be sufficient to give clear and unambiguous answers to important practical questions. As an empirical hypothesis, the size principle is not immune from such difficulties when, for example, the question is raised of how long an oversized coalition may be expected to persist. To respond that this will depend on how imperfect information is, or even how over-riding ideological cleavages are, is not helpful when time-dependent functional relationships between these variables and the size principle are unspecified. Will the excess majority of an oversized coalition be dissipated in weeks, years, or decades?

Riker tries indirectly to meet these difficulties by developing the out-lines of a dynamic model based on considerations that govern the exchange of side payments. This model, however, does not really answer the question of what time-dependent relationships govern these exchanges. Neither does his discussion of strategy in coalition building, though it does lead to the important insight that smaller *protocoalitions* (i.e., coalitions not of sufficient size to be winning) are more frequently favored in the forma-tion of minimal winning coalitions than one might expect on the basis of their size alone. The inherent instability and lack of equilibrium in zero-sum, coalition politics that encourage participants to make decisions which lead toward the elimination of some suggests a dynamic that upsets any balance among forces that might temporarily exist. But Riker does not spell out the development of this process in a formal way. To establish the existence of disequilibrium and necessary and sufficient conditions for its occurrence does not require that one specify a mechanism that charts the operation and movement of destabilizing forces over time.

The mostly static nature of Riker's model, at least in its formal aspects, is not the fault of his analysis, which as we have seen leads to some impor-tant and provocative findings. These findings are limited, however, by the questions Riker asked, which were primarily related to the nature of the outcomes, and to a lesser extent the strategies of the participants, in coalition-building processes. (This is perhaps one consequence of using a game-theoretic model, since game theory is essentially a static theory.[61])

---

60. A similar argument is made by Eric C. Browne, who is particularly critical of models based on the minimization of size and favors instead spatial models like DeSwann's based on the minimization of policy differences among actors. See Eric C. Browne, *Coalition Theories: A Logical and Empirical Critique*, Sage Professional Paper in Comparative Politics, vol. 4, no. 01-043 (Beverly Hills, Calif.: Sage Publications, 1973).

61. A dynamic component to the theory has recently been introduced in the form of "differential games," whose study has become a subfield of game theory over the past decade. See Rufus Isaacs, *Differential Games: A Mathematical Theory with Applications to Warfare and Pursuit, Control and Optimization* (New York: John Wiley & Sons, 1965);

Riker seems to have sensed these limitations when he candidly remarks:

> The model itself is at this point quite vague simply because we cannot easily abstract a pattern from the rich complexity of events in the growth of protocoalitions. It may be regarded as the main task of a dynamic theory of coalitions to specify the pattern of growth and the strategic considerations involved in the process by which a proto-coalition passes from the . . . stage of competing protocoalitions to the final stage when some coalition enforces a decision.[62]

To probe more deeply into the dynamic nature of transitions from stage to stage, we must postulate and develop the implications of different goals to which rational actors might aspire. We shall consider one such goal in the next section and sketch what implications its satisfaction has on time-dependent, sequential aspects of coalition-formation processes.

6.11. AN ALTERNATIVE GOAL: MAXIMIZING ONE'S SHARE OF SPOILS

Our review of the literature has so far indicated that the study of coalition behavior has been vigorously pursued in recent years. Yet little attention has been devoted to the construction of theoretical models of coalition-formation processes that occur over time. Rather, studies of coalition behavior have focused more on *static outcomes* than on the *dynamic processes* that produced them *prior* to the point at which one coalition has gone on to win. Even in the relatively well-structured context of voting bodies, to which we shall confine our subsequent analysis, the development of dynamic models has only recently been initiated.

The existence of this gap in the literature is due in considerable part to the failure of theorists to postulate goals for rational actors that raise questions about the *timing* of their actions—and therefore about political processes that occur over time. Later we shall offer one such goal, but first we shall outline a verbal model of coalition-formation processes in voting bodies. We restrict our analysis to the study of coalition-formation processes involving as active opponents only two protocoalitions—coalitions with too few members to be decisive alone—that vie for the support of uncommitted members in a voting body in order that they can become winning coalitions (which are decisive, given some decision rule). As previously, we shall distinguish in the subsequent analysis all winning coalitions from the subset of winning coalitions that are minimal winning —coalitions in which the subtraction of a single member reduces them to (nonwinning) protocoalitions.

Avner Friedman, *Differential Games and Related Topics*, ed. H. W. Kuhn and G. P. Szego (New York: American Elsevier Publishing Co., 1971); *Topics in Differential Games*, ed. Austin Blaquière (Amsterdam: North-Holland Publishing Co., 1973); and Avner Friedman, *Differential Games* (Providence, R.I.: American Mathematical Society, 1974).

62. Riker, *Theory of Political Coalitions*, pp. 107–108.

Our main interest is in studying the formation of winning coalitions including one or the other of the two protocoalitions, but not both. We assume that the two protocoalitions are totally at odds with each other and seek victory only through securing the commitment of *uncommitted members*, not through the switching of members' commitments from one protocoalition to the other. We thus preclude members of the two proto-coalitions from combining with each other to form a winning coalition, as Luce and Rogow precluded potential defectors in their legislative model of a two-party system from coalescing (see section 6.3).

Although we attach no magic to the number 2, our assumption of two opposed protocoalitions provides a useful starting point for analysis. Empirically, it seems compatible with many winner-take-all electoral systems, which tend to reduce conflicts to those involving only two opponents. At a theoretical level, the assumption in the characteristic-function form of n-person game theory that a coalition will inspire the formation of a countercoalition, whose members can get at least as much and possibly more by banding together, provides some justification for viewing even n-person games as reducible to two-person games, where the players are protocoalitions of actors.

Our purpose in making these simplifying assumptions is to focus attention on the conflict between the two protocoalitions both striving to enlist the support of uncommitted members in a zero-sum environment. We are particularly interested in exploring the possible calculations that leaders of the protocoalitions may make in trying to determine how much to offer the uncommitted members to join, and, conversely, the possible calculations that the uncommitted members may make in trying to decide what a commitment is objectively worth.[63]

To be sure, a host of situational factors—including "uncommitted" members predisposed to one or the other of the two protocoalitions, cross-cutting affiliations of members of the two protocoalitions that encourage switching from one to the other, third protocoalitions, and so on—will upset any strict correspondence that this model might have to coalition formation in real voting bodies. Nevertheless, as the following example illustrates, the model seems to provide a sufficiently good approximation of the manner in which coalitions form in some actual voting bodies so as to render its calculations useful in analyzing the dynamics of coalition-formation processes.

---

63. For a fruitful exploration of this question using the methods of economic analysis, see James S. Coleman, "The Marginal Utility of a Vote Commitment," *Public Choice*, 5 (Fall 1968), pp. 39–53. Coleman specifically dismisses game theory as a not very useful model for the study of collective decisions and instead uses probability theory as the basis for his formal analysis of voting decisions. The probabilistic models that we shall illustrate in this section are closely related to those of Coleman, but we shall also indicate how a game-theoretic element can be explicitly introduced. For a further elaboration of Coleman's probabilistic approach, see James S. Coleman, "The Benefits of Coalition," *Public Choice*, 8 (Spring 1970), pp. 45–61, in which a probabilistic model is used to explore gains and losses associated with joining or not joining a coalition with and without vote trading. See also James S. Coleman, *The Mathematics of Collective Action* (Chicago: Aldine Publishing Co., 1973).

To support this claim with a real-life example, consider the final stage of the coalition-formation process in the U.S. House of Representatives prior to the election of John Quincy Adams as president in 1825 (see section 6.9). Recall that the standing of the four candidates before voting began was

Adams: 10 votes (states),

Jackson: 7 votes (states),

Crawford: 4 votes (states),

Clay: 3 votes (states),

which would have forced the elimination of Clay under the Twelfth Amendment, which allows only three runoff candidates. Realizing that the votes Clay controlled were open for bidding, Crawford's managers, it seems fair to suppose, might also have considered selling their votes since their candidate ranked well below the two front-runners.[64] If we assume that the two leading candidates, Adams and Jackson, represent the two protocoalitions in the model we have described, and the "uncommitted" votes are the blocs controlled by Clay and Crawford, it is simple to determine how each of the two leading candidates could have won (the decison rule is 13 out of 24 states):

Adams, with the support of (1) Clay, (2) Crawford, or (3) both Clay and Crawford;

Jackson, with the support of (4) both Clay and Crawford.

If we are given no additional information about coalitions that might form in this situation, it is reasonable to assume that each of these four outcomes is equally likely (which is an assumption we shall modify later in light of the size principle). Hence, we can calculate the complementary probabilities (P's) that each of the two leading candidates (i.e., protocoalitions) would become winning:[65]

$$P(\text{Adams}) = \tfrac{3}{4} = 0.75; \ P(\text{Jackson}) = \tfrac{1}{4} = 0.25.$$

In fact, of course, Clay threw his support to Adams as a result of the "corrupt bargain," making Adams the winner (with probability equal, in

---

64. If a deadlock had developed, this appears indeed to have been their intention. See John Spencer Bassett, *The Life of Andrew Jackson*, 2d ed. (New York: Macmillan Co., 1925), p. 368.

65. Probabilities like these based on equiprobability assumptions are common in the natural sciences. For their use in the Maxwell-Boltzman and Bose-Einstein statistics in physics, and analogous applications in political science, see William Feller, *An Introduction to Probability Theory and Its Applications*, 2d ed. (New York: John Wiley & Sons, 1957), pp. 20–21. 38–40; and Steven J. Brams, "The Search for Structural Order in the International System: Some Models and Preliminary Results," *International Studies Quarterly*, 13 (Sept. 1969), pp. 254–80.

effect, to 1.00). If Clay had instead supported Jackson, then Jackson and Adams would have tied with ten votes each. If, at this hypothetical juncture, Crawford had been equally likely to have given his support to either leading candidate, then each would have had a probability equal to 0.50 of going on to win.

The two choices open to Clay—support Adams or support Jackson— are pictured in the accompanying diagram, with the probabilities that each of the main contenders, Adams and Jackson, would win before and after Clay's commitment to one of them:[66]

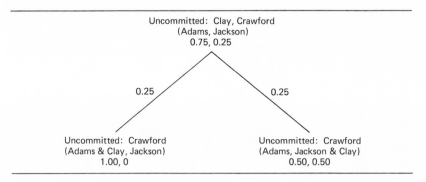

Uncommitted: Clay, Crawford
(Adams, Jackson)
0.75, 0.25

0.25

0.25

Uncommitted: Crawford
(Adams & Clay, Jackson)
1.00, 0

Uncommitted: Crawford
(Adams, Jackson & Clay)
0.50, 0.50

Along each branch are shown the *probabilistic contributions* that Clay's support makes to raising the probability of winning of Adams (from 0.75 to 1.00, or an increment of 0.25) and to Jackson (from 0.25 to 0.50, or an increment of 0.25). The fact that these probabilistic contributions are both equal to 0.25 means that no advantage would accrue to Clay from supporting one candidate over the other if his goal were to maximize his probabilistic contribution to a contender.[67] The goal we postulate, however, is that uncommitted actors seek to maximize their *share of spoils* (*SS*), which is defined as the probabilistic contribution (*PC*) an uncommitted actor makes to a protocoalition times the probability (*P*) that, with this contribution, the protocoalition will go on to win.[68] Thus,

$$SS(\text{Clay} \rightarrow \text{Adams}) = PC(\text{Clay} \rightarrow \text{Adams})P(\text{Clay} \rightarrow \text{Adams})$$
$$= (0.25)(1.00) = 0.25,$$
$$SS(\text{Clay} \rightarrow \text{Jackson}) = PC(\text{Clay} \rightarrow \text{Jackson})P(\text{Clay} \rightarrow \text{Jackson})$$
$$= (0.25)(0.50) = 0.125,$$

66. We have assumed that Clay makes the first commitment not only because as low man he was eliminated from the contest in the House but also because Crawford's supporters felt bound to vote for their candidate on the first ballot. Bassett, *Life of Andrew Jackson*, p. 363.

67. If an uncommitted actor's probabilistic contribution can be taken as a measure of the resources he brings to a protocoalition, then Gamson predicts that payoffs after winning will be distributed according to these contributions. See Gamson, "A Theory of Coalition Formation."

68. A similar calculation is suggested in Norman Frohlich, Joe A. Oppenheimer, and Oran R. Young, *Political Leadership and Collective Goods* (Princeton, N.J.: Princeton University Press, 1971), p. 89. See also Anthony Downs, *An Economic Theory of Democracy* (New York: Harper and Row, 1957), pp. 47–50, 159.

where the actor before the arrow indicates the uncommitted "giver" and the actor after the arrow the protocoalition "receiver." Assuming Clay's goal was to maximize his share of spoils, then it was rational by this calculation for him to support Adams rather than Jackson (as he in fact did), since the former course of action would yield twice as much in "spoils" as the latter.

In the share-of-spoils calculation, we assume that an uncommitted actor desires first to maximize his probability of being in a winning coalition, and second to maximize the portion of the benefits he derives from that winning coalition. As Anthony Downs has shown, however, a benefit-maximizing strategy on the part of individuals may not be consistent with their acting together as a group to form a winning coalition,[69] so it is convenient to combine these two goals in an expected-value calculation, as we have done in the share-of-spoils concept. When uncommitted actors maximize their share of spoils, they are not necessarily choosing the protocoalition with the greatest probability of winning, or the proto-coalition to which they can make the largest probabilistic contribution (and from which, presumably, they can derive the greatest benefits should it win). Instead, they are choosing that combination of "winningness" and benefits that on the average gives them the greatest "spoils." It is this bundle of goods which we assume uncommitted actors, acting rationally, seek to maximize. This payoff may take the form of not only material and divisible benefits like patronage or even graft (with which the term "spoils" has often been linked), but also less divisible collective benefits that flow from policies and actions (e.g., related to the prevention of war) that have repercussions on all members of a political system.

In the preceding example, we computed the probabilities that the two leading candidates would become *any winning* coalitions—that is, coalitions with *at least* a simple majority of members—before and after the commitment of Clay. If we restrict coalitions to those that are *minimal winning* with *exactly* a simple majority of members, then consonant with the size principle there is only one way that Adams can win with exactly thirteen of the twenty-four votes (i.e., by gaining the support of Clay), but no way Jackson can win with this bare majority. Since the Adams-Clay alliance is the only possible minimal winning coalition, it will become minimal winning with probability equal to one *before* Clay's commitment, given that only minimal winning coalitions form. *After* Clay's commitment, the coalition will be winning (with probability equal, in effect, to one), so Clay's probabilistic contribution, and share of spoils that he can expect to receive from Adams, is zero.

This may seem a rather artificial result, especially in light of the fact that Clay became Secretary of State under Adams and was importuned by all parties for support before the House vote.[70] Apparently, not everybody

69. Downs, *Economic Theory of Democracy*, p. 159.
70. Bassett, *Life of Andrew Jackson*, pp. 350–51.

was aware of the size principle, for if they had been, all except Adam's supporters should have given up hope!

Despite its artificiality, however, this result points up one nonobvious fact about coalition formation: To the extent that an uncommitted actor is constrained in the range of choices open to him, his influence will be diminished. Indeed, in the extreme case where he is a captive of one proto-coalition, his probabilistic contribution and share of spoils are effectively zero since he is not truly uncommitted. This was the case in our example, for if Clay adhered to the size principle, he could support only Adams. In other words, the size principle, like friendship, ideology, and other constraints on free association, may cut down an actor's range of choices and force his hand before all alternatives but one are closed to him. In fact, one month before the actual balloting in the House, Clay confined to intimates that he would support Adams,[71] which became public knowledge two weeks before the election.[72] But there was still great uncertainty about what the outcome would be because the New York delegation (counted earlier in Adam's total) was deadlocked up to the day of voting. The election eventually turned on the vote of one of the thirty-four representatives from New York, who on the day of voting broke the deadlock and gave New York to Adams.[73]

How do these historical details bear on models of coalition formation? First, we have shown that an alternative goal—maximization of share of spoils based on the any winning model—also retrodicts that Clay would support Adams over Jackson. In the example we have been considering, this goal is perhaps more realistic than the goal of winning because it directly incorporates the notion of private and divisible benefits that uncommitted actors may realize in addition to the public and indivisible benefits of winning that all members of the winning coalition share. Second, we have shown some curious implications of the share-of-spoils calculation when the size principle is invoked. If all but minimal winning coalitions are disallowed, the share-of-spoils calculation suggests that Clay's votes were not essential to Adams, for the size principle precludes their going to anybody else. Manifestly, the situation was not viewed this way by the participants, although, as we have seen, Clay's votes were committed before the day of the election and it fell to somebody else to make the critical choice on that day. This leads us to ask: What makes a player's choices critical?

Choices of players are *critical*, it seems, if (and only if) they are preceded by prior choices of other players that make them decisive in the

---

71. Bassett, *Life of Andrew Jackson*, p. 352.

72. Marquis James, *Andrew Jackson: Portrait of a President* (New York: Grosset and Dunlap, 1937), p. 120.

73. Riker views skeptically an interesting (and possibly apocryphal) story about the intense pressure exerted on this representative and how he made his decision. Riker, *Theory of Political Coalitions*, pp. 155–57.

determination of outcomes. For this reason, it is useful to view coalition-formation processes as *sequences of moves* on the part of players. In one sense, game theory takes account of such sequences in the concept of a "strategy" (see section 1.2), which is a plan that prescribes choices for every contingency that might arise. But in most real coalition-building situations, it is no easy task to abstract any set of general contingencies.[74] Even in a particular situation, the possible alignments, and the ways in which they can form, are manifold. What we can offer, however, is a prescription of how protocoalitions should grow *at all stages* in the coalition-formation process in order to maximize their attractiveness to uncommitted actors.

Intuitively, it seems clear that uncommitted actors who wish to maximize their share of spoils in the previously described model would be interested in joining the protocoalition with the greater probability of winning. Otherwise, their probabilistic contributions would be cut by more than half in the share-of-spoils calculation. On the other hand, if the probability that one of the two protocoalitions would become winning were overwhelming (i.e., close to one), the probabilistic contribution that an uncommitted actor would make by joining it would be necessarily small, even though the contribution itself would not be heavily discounted in the share-of-spoils calculation. It turns out that if there are two protocoalitions, one of which grows faster than the other, an uncommitted actor should join the one he thinks is growing faster—as measured by its larger size at any stage—when its probability of becoming *minimal winning* is two-thirds.[75] That is, he should wait until the probability that one protocoalition becomes minimal winning is exactly twice that of the other before committing himself to what he perceives to be the faster-growing protocoalition. This advice is applicable whether this point (or points) is reached early or late in the coalition-formation process.

From the viewpoint of the coalition leaders, an awareness of this strategy provides no guarantee of victory. For example, a protocoalition may simply not have the resources to pull ahead of another protocoalition so that it is perceived as the two-to-one favorite. Given that it is able to

---

74. Although the calculations are rather complicated, some attempt at generalization has been made through the utilization of lattice structures for arraying the step-by-step buildup of protocoalitions in ten-member voting bodies, and in larger bodies through computer-simulation techniques. See Steven J. Brams and William H. Riker, "Models of Coalition Formation in Voting Bodies," in *Mathematical Applications in Political Science, VI*, ed. James F. Herndon and Joseph L. Bernd (Charlottesville, Va.: University Press of Virginia, 1972), pp. 79–124; Steven J. Brams, "A Cost/Benefit Analysis of Coalition Formation in Voting Bodies," in *Probability Models of Collective Decision-Making*, pp. 101–24; and Steven J. Brams and John G. Heilman, "When to Join a Coalition, and with How Many Others, Depends on What You Expect the Outcome to Be," *Public Choice*, 17 (Spring 1974), pp. 11–25.

75. Steven J. Brams and José E. Garriga-Picó, "Bandwagons in Coalition Formation: The 2/3's Rule," *American Behavioral Scientist*, 7 (March–April 1975) (forthcoming).

accomplish this feat, however, the "2/3's rule" prescribes that it should not attempt to increase these odds in its favor still further, for then it will become less attractive to uncommitted actors who desire to maximize their share of spoils. We would expect, therefore, that a protocoalition will maximize the commitments it receives from uncommitted actors when it enjoys a 2:1 probabilistic advantage over the other protocoalition.[76]

Of course, an opposition protocoalition would be attempting to achieve the same kind of effect, which means that neither protocoalition has a surefire winning strategy unless additional assumptions are made about the total resources possessed by each side, the manner in which they are allocated, their effects, and so on. By the same token, uncommitted actors cannot anticipate if and when optimal commitment times will occur in the absence of additional assumptions. The "2/3's rule" as such says only that there exist optimal commitment times, but it says nothing about instrumental strategies for realizing them.

Still, this rule offers a precise quantitative statement of when bandwagons would be expected to develop and provides, in addition, a rationale for their existence based on the assumptions of the model from which it is derived (which we shall not try to develop rigorously here). Although it may be thought of more as a benchmark for developing strategies than a strategy itself, we now present one extremely simplified example of how the "2/3's rule" might be exploited in a coalition-formation situation.

In a 101-member voting body, assume uncommitted members join each of two protocoalitions until one acquires 51 members and becomes minimal winning. The rule says that if one protocoalition has 50 members, it is more attractive to the remaining uncommitted members if the other protocoalition has 49 members rather than any smaller number. The reason is that there are exactly two ways in which the 50-member protocoalition can become minimal winning with 51 members (i.e., if either of the two uncommitted members joins it), but only one way in which the 49-member protocoalition can become minimal winning (i.e., if both of the uncommitted members join it), which gives the 50-member protocoalition a 2:1 probabilistic advantage if all minimal winning outcomes are considered equally likely. Therefore, it would actually be to the advantage of a 50-member protocoalition to "allow" a smaller protocoalition to expend resources to attract uncommitted members and increase its membership to 49, for then the larger protocoalition's probability of becoming minimal winning will be two-thirds and the share of spoils it can offer to the uncommitted member who makes it minimal winning is maximal.

---

76. The surge in support that the expected winner receives under such circumstances is often referred to as the "bandwagon effect." For an empirical test of the relationship between the share-of-spoils goal and bandwagons, see Steven J. Brams and G. William Sensiba, "The Win/Share Principle in National Party Conventions" (Unpublished paper, New York University, 1970).

Even in a game of perfect information wherein the smaller protocoalition could anticipate this strategy of the larger protocoalition, there is little it could do about it because there is no way it could make its probability of becoming minimal winning two-thirds (given that there are no defections), which gives the larger protocoalition a dominant strategy (see section 1.4). This would not be true, however, if the larger protocoalition had fewer than 50 members.

It is in such situations, it would seem, that game theory has something to say about coalition processes as well as coalition outcomes, although practically no research has been done in this area. The main reason seems to be that goals like winning, which apply to a coalition as a whole, tend to shut off inquiry into the dynamic analysis of what offers protocoalitions are likely to make, and uncommitted actors accept, at particular times in the formation and disintegration of winning coalitions. Simply to say that uncommitted actors seek to obtain the benefits of winning does not by itself entail the choice of a particular strategy on their part. By contrast, postulating the goal of not just obtaining the benefits from winning, but maximizing one's portion of these benefits[77]—as measured by the share of spoils—directs one's attention to the optimal timing of strategic choices (e.g., when to join a protocoalition), and hence the unfolding of events (e.g., bandwagons) over time.

While there has been some work done on games of timing (e.g., when to fire the shot—or shots—in a duel),[78] these do not appear to have much relevance for coalition games and, more generally, questions of timing in politics. Some of the most interesting tactical and strategic questions in politics relate to the timing of actions in an essentially zero-sum environment (e.g., when to initiate war, announce one's candidacy, or take a position on an issue), and such questions deserve much more study than they have so far received.

6.12. SUMMARY AND CONCLUSION

The study of n-person cooperative games, where communication and bargaining among players are allowed, reduces largely to the study of coalition formation in these games if no single player can alone dictate the outcome. Treating power as control over outcomes, we began our analysis by reviewing the relationship between power and coalitions, as manifested in voting and vote-trading games discussed in previous chapters. Specifically, we contrasted (1) the aggregation of players into coalitions, which proceeds from a coincidence of their preferences and their

---

77. This idea is incorporated in Adrian and Press's reformulation of the size principle. See Charles H. Adrian and Charles Press, "Decision Costs in Coalition Formation," *American Political Science Review*, 62 (June 1968), p. 557.

78. See Melvin Dresher, *Games of Strategy: Theory and Applications* (Englewood Cliffs, N.J.: Prentice-Hall, 1961), pp. 128–44.

ability to enforce an agreement (Chapters 2 and 4) ,with (2) the disaggregation of the value of coalitions into the voting power of players, which depends upon the bargaining positions of players in coalitions (Chapter 5). We then showed how certain restrictions could be imposed on coalition formation to locate actors in positions of power and predict real-world coalition outcomes.

These restrictions took two different forms. One kind of restriction permitted only certain realignments of a given coalition structure from forming. In the case of Luce and Rogow's model of a stable two-party legislative system, they permitted the formation of coalitions across party lines, but not for the purpose of forming a third party. The imputations and coalition structures that were ψ-stable—that is, that were incapable of being disrupted by an admissible coalition—enabled them to determine actors in "locations of power" in various situations. As it turned out, the actors who were in positions of power in each situation were those whose presence was necessary in every winning coalition allowed by admissible changes in the coalition structure.

Probably the most interesting result revealed in this analysis is that the president is not in a position of power whenever two-thirds majorities control Congress, even if they are of his own party. A check on vetoes cast by Franklin Roosevelt and Lyndon Johnson, the only presidents since the popular election of senators to have had both simple and two-thirds majorities of their own parties in Congress during their administrations, showed that each president cast considerably more vetoes on the average in his two-thirds majority Congresses than in his simple majority Congresses. This finding would seem to substantiate the theoretical consequence of the Luce-Rogow model that the president lacks power when two-thirds majorities of his own party control Congress—although it was pointed out that vetoes are more a measure of the negative power to prevent action than the positive power to initiate action.

The second set of restrictions discussed was that imposed by Riker on the characteristic function. In addition to certain mathematical constraints, he assumed it had positive value only if a coalition were winning, that all the members of a winning coalition received positive payoffs, and that the winning coalition had control over its membership. Given these restrictions, we sketched Riker's proof of the size principle that in n-person zero-sum games of perfect information where side payments are allowed, only minimal coalitions will form.

We reviewed historical evidence from American and international politics adduced in support of the size principle and suggested that the disintegration of overwhelming majorities will tend to be associated with "soft" side payments, the formation of minimal majorities with "hard" side payments. Oversized coalitions will also tend to form in situations of imperfect information, which induce miscalculations by leaders and lead to their eventual replacement by minimal winning coalitions. Although the size principle has received its greatest empirical support in situations wherein ideological considerations are not paramount, recent attempts to develop models that take account of the effects of ideology on coalition outcomes have tended to be based on circular or ad hoc reasoning.

Riker originally proposed winning as a goal of political actors when a tractable definition of power proved elusive. As an alternative to the goal of winning, which does not specify how the benefits of winning will be divided among members of a winning coalition, we proposed that uncommitted actors in a voting body would attempt to maximize their share of spoils, which is the probabilistic contribution that they make to a protocoalition discounted by the protocoalition's probability of going on to win after the contribution. (We assumed an equivalence between the payoff a member of a protocoalition receives if the protocoalition becomes winning and the size of the discounted contribution that he makes to its becoming winning, which will depend on what stage in the coalition-formation process he joins it.) We discussed the application of this goal to the United States presidential election of 1824–25, assuming that either minimal winning or any winning coalitions could form. Its relevance to the dynamics of coalition-formation processes, and the timing of political actions generally, was also discussed, and game-theoretic strategies based on this goal were outlined.

7

ELECTION GAMES

7.1. INTRODUCTION

In this chapter we shall develop two different game-theoretic models for the allocation of resources in a United States presidential campaign. In the first model, we assume that the goal of a candidate is to maximize his popular vote (or more accurately, expected popular vote, since we assume a probabilistic decision rule on the part of each voter), and the candidate who receives the greatest number of popular votes wins the election. In the second model, we assume that a candidate desires to maximize his expected electoral vote, which takes account of the fact that under the rules of the Electoral College the popular-vote winner in each state wins all the electoral votes of that state.

In both models, we assume that a presidential campaign influences only the voting behavior of voters uncommitted to a candidate prior to the start of the campaign. For these uncommitted voters, the amount of resources (e.g., time, money, and media advertising) that a candidate allocates to each state—relative to that allocated by his opponent(s)—determines the probability that an uncommitted voter will cast his vote for him in the election. Since the total resources of each candidate are assumed to be fixed, the question of how to allocate these resources to each state in a campaign assumes great importance in close races that may be decided by the choices of relatively few uncommitted voters.

For convenience, we shall assume in the initial development of the models that there are only two candidates (a Republican and a Democrat), and that the numbers of voters committed to each candidate prior to the

This chapter is based largely on Steven J. Brams and Morton D. Davis, "Models of Resource Allocation in Presidential Campaigning: Implications for Democratic Representation," *Annals of the New York Academy of Sciences* (*Conference on Democratic Representation and Apportionment: Quantitative Methods, Measures and Criteria*, ed. L. Papayanopoulos), 219 (New York: New York Academy of Sciences, 1973), pp. 105–23; and Steven J. Brams and Morton D. Davis, "The 3/2's Rule in Presidential Campaigning," *American Political Science Review*, 68 (March 1974), pp. 113–34; the permission of the American Political Science Association to adapt material from the latter article is gratefully acknowledged.

campaign are evenly divided between the two candidates in each state. The choices of the uncommitted voters will therefore be decisive to both the collective choice of a majority of voters in each state and to the collective choice of a majority of voters nationwide. As we shall show, however, the resource-allocation strategy that maximizes a candidate's expected electoral vote, which is based on the probabilities that majorities of voters in each state favor a particular candidate, differs markedly from the strategy that maximizes a candidate's expected popular vote, which depends only on the probabilities that individual voters in each state favor a particular candidate and not on the majority choice in each state.

The main conclusion we shall draw from this analysis is that the winner-take-all feature of the Electoral College induces candidates to allocate campaign resources roughly in proportion to the 3/2's power of the electoral votes of each state. This creates a peculiar bias in presidential campaigns that makes the largest states the most attractive campaign targets of the candidates—even out of proportion to their size—which would not be the case if the Electoral College were abolished and presidents were elected by direct popular vote. On the basis of the 1970 census and the electoral votes of each state through 1980, we shall indicate a presidential candidate's "optimal" expenditures of resources in all fifty states and the District of Columbia under both the Electoral College system and the popular-vote alternative. In the case of the Electoral College, the optimum we shall derive will be shown to be unstable: Any allocation of resources can be "beaten" under this system. This is not the case for the popular-vote alternative, which, because of the stability of its optimum, would tend to relieve the candidates of the necessity of making some of the manipulative strategic calculations that are endemic to the present system.

We shall conclude our analysis by comparing the ability of the two different systems to translate the attention (in time, money, and other resources) that presidential candidates—and after the election, incumbent presidents looking to the next election—pay to their state constituencies as a function of their size. Our comparison will reveal that the non-egalitarian bias of the Electoral College, which makes a voter living in one state as much as three times more attractive a campaign target as a voter living in another state, would be eliminated if the president were elected by direct popular vote.

7.2. THE NEED FOR MODELS TO ASSESS THE CONSEQUENCES OF ELECTORAL REFORM

Probably the most important reason that the structural reform of major political institutions is so controversial is because reforms often produce shifts in the distribution of power among political actors. When the precise effects of these shifts are uncertain, confusion tends to beset and compound controversy. As Senator Birch Bayh of Indiana said on the floor of the

U.S. Senate on September 8, 1970 about hearings that had been held on the reform of the Electoral College:

I must say, sitting through two or three volumes of hearings over the last 4 or 5 years was not at all times an inspirational experience. Some of the testimony was repetitive. Nevertheless, as chairman of the Subcommittee on Constitutional Amendments, I sat there. I thought it amusing, if not ironic, that on the last day—and I am not going to name individuals or organizations—and after 4 years of study, the last two witnesses appeared before our committee. One witness came before our committee suggesting the present [electoral] system should be maintained because it gave an advantage to the larger States and the next witness suggested the present system should be maintained because it gave an advantage to the small States.[1]

After several thousand pages of testimony before both Senate and House committees and subcommittees in the past few years,[2] there remains today a good deal of confusion and controversy about the possible effects of various proposed changes in the Constitution—relating to the election of a president—on the creation of new parties and minor candidates, the political influence of small and large states and groups and individuals within these states, governmental stability, and a host of other aspects of the electoral process. This is true despite the plethora of proposals for electoral reform that have been extensively discussed, if not analyzed, in congressional hearings and in numerous books and articles.

This discussion and analysis has in many cases been shallow, however, producing controversy based not on genuine differences of opinion but rather on a confused understanding on the part of different analysts of what consequences would follow from various changes in the electoral system. The main reason for this confusion seems not to stem from any paucity of factual information on national elections. Rather, there has been a lack of rigorous deductive models which can be used to explore the logical and quantitative consequences of different electoral systems, particularly as manifested in their effects on competition among candidates.

The consequences that would flow from electoral reform cannot be assessed simply by asking how hypothetical changes in the present system —specifically, in procedures for aggregating popular votes that produce a winner in presidential elections—would have affected previous election outcomes, as has been frequently done in the past. When an alternative procedure would be likely to have changed the campaign strategies of candidates in these elections to produce returns in states different from

---

1. *Congressional Record*, Sept. 8, 1970, p. 30813.

2. In the Senate, the most recent hearings have been *Electoral College Reform*, Hearings before the Committee on the Judiciary, United States Senate, 91st Congress, Second Session (1970); in the House, *Electoral College Reform*, Hearings before the Committee on the Judiciary, United States House of Representatives, 91st Congress, First Session (1969).

those that actually occurred, it is evident that the past returns cannot be held constant, with only the hypothetical procedure for aggregating them being allowed to vary, for purposes of estimating what consequences alternative aggregation procedures *would have had* on previous outcomes. As Alexander M. Bickel argued, "Any [major] change in the system . . . may induce subtle shifts in electoral strategies, rendering prediction based on past experience hazardous."[3] Curiously, it is just such "hazardous" predictions that Bickel conjures up to support his argument for retention of the Electoral College.

Toward improving this state of affairs, we shall outline and test some game-theoretic models of the presidential campaign process that will help us assess the consequences produced by the present electoral system and its most prominent alternative, direct popular-vote election of the president.[4] We have chosen to focus on the direct popular-vote alternative, and not proportional, district, and other plans for electing a president, because it has been the most widely discussed of the proposed alternatives to the Electoral College.[5] With a provision for a runoff election between the two top candidates if neither secures as much as 40 percent of the popular vote in the initial election, this alternative was approved by the U.S. House of Representatives on September 18, 1969 by a vote of 338 to 70, considerably more than the two-thirds majority required for the proposal

---

3. Alexander M. Bickel, *Reform and Continuity: The Electoral College, the Convention, and the Party System* (New York: Harper & Row, 1971), p. 35.

4. From a critical review of recent literature on campaigning, Gerald Pomper concludes that "in future research, more attention needs to be directed to the effects, rather than the characteristics, of campaigns." See Gerald M. Pomper, "Campaigning: The Art and Science of Politics," *Polity*, 2 (Summer 1970), p. 539. Toward this end, mathematical models of the campaign process are developed in John Ferejohn and Roger Noll, "A Theory of Political Campaigning" (Paper delivered at the 1972 Annual Meeting of the American Political Science Association, Washington, D.C., Sept. 5–9); and Gerald H. Kramer, "A Decision-Theoretic Analysis of a Problem in Political Campaigning," in *Mathematical Applications in Political Science, II*, ed. Joseph L. Bernd (Dallas: Southern Methodist University Press, 1966), pp. 137–60. In the Kramer article, a resource-allocation model is used to analyze the effects of different canvassing techniques on turnout and voting from the vantage point of one candidate—and not his opponent(s) directly, whose possible strategies our later game-theoretic models explicitly take into account; for an empirical test of the effects of canvassing in recent elections, see Gerald H. Kramer, "The Effects of Precinct-Level Canvassing on Voter Behavior," *Public Opinion Quarterly*, 34 (Winter 1970–71), pp. 560–72. The most explicit treatment of different strategic factors in a presidential campaign, developed from a coalition-theoretic perspective, is John H. Kessel, *The Goldwater Coalition: Republican Strategies in 1964* (Indianapolis: Bobbs-Merrill Co., 1968). A useful compilation of material on techniques of campaign management and communication and their effects on the electorate can be found in Dan Nimmo, *The Political Persuaders: The Techniques of Modern Election Campaigns* (Englewood Cliffs, N.J.: Prentice-Hall, 1970); and for a collection of articles on new campaign methods, see *The New Style in Election Campaigns*, ed. Robert Agranoff (Boston: Holbrook Press, 1972). A study of the use of electoral propaganda in Great Britain, whose practices are compared with those of the United States, is given in Richard Rose, *Influencing Voters: A Study of Campaign Rationality* (London: Faber & Faber, 1967).

5. A recent summary of different proposals, and a biased assessment (in favor of the present Electoral College) of their likely impact, considered especially in light of the three-way presidential contest in 1968, is given in Wallace S. Sayre and Judith H. Harris, *Voting for President: The Electoral College and the American Political System* (Washington, D.C.:

of constitutional amendments.[6] This plan fared less well in the U.S. Senate and eventually became the victim of a filibuster by southern and (strangely, as we shall see) small-state senators; cloture motions on September 17 and 29, 1970 won the approval of a majority of senators but failed to receive the required two-thirds endorsement needed to cut off debate.[7] This plan, nevertheless, has been strongly supported by the American public, receiving 66 to 19 percent approval (15 percent undecided) prior to the three-way 1968 presidential election, 81 to 12 percent approval (7 percent undecided) after that election.[8]

7.3. PRESIDENTIAL CAMPAIGNS AND VOTING BEHAVIOR

If there is anything that has emerged from research on electoral behavior over the past forty years, it is that most people make up their minds about whom they will vote for in a presidential election well before the onset of the campaign—at least the final campaign between the two major party nominees, which traditionally commences at the beginning of September in a presidential election year. Yet, although the campaign changes few minds, it does serve the important function of reinforcing choices already made, as many studies have documented.

On the other hand, for the typically 20 to 40 percent of the electorate undecided about their choice of a candidate at the start of a presidential campaign,[9] the campaign will not only be decisive to their individual

---

Brookings Institution, 1970). In view of the controversial aspects of Electoral College reform alluded to in the text, Sayre and Parris's belief (p. 43) that "the political effects of the electoral college system are as clear as any in the nonexact science of American politics" is difficult to accept, their "nonexact science" qualification notwithstanding. Other summaries that reflect a similar bias in favor of the Electoral College include Bickel, *Reform and Continuity*, pp. 4–36; and Nelson W. Polsby and Aaron Wildavsky, *Presidential Elections: Strategies of American Electoral Politics*, 3d ed. (New York: Charles Scribner's Sons, 1971), pp. 258–71. On the other side, a report by the American Bar Association's Commission on Electoral College Reform has called the popular-vote plan "the most direct and democratic way of electing a President." See American Bar Association (ABA), *Electing the President: A Report of the Commission of Electoral College Reform* (Chicago: ABA, 1967). Also supportive of the direct-vote plan is Neal R. Pierce, *The People's President: The Electoral College in American History and the Direct-Vote Alternative* (New York: Simon and Schuster, 1968); Lawrence D. Longley and Alan G. Braun, *The Politics of Electoral College Reform* (New Haven, Conn.: Yale University Press, 1972); and Harvey Zeidenstein, *Direct Election of the President* (Lexington, Mass.: Lexington Books, 1973). See also John H. Yunker and Lawrence D. Longley, "The Biases of the Electoral College: Who Is Really Advantaged?" and Max S. Power, "Logic and Legitimacy: On Understanding the Electoral College Controversy," both in *Perspectives on Presidential Selection*, ed. Donald R. Matthews (Washington, D.C.: Brookings Institution, 1973), pp. 172–203, 204–37.

6. *Congressional Record*, Sept. 18, 1969, pp. 26007–26008.

7. *Congressional Quarterly Almanac*, vol. 26, 91st Congress, Second Session (Washington, D.C.: Congressional Quarterly, 1971), p. 840. For further details, see Alan P. Sindler, "Basic Change Aborted: The Failure to Secure Direct Popular Election of the President, 1969–1970," in *Policy and Politics in America*, ed. Alan P. Sindler (Boston: Little, Brown, and Co., 1973), pp. 30–80.

8. Sayre and Parris, *Voting for President*, p. 15.

9. William H. Flanigan, *Political Behavior of the American Electorate*, 2d ed. (Boston: Allyn and Bacon, 1972), Table 5.3, p. 109.

voting decisions but also will often prove decisive to the choice of a candidate by a majority or a plurality of the electorate. This 20 to 40 percent minority of the electorate is usually more than sufficient to change the outcome of almost all presidential elections, which is why most campaigns are waged to make only marginal changes in the distribution of voter preferences. Indeed, when a presidential candidate does succeed in capturing as much as 55 percent or more of the popular vote, his victory is considered a landslide.

If presidential campaigns are decisive principally for the minority of undecided or uncommitted voters who will be crucial in determining the election outcome, then a candidate's ability to project favorably his personality and positions on issues during the campaign assumes great importance. Recent research in voting behavior has suggested the importance of issue-oriented aspects of elections first surfaced by V. O. Key, Jr., some years ago.[10] Moreover, since the pioneering work of Anthony Downs, the development of spatial models of party competition, which also make the positions that candidates and parties take on issues the focal point of the analysis, has proceeded apace.[11]

The campaign models we shall describe in the sections that follow have much in common with the logical structure, though not the substantive assumptions, of the party-competition models. As in these models, for example, we define optimal strategies to be those strategies that maximize

---

10. V. O. Key, Jr., with the assistance of Miltion C. Cummings, Jr., *The Responsible Electorate: Rationality in Presidential Voting, 1936–1960* (Cambridge, Mass.: Belknap Press, 1966). For a discussion of the role of issues in presidential elections, see the articles, comments, and rejoinders by Gerald M. Pomper, Richard W. Boyd, Richard A. Brody, Benjamin I. Page, and John H. Kessel, *American Political Science Review*, 66 (June 1972), pp. 415–70. The voting decisions of ticket-splitters (those who vote for a candidate of one party for one office, a candidate of another party for a different office), in particular, who constituted 54 percent of the American electorate in the 1968 election, are influenced primarily by the candidates and issues, and only secondarily by party identification and other group affiliations; see Walter De Vries and Lance Tarrance, Jr., *The Ticket-Splitter: A New Force in American Politics* (Grand Rapids, Mich.: William B. Erdemans Publishing Co., 1972).

11. Anthony Downs, *An Economic Theory of Democracy* (New York: Harper & Row, 1957). For a review of the more recent literature on party-competition models, see William H. Riker and Peter C. Ordeshook, *An Introduction to Positive Political Theory* (Englewood Cliffs, N.J.: Prentice-Hall, 1973), chaps. 11 and 12; and Kenneth A. Shepsle, "Theories of Collective Decision-Making," in *Political Science Annual, V: Collective Decision-Making,* ed. Cornelius P. Cotter (Indianapolis: Bobbs-Merrill Co., 1974), pp. 4–77. Generally speaking, work on party-competition models has not utilized the mathematics of game theory, but rather classical optimization techniques that do not explicitly assume the existence of an opponent. Recent exceptions include John Chamberlin, "A Game-Theoretic Analysis of Party Platform Selection," in *Social Choice,* ed. Bernhardt Lieberman (New York: Gordon and Breach, 1971), pp. 409–24; James S. Coleman, "The Positions of Political Parties in Elections," in *Probability Models of Collective Decision Making,* ed. Richard G. Niemi and Herbert F. Weisberg (Columbus, Ohio: Charles E. Merrill Publishing Co., 1972), pp. 332–57; Melvin J. Hinich, John O. Ledyard, and Peter C. Ordeshook, "A Theory of Electoral Equilibrium: A Spatial Analysis Based on the Theory of Games," *Journal of Politics* (forthcoming); and Richard D. McKelvey, "Policy Related Voting and Electoral Equilibrium," *Econometrica* (forthcoming).

(or minimax) some objective function.[12] Unlike these models, however, we ignore the positions that candidates adopt on issues. Although these positions are central to the determination of optimal strategies in party-competition models, we instead take the positions (and personality) of a presidential candidate as given and ask how he should allocate his total resources among the fifty states and the District of Columbia in order to convey as favorable an image as possible to the voters. An optimal strategy in our models is thus a set of resource allocations to each state rather than a specification of issue positions of candidates. We shall show later how these allocations can be derived from the maximization of different objective functions and discuss the different goals that these functions embody.

Although expenditures beyond a certain point in some campaigns may become counterproductive, we assume that a positive correlation in general exists between the amount of resources a candidate spends in a state—in relationship to that spent by his opponent—and the favorable-ness of the image he projects to voters in that state. We further assume that the more favorable a candidate's image, the more likely previously un-committed voters will be to vote for him.

Given this connection between campaign spending and voting behavior, the major strategic problem a candidate faces is how best to allocate his total resources among the states to win over that portion of the electorate (the uncommitted voters) who will prove decisive in most states—without in the process alienating voters already predisposed to his candidacy. His problem is rendered even more difficult by the fact that his opponent(s) will tend to allocate his (their) resources in such a way as to exploit any mistakes he might make in his allocations. It is this competitive aspect of presidential campaigns that the mathematics of game theory will prove helpful in illuminating.

We shall not consider here second-order strategic and tactical questions relating to how a candidate should spend his campaign resources *within* each state (e.g., on mass media advertising versus canvassing). Neither shall we consider the question of what portion of a candidate's resources should be devoted to *nonstate*-oriented campaign activities (e.g., nation-wide TV broadcasting).

7.4. THE GOALS OF CANDIDATES

The basic assumption of our analysis is that both voters and candidates are rational individuals who seek to maximize the attainment of certain

---

12. To avoid always speaking of both "maximizing" and "minimaxing" strategies in the subsequent discussion, we shall usually use only the term *maximization*; in describing the different solutions of each model, however, we shall distinguish game-theoretic minimax solutions from nongame-theoretic maximization solutions.

goals. To be sure, these goals may be the products of sociological, psycho-logical, and other conditioning forces in their lives, but this does not invalidate the assumption that, whatever their goals, candidates and voters seek the most rational means to achieve them. In the 1964 presidential election, for example, there is strong evidence to support the contention that the Republican nominee, Senator Barry Goldwater, did not so much desire to win as to present voters with "a choice, not an echo." If this is true, then his apparently aberrant campaign behavior, at least as measured against the normal canons of presidential campaigning, may have been quite rational, given his principal goal of espousal of a conservative ideology rather than winning.[13] Provided that we can impute plausible—if not totally realistic—goals to presidential candidates and voters, the test of rationality involves determining whether they behave *as if* they order their alternative courses of action and choose that which is most preferred, consistent with the postulated goals.

A goal one might postulate initially for voters in each state is that they vote for a presidential candidate solely on the basis of how much time (and, in principle, other resources) he spends in each state, as compared with that spent by his opponent. Clearly, this assumption is wildly unrealistic for the majority of voters already predisposed to one candidate before the start of a campaign. It is even a radical simplification for uncommitted voters, whom we argued earlier are usually decisive to the outcome of most presidential elections. Yet this assumption, which we *shall* apply to un-committed voters in our models, does offer a means for capturing one salient aspect of the campaign—how candidates view the relationship between their expenditures and the potential voting behavior of uncom-mitted voters in each state—from which we can derive prescriptions of how much time the candidates should allocate to each state. Whether in fact this assumption is plausible is an empirical proposition that we shall test not directly but rather indirectly through corroboration of the con-sequence deducible from it of how much time the candidates would spend in each state if their goals were to maximize their expected electoral or popular vote.

The goal of maximizing one's expected electoral vote under the present system, and one's expected popular vote under a system allowing for popular-vote election of a president, seem plausible goals to ascribe to

---

13. As another possible goal, Stanley Kelley, Jr., has suggested that ". . . at least some of the Goldwater inner circle set control of the Republican party—not winning the Presidency—as their principle objective in 1964. That is the implication, certainly, of Senator Goldwater's statement that the conservative cause would be strenghtened if he could win as much as 45 percent of the vote." Stanley Kelley, Jr., "The Presidential Campaign," in *The National Election of 1964*, ed. Milton C. Cummings, Jr. (Washington, D.C.: Brookings Institution, 1966), p. 58. Furthermore, there is evidence that Goldwater, who was well aware of his impending defeat from polls commissioned by the Republican National Committee, did little to try to stem its magnitude in the latter half of his campaign (which would be consistent with the goal of winning), but instead tried to rationalize the conduct of his campaign and the anticipated action of the voters. See Stephen C. Shadegg, *What Happened to Goldwater?* (New York: Holt, Rinehart and Winston, 1965), p. 241.

most presidential candidates, the case of Senator Goldwater notwith-standing. Both goals, based on probabilistic calculations, incorporate the idea that presidential campaigning is shot full of uncertainties and that there is no surefire campaign strategy that can guarantee victory. For the models we shall develop (but not necessarily for others), the goal of maximizing one's expected popular-vote will always be synonymous with maximizing one's probability of winning under the popular-vote system, though under the Electoral College system maximization of one's expected electoral vote may under certain circumstances be inconsistent wtih maximizing one's probability of winning. We shall point out some implica-tions of these different maximization goals for the Electoral College in our later analysis.

7.5. THE POPULAR-VOTE MODEL

To point up the need for models of the campaign process, consider first Richard Goodwin's testimony before the Judiciary Committee of the Senate:

> Today, nearly every State has a swing vote which, even though very small, might win that State's electoral vote. Thus, nearly every State is worth some attention. If the focus shifts to numbers alone [under a system of direct popular vote], then the candidate will have to con-centrate almost exclusively on the larger States. That is where the people are, and where the most volative [*sic*] vote is to be found. . . . What does this mean? It means that direct election would greatly intensify the attention given to the largest States.[14]

Now compare Goodwin's statement to Senator Bayh's response:

> The record will show that the major party candidates spend con-siderably more time [under the Electoral College system] in States that have large blocks of electoral votes. . . . It seems to me you have to be rather naive to overlook the fact that today the whole emphasis of the campaign is in . . . major states.[15]

If these contrary assertions leave one perplexed, what is one to say after several further pages of testimony—interspersed with inconclusive evidence presumably supporting each of these diametrically opposed viewpoints—about which system will force candidates to spend a disproportionately large portion of their time in the largest states? A recognition that the rules of politics are not neutral does not necessarily produce an immediate understanding of what biases they create and which contestants they favor.

---

14. Senate Hearings, *Electoral College Reform* (1970), p. 82.
15. Senate Hearings, *Electoral College Reform* (1970), pp. 82–83.

To develop game-theoretic models that may help resolve such a question as that discussed above, we shall consider first the popular-vote plan. We begin by assuming that the probability that a randomly selected *uncommitted voter* in state i votes for the Republican candidate is

$$p_i = \frac{r_i}{r_i + d_i},$$

where r_i is the amount (of time, money, or other resources) spent by the Republican candidate in state i and d_i is the amount spent by the Democratic candidate over the course of the campaign. In addition, we assume that all uncommitted voters vote. (These are the same assumptions we shall make in the case of the electoral-vote model.)

If n_i is the number of uncommitted voters in state i, then to maximize his *expected popular vote* among the uncommitted voters in all fifty states, the Republican candidate should maximize the quantity W_p, which is defined below:[16]

$$W_p = \sum_{i=1}^{50} n_i p_i \qquad r_i, d_i, n_i > 0,$$

where

$$\sum_{i=1}^{50} r_i = R, \qquad \sum_{i=1}^{50} d_i = D, \qquad \text{and} \qquad \sum_{i=1}^{50} n_i = N.$$

The term "expected" is used to signify the fact that W_p, the sum of the number of uncommitted voters in each state times the probability of their voting Republican, is not a certain quantity but instead the *average* Republican share of the total uncommitted vote for given allocations r_i and d_i by both candidates in all states.

Given that neither candidate has any information about the allocations made by his opponent(s), one can show that the optimal strategy for each candidate consists of allocating funds in proportion to the number of uncommitted voters in each state.[17] That is,

$$r_i = \left(\frac{n_i}{N}\right) R, \tag{7.1}$$

and

$$d_i = \left(\frac{n_i}{N}\right) D, \tag{7.2}$$

---

16. This model leads to the same results as one in which the quantity maximized is the *expected plurality* of uncommitted voters:

$$\sum_{i=1}^{50} n_i \left(\frac{r_i - d_i}{r_i + d_i}\right) = \sum_{i=1}^{50} n_i \left(\frac{2r_i}{r_i + d_i} - 1\right) = 2 \sum_{i=1}^{50} n_i p_i - N.$$

See Richard A. Epstein, *The Theory of Gambling and Statistical Logic* (New York: Academic Press, 1967), pp. 121–23. Note that the summations which follow range only over the fifty states, though in presidential elections beginning in 1964 the District of Columbia must also be included.

17. Because the mathematical derivation of this result and others described in subsequent sections involve more advanced methods than are assumed in this book, they will not be given here. For derivations and proofs of this and subsequent mathematical results, see Brams and Davis, "Models of Resource Allocation in Presidential Campaigning."

for all states i. For these allocations, the expected number of uncommitted voters that the Republican candidate can assure himself of from the entire pool of uncommitted voters is

$$W_p = \left(\frac{R}{R + D}\right) N.$$

If $R = D$ (i.e., the total resources of the Republican and Democratic candidates are equal), then

$$W_p = \frac{N}{2}.$$

That is, the two candidates would split the total uncommitted vote.

To this vote, of course, must be added the votes of previously committed Republican and Democratic voters—on whom we assume the campaign has no effect—to get the total number of votes that each candidate receives. (We shall show later how these committed voters can be incorporated directly into the resource-allocation calculations.) For now, it will be convenient to assume that there are only two candidates (or parties) in each state, though this assumption will later be abandoned. Also, we shall assume that the committed voters are split evenly between the parties in each state so that the winner of the uncommitted vote in each state captures a majority of popular votes in that state.

The allocations just specified are optimal in the sense that if either candidate adopts the proportional-allocation strategy, the other cannot gain by deviating from such a strategy. This sense of optimal is game-theoretic in nature: Neither candidate has an incentive to depart from this strategy because he might fare worse if he did. Thus, the strategies are in equilibrium.

Any departure from an equilibrium strategy by one candidate can be exploited by his opponent. If the Republican candidate, for example, were able to obtain information about deviations by the Democratic candidate from his equilibrium strategy, he could act on this information (i.e., the nonoptimal allocations of the Democratic candidate) in distributing his own resources so as to capitalize on his opponent's mistakes. It can readily be demonstrated that, *knowing* the d_i, the Republican candidate can maximize his expected popular vote among the uncommitted voters by allocating his own resources in the following way:

$$r_i = \frac{\sqrt{n_i d_i}}{\sum\limits_{i=1}^{50} \sqrt{n_i d_i}} (R + D) - d_i. \tag{7.3}$$

Similarly, for the Democratic candidate,

$$d_i = \frac{\sqrt{n_i r_i}}{\sum\limits_{i=1}^{50} \sqrt{n_i r_i}} (R + D) - r_i. \tag{7.4}$$

TABLE 7.1 ALLOCATION OF RESOURCES

State i	Number of voters n_i	Nonoptimal Democratic allocations d_i	Optimal Republican allocations r_i	Expected vote W_{P_i}	Minimax allocations r_i and d_i
1	20	100	143	11.8	200
2	30	200	247	16.6	300
3	40	400	404	20.1	400
4	50	600	553	23.4	500
5	50	600	553	23.4	500
Sum	190	1,900	1,900	95.3	1,900

If the Democratic candidate pursues an optimal strategy of proportional allocations according to equation 7.2, then for the Republican candidate equation 7.3 reduces to equation 7.1—that is, his best response is to allocate his resources proportionally, too, which is the minimax solution in pure strategies for both candidates in this two-person, constant-sum, infinite game. The game is constant-sum because what uncommitted votes one candidate wins the other candidate necessarily loses (the game is zero-sum if it is conceptualized as a plurality maximization game; see note 16 above), and it is infinite because each player has a choice among infinitely many possible expenditure levels in each state. The outcome is a saddlepoint when the pure strategies (i.e., sets of resource allocations to the states) prescribed by equations 7.1 and 7.2 are played, for if either candidate plays his optimal (minimax) strategy, his opponent can do no better than play his own.

To see how the Republican candidate could do better than his minimax strategy, given nonoptimal allocations by his Democratic opponent, consider the nonoptimal allocations d_i of the Democratic candidate shown in Table 7.1 (which we assume for convenience total ten times the total number of voters). Note that the Democrat's allocations are less than the (proportional) minimax allocations for the two smallest states (states 1 and 2) and greater than these minimax allocations for the two largest states (states 4 and 5). Assuming that the Republican candidate is privy in advance to these (planned) allocations by his Democratic opponent, his best response according to equation 7.3 is to outspend his opponent in the three smallest states and underspend him in the two largest states, as shown in Table 7.1.

It can be observed that the Republican candidate, knowing the Democratic candidate's allocations, usually responds with an allocation somewhere between his opponent's allocations and the minimax allocations. Yet of 190 votes, this best response garners the Republican candidate only 0.6 votes (95.3 for Republican to 94.7 for Democrat) more than his opponent when their total resources are equal, illustrating the relative insensitivity of the popular-vote model to nonoptimal allocations. In sharp contrast, it can be shown that allocations under the Electoral College system are extremely sensitive in the range where the two candidates about match each other's expenditures in states (i.e., $p_i \approx 0.50$ for voters in state i).

7.6. THE ELECTORAL-VOTE MODEL

Under the present Electoral College system, the geographic origin of the vote is salient. Because all the electoral votes of each state—equal to the total number of its senators plus representatives—are awarded to the majority (or plurality) winner in that state, such a winner-take-all decision rule is often referred to as "unit rule." In this section we shall outline some of the difficulties connected with the concept and interpretation of "optimal strategy" under this system, and in section 7.7 we shall develop an intriguing, but somewhat fragile, solution to the resource-allocation problem for this system.

For purposes of comparison with the popular-vote model, assume that v_i is the number of electoral votes of state i, where $\sum_{i=1}^{50} v_i = V$. Then, to maximize his *expected electoral vote* among the uncommitted voters in all fifty states, the Republican candidate should maximize the quantity W_e:

$$W_e = \sum_{i=1}^{50} v_i \pi_i, \qquad (7.5)$$

where, assuming for convenience an even number of voters n_i,

$$\pi_i = \sum_{k=\frac{n_i}{2}+1}^{n_i} \binom{n_i}{k} p_i^k (1 - p_i)^{n_i-k} \qquad (7.6)$$

is the probability that the Republican candidate obtains *more* than 50 percent of the uncommitted voters in state i—that is, the probability that state i goes Republican if the committed voters are split fifty-fifty. We assume here that the voting of uncommitted voters *within* each state is statistically independent.[18]

---

18. For the reader unfamiliar with the binomial distribution, equation 7.6 simply sums the probabilities that the Republican candidate obtains exactly

$$\frac{n_i}{2} + 1 \qquad \text{or} \qquad \frac{n_i}{2} + 2 \qquad \text{or} \qquad \dots n_i$$

uncommitted votes in state i and thereby wins the state. To use the binomial distribution in this case requires that we assume that the votes cast by each uncommitted voter have no effect on (i.e., are statistically independent of) the votes cast by other uncommitted voters in state i.

Unfortunately, unlike the maximization of W_p in the popular-vote model according to equation 7.3 (or the minimaxing of W_p according to equation 7.1), the maximization of W_e in the electoral-vote model does not yield a closed-form solution for the optimal values of r_i (d_i) when the d_i (r_i) are known.[19] This makes problematic comparisons with the explicit solution to the popular-vote model (equations 7.3 and 7.4).

When the allocations of an opponent are not known, the minimax solution to the optimization problem in general does not take the form of pure strategies in a two-person, constant-sum, infinite game, where the players are the two candidates. Like the popular-vote model, the contest is a game because it depends on the strategy choice of the other player (i.e., an interdependent decision situation exists). Unlike the popular-vote model, this is not a game of perfect information and therefore has no saddlepoint (see section 1.3). This means that the choice by a candidate of a best strategy will depend on his opponent's choice. To keep his opponent from discovering his choice, each candidate will leave his choice to chance by choosing from a set of pure strategies at random, with only the probabilities determined. That is, he will use a mixed strategy, or a probability distribution over a set of pure strategies.

(Only under special circumstances does such a game have a pure-strategy solution. To wit, when one candidate has more than twice the resources of the other candidate, it is easy to show that the richer candidate can assure himself, at least with a high probability, of a majority in the Electoral College. He can accomplish this by pursuing the pure strategy of distributing his greater campaign resources among the various states in proportion to their electoral votes v_i, which is analogous to the pure minimax strategy of both candidates in the popular-vote model. Since this is a pure strategy, the poorer candidate will know what this strategy is, and in particular will know the richer candidate's expenditure levels in various states. But even if the poorer candidate spends only slightly more than the richer candidate in several states, the electoral votes of these states can never equal a majority because the resources of the poorer candidate are less than half those of the richer candidate. More specifically, if the poorer candidate concentrates all his resources on a set of states with a bare majority of electoral votes, he may capture most of these states by spending just a little more than the richer candidate spends in them, but the richer candidate will still be able to outspend him in at least one state, which will deny the poorer candidate his majority with almost certainty.)

In the more plausible case in which the two candidates' campaign resources are more evenly balanced, the richer candidate cannot use a pure strategy because this would enable the poorer candidate to outspend him in a sufficient number of states to achieve a majority in the Electoral College. In other words, the richer candidate must keep secret how much

---

19. Roughly speaking, a closed-form (or analytic) solution is one in which an explicit expression, or formula, can be given for the unknowns (i.e., r_i and d_i) in terms of the other variables.

he is going to spend in each state, and so, for similar reasons, must the poorer candidate. (In game-theoretic language, neither candidate has a pure minimax strategy since such a strategy on the part of either candidate would be known and could be "beaten" by the other candidate.) The solution for both candidates, therefore, will be in mixed strategies, but the mixed-strategy solution to this game is not known.[20]

It can be shown that if $p_i = \dfrac{r_i}{r_i + d_i}$ is *defined* to be the probability that a majority of uncommitted voters (rather than a randomly selected uncommitted voter) in state i vote Republican, i.e., if $\pi_i = p_i$, then the game will have a solution in pure strategies for both candidates. Analogous to the solution in the popular-vote model (with v_i and V now substituted for n_i and N, respectively), the (pure) minimax strategy of each candidate is to allocate resources in proportion to the number of electoral votes of each state:

$$r_i = \left(\frac{v_i}{V}\right) R; \quad \text{and} \quad d_i = \left(\frac{v_i}{V}\right) D.$$

Because this consequence is derived from the assumption that resource expenditures under the Electoral College system affect directly a majority—and not single individuals—in each state, it is not directly comparable to the proportional-allocation rule in the popular-vote model, which is derived from the individualistic assumption. Although it is hard to entertain the belief that aggregates, rather than individuals, respond to the actions of the candidates, we shall nonetheless compare the empirical correspondence of the proportional-allocation strategy under the Electoral College with an allocation strategy based on the voting behavior of individuals (to be described in the next section).

Whatever its exact form, a mixed-strategy solution to the electoral-vote model based on the individualistic assumption will be difficult to interpret, let alone to verify empirically. To exorcise this strategy of its random element, we shall show in section 7.7 how an additional assumption can be made that invests optimal strategies under the Electoral College system with greater determinateness.

7.7. THE 3/2's ALLOCATION RULE

The assumption we shall make is that the two presidential candidates match each other's resource expenditures in each state. This assumption seems

---

20. In so-called Colonel Blotto games—where the candidate who outspends his opponent in a state by whatever amount wins that state with certainty, in contrast to the probabilistic relationship that we assume here between expenditures and winning—a minimax solution in mixed strategies has been found when all states have the same number of electoral votes. See David Sankoff and Koula Mellos, "The Swing Ratio and Game Theory," *American Political Science Review*, 56 (June 1972), pp. 551–54; and for a discussion of related Colonel Blotto games, see Lawrence Friedman, "Game Theory Models in the Allocation of Advertising Expenditures," *Operations Research*, 6 (Sept.–Oct. 1958), pp. 699–709.

reasonable in light of the fact that the candidates tend to agree on what states (usually large and heterogeneous) are the most attractive campaign targets. Even Senator Goldwater decided to abandon early plans in 1964 to write off the big industrial northeastern states and instead "go shooting where the ducks are." Of the seven states he visited most frequently, and which collectively claimed more than half of his time, three (New Jersey, New York, and Pennsylvania) were in the Northeast.[21]

As further support for the matching assumption, the product-moment correlation coefficients between the combined appearances (to be defined later) of the Republican and Democratic presidential and vice-presidential candidates in all fifty states are .92 in the 1960 campaign, .83 in the 1964 campaign, .90 in the 1968 campaign, and .74 in the 1972 campaign. Such strong empirical relationships between appearances of the two slates lend support to the proposition that the goal of maximizing one's expected electoral vote that we postulated for the candidates is generally consistent, at least in four recent presidential campaigns, with concentrating one's time in the same set of states.

In the statement of the Electoral College model, we can incorporate this matching assumption by assuming that $r_i = d_i$ (and necessarily $R = D$) for all states i in solving equation 7.5 for a maximum. This assumption enables us to obtain an explicit expression for r_i that maximizes the Republican's expected electoral vote:

$$r_i = \left(\frac{v_i \sqrt{n_i}}{\sum_{i=1}^{50} v_i \sqrt{n_i}} \right) R. \tag{7.7}$$

Since we assumed that $d_i = r_i$ (and $D = R$) for all i, the Democrat's optimal resource allocation,

$$d_i = \left(\frac{v_i \sqrt{n_i}}{\sum_{i=1}^{50} v_i \sqrt{n_i}} \right) D, \tag{7.8}$$

is the same as his Republican opponent's.

It can be shown that these allocations are in equilibrium: Any *arbitrarily small* (but finite) deviation from these strategies on the part of either candidate will be nonoptimal. A *sufficiently large* deviation by one candidate could, however, prove profitable. If, for example, the Republican candidate reallocated all the resources he spends in one state, as prescribed by equation 7.7, to a larger state, he would lose the smaller state with certainty but almost surely win a majority of popular votes (and all the electoral votes) in the larger state. Thus, instead of exactly splitting his expected electoral vote with his opponent—by matching expenditures

---

21. Kelley, "The Presidential Campaign," pp. 50–51, 75.

with him in all states according to equations 7.7 and 7.8—he could, by such a unilateral deviation, easily win a majority of the electoral votes, holding constant his total expenditures.

We cannot readily specify *how small* the deviation must be so as to be nonoptimal for a candidate. As we have shown by the above example, however, the equilibrium is only *local* and not a minimax solution, for *some* unilateral deviation can secure a candidate a more-preferred outcome (i.e., a higher expected electoral vote). The matching strategies prescribed by equations 7.7 and 7.8 are therefore not optimal in a *global* sense. To be optimal in this sense, the strategies would have to be such that, whenever the candidates spend the same amount in all states, there would be an incentive for both to move toward the particular matching expenditures defined by equations 7.7 and 7.8. There is, however, no such incentive. Only when the candidates' expenditures are near or at this point is there an incentive *not to deviate* from these expenditures by a small amount.

Because the allocations defined by equations 7.7 and 7.8 constitute only a local equilibrium point, it is unstable.[22] Yet, despite its instability, it is suggestive of a possible reference point in the calculations of candidates that does have an interesting interpretation. This interpretation can most easily be grasped if we make the simplifying assumption that the electoral votes of state i, v_i, are proportional to the number of uncommitted voters, n_i, in that state. Given this assumption, the terms in parentheses in the numerator, and in the summation of the denominator, of equations 7.7 and 7.8, $v_i \sqrt{n_i}$, will be proportional to $v_i \sqrt{v_i} = v_i^{\frac{3}{2}}$. The (local) equilibrium strategies of the Republican and Democratic candidates who match each other's resource expenditures in each state will then be, respectively,

$$ r_i = \left(\frac{v_i^{\frac{3}{2}}}{\sum\limits_{i=1}^{50} v_i^{\frac{3}{2}}} \right) R, $$

and

$$ d_i = \left(\frac{v_i^{\frac{3}{2}}}{\sum\limits_{i=1}^{50} v_i^{\frac{3}{2}}} \right) D. $$

In other words, to maximize his expected electoral vote, each candidate should allocate his total resources, which we assumed are the same for the Republican and Democratic candidates ($R = D$) in making the matching assumption ($r_i = d_i$ for all i), in proportion to the 3/2's power of the number of electoral votes of each state.

This is what is meant by the "3/2's allocation rule" in the heading of this section. It implies that both candidates, in matching each other's

---

22. Whenever we use the term *unstable*, it means "globally unstable," for this point is stable locally (i.e., is impervious to small deviations).

resource expenditures, should not simply allocate on the basis of the electoral votes of each state but rather should allocate decidedly more than proportionally to large states than to small states. For example, if one state has 4 electoral votes, and another state has 16 electoral votes, even though they differ in size only by a factor of four, the candidates should allocate eight times as much in resources to the larger state because

$$\frac{16^{\frac{3}{2}}}{4^{\frac{3}{2}}} = \frac{64}{8} = 8.$$

This allocation rule thus favors large states with more electoral votes, even out of proportion to their size. It is a strikingly simple and non-obvious consequence of the postulated goal that candidates seek to maximize their expected electoral vote, given that they match each other's resource expenditures.[23] Although simple in form, however, it is not immediately obvious why large states are so advantaged, apart from the commonsensical observation that voters in large states have greater potential influence over the disposition of large blocs of electoral votes.

7.8. WHY THE LARGE STATES ARE FAVORED

To give greater insight into the quantitative dimensions of this "potential influence," it is useful to calculate the "expected minimum number of voters sufficient to change the outcome" in a state. This quantity is simply the sum of the probability of an exactly even split (i.e., a tie, assuming an even number of voters) times 1 (one "swing voter" is minimally sufficient to change the outcome—a deadlock—by changing his vote), the probability of a 1-vote victory by either candidate ($k = +1$ or $k = -1$ below) times 2 (two "swing voters" are minimally sufficient to change the outcome by changing their votes), and so on. If we assume for convenience that there are $2n$ voters (i.e., an even number) in a state, then this expected minimum number, which we shall refer to as the *expected number of decisive voters*, is

$$N(D) = \sum_{k=-n}^{n} \binom{2n}{n+k} p^{n+k}(1-p)^{n-k}(|k|+1),$$

where p is equal to the probability that a randomly selected uncommitted voter in the state votes for one candidate and $(1-p)$ the complementary probability that such an uncommitted voter votes for the other candidate. When $p = (1-p) = 1/2$, it can be shown that

$$N(D) = \left(\frac{1}{2}\right)^{2n}\left[n\binom{2n}{n} + 2^{2n}\right].$$

---

23. If the total resources of the candidates are not equal (i.e., $R \neq D$), it is not difficult to show that a "proportional matching" of expenditures, wherein the candidates spend the same proportion (or percentage) of their total resources in each state, yields a somewhat more complicated expression for a local maximum.

Since $\left(\dfrac{1}{2}\right)^{2n}\dbinom{2n}{n} \approx \sqrt{\dfrac{1}{\pi n}}$, the central term of the binomial distribution,[24]

$$N(D) = \sqrt{\frac{n}{\pi}} + 1. \tag{7.9}$$

That is, when the probability is fifty-fifty that the voters will vote for one candidate or the other—and thereby enhance the likelihood of a close outcome—the expected number of decisive voters in a state varies with the square root of its size, the first term on the right-hand side of equation 7.9.

The number of decisive voters *per voter* in a state, or what might be called the *decisiveness* of an average voter, is found by dividing $N(D)$ by $2n$:

$$D(2n) = \frac{1}{2\sqrt{\pi n}} + \frac{1}{2n}.$$

To illustrate this measure of decisiveness, in a state with $2n = 100$ voters,

$$D(100) = 0.040 + 0.010 = 0.050,$$

whereas in a state with $2n = 400$ voters,

$$D(400) = 0.020 + 0.003 = 0.023.$$

We see that although the larger state has four times as many voters as the smaller state, the decisiveness of an individual voter in the larger state decreases only by a factor of about two. Thus, even though an individual voter has a reduced chance of influencing the outcome in a large state because of the greater number of people voting, this reduction is more than offset by the larger number of electoral votes he can potentially influence. Hence, despite the apparent dilution of his vote under a winner-take-all system like the Electoral College, a voter in a large state has on balance greater potential voting power to affect the outcome of a presidential election than a voter in a small state.[25]

---

24. James S. Coleman has developed a measure of power based on this term, which here gives the probability of an exactly even split of the uncommitted voters in a state. (Note that π in the expressions in this section refers to the number 3.14159, not the probability we defined earlier.) See James S. Coleman, "Loss of Power," *American Sociological Review*, 38 (Feb. 1973), pp. 1–17.

25. Cf. the similar findings in John P. Banzhaf III, "One Man, 3.312 Votes: A Mathematical Analysis of the Electoral College," *Villanova Law Review*, 13 (Winter 1968), pp. 304–32. Banzhaf's analysis of the voting power of voters in a state is based on the concept of a "critical vote," which can occur only if there is a fifty-fifty split and is similar to that used in Coleman, "Loss of Power." By contrast, the concept of "decisiveness" given in the text takes into account other possible divisions of the vote. It should be noted that Banzhaf combines the concept of a critical vote of a voter in a state with his index of voting power (see section 4.3) applied to states in the Electoral College to obtain a composite measure of the power of a voter in each state to affect the outcome in the Electoral College. This measure is not only rather ad hoc but, interpreted as a probability measure, is based on the assumption that the power of a voter to affect the popular-vote outcome in a state is independent of the power of the state to affect the electoral outcome in the Electoral College. This assumption is not justified in Banzhaf's analysis and seems, in fact, to be false. Incidentally, the correctness of

It is precisely this greater potential voting power of voters in large states which makes them more attractive as campaign targets to the candidates. Is it any wonder why, then, the candidates view the large states as more deserving of their attention—even on a per capita basis—than small states?

This result, though not directly tied to our earlier resource-allocation models, sheds considerable light on the strategic advantage of voters in large states. Yet, because the 3/2's rule that favors large states is only a local equilibrium point when candidates match each other's resource allocations, it is highly vulnerable when the matching assumption is discarded. In fact, it can be shown that when a candidate knows the allocations of his opponent under the Electoral College system, the most devastating strategy he can generally use against him is to spend nothing in the smallest state and instead use these extra resources (assuming the total resources of both candidates are equal) to outspend his opponent in each of the other states.

In general, adaptive strategies which exploit the commitments of an opponent demand a flexibility in responding to an opponent's allocations and recommitting one's own resources that does not seem to accord with the campaign realities of advance scheduling, purchase of future broadcast time, and so forth. It is for this reason that the 3/2's rule, despite its instability, may better reflect a fixed, if intuitive and not well understood, point of reference for the major candidates than do allocation strategies directed only to responding to an opponent's commitments. Although admittedly most candidates, and especially successful ones, are probably incrementalists in the way they define and quickly respond to issues in a campaign,[26] it nevertheless appears that decisions about resource allocations are inherently less fluid.

Having already justified the assumptions and logic that generate the 3/2's rule, and posted warnings about its fragile nature, we shall report in the next section on its empirical validity. Since by election time the candidates necessarily have made a set of choices that fixes their allocations, we can check these against the 3/2's and other rules without inquiring into what determines each and every choice. In this analysis, we shall refine the 3/2's rule by assuming that the number of uncommitted voters in each state is directly proportional to the population, and not to the number of electoral votes, of that state. Especially in small states, where the two-senator bonus in the Electoral College greatly magnifies

Banzhaf's numerical calculations of the voting power of voters in the Electoral College has recently been contested in Lawrence D. Longley and John H. Yunker, "The Changing Biases of the Electoral College" (Paper delivered at the 1973 Annual Meeting of the American Political Science Association, New Orleans, Sept. 4–8), though both Banzhaf, and Longley and Yunker, conclude that citizens in the largest states are considerably advantaged. This accords pretty much with our conclusions here, as will be shown in detail later.

26. See Karl A. Lamb and Paul A. Smith, *Campaign Decision-Making: The Presidential Election of 1964* (Belmont, Calif.: Wadsworth Publishing Co., 1968).

their per capita electoral-vote representation, there is no reason to assume that the proportion of uncommitted voters will be so magnified. It seems far more reasonable to tie estimates of uncommitted voters directly to the population, thus retaining the original form of the 3/2's rule, $v_i \sqrt{n_i}$, where n_i is assumed proportional to the population.[27]

7.9. TESTING THE MODELS

Political campaigns in the United States have grown enormously expensive in recent years. Herbert E. Alexander, director of the Citizens' Research Foundation, estimated that political spending on all levels during the 1968 campaign ran to $300 million—up 50 percent from the total cost of campaigns in 1964—and that it cost $100 million to elect a president that year.[28] The *New York Times* estimated that a total of $400 million was spent on political campaigns in 1972,[29] though a final accounting of expenditures at the presidential level has been complicated by the maelstrom of Watergate.

The per capita costs of presidential campaigns have also soared, with expenditures per potential voter increasing by 71 percent between 1964 and 1968.[30] Only recently enacted legislation promises some relief from these spiraling costs by limiting the campaign spending of presidential candidates for radio, television, and other media advertising, as well as paid telephone solicitation, to 10 cents per voter, with no more than 60 percent to be spent on broadcast advertising.[31]

Unfortunately, reliable data on the financial expenditures of the presidential candidates in each state are not generally available. Although the Federal Communications Commission has published data on the campaign expenditures of the major political parties for radio and television advertising in recent presidential election years,[32] these figures are not disaggregated for presidential races. Despite the wealth of data that has been collected on the financial contributions and expenditures of

---

27. The population of a state, of course, is not an exact reflection of the proportion of the voting-age population who are registered and actually do vote in a presidential election. Since it is not at all clear whether and how the proportion of uncommitted voters in a state is related to differences in registration and turnout among the states, we have taken the simplest course of using population as a first-approximation estimate of the proportion of voters likely to be uncommitted at the start of a campaign. This assumption can, of course, be modified at a later time if found to be deficient.

28. *New York Times*, Jan. 31, 1972, p. 48.

29. *New York Times*, Nov. 19, 1972, p. 1.

30. Computed from figures given in Delmer D. Dunn, *Financing Presidential Campaigns* (Washington, D.C.: Brookings Institution, 1972), Figure 1, p. 33.

31. *New York Times*, April 9, 1972, p. 44.

32. Federal Communications Commission (FCC), *Survey of Political Broadcasting: 1960, 1964, 1968* (Washington, D.C.: FCC, April 1961, July 1965, August 1969), which includes data on both primary and general election campaigns.

candidates,[33] its completeness and reliability have been hampered by unsystematic reporting and the lack of effective governmental controls on contributions and expenditures. The Federal Election Campaign Act of 1971, which went into effect on April 7, 1972, tightens up reporting procedures and should improve this situation.

In the absence of reliable state-by-state data on financial expenditures of presidential candidates, we have turned to the one resource which imposes the same implacable restraints on the campaign behavior of all candidates—time. With a finite amount of it to spend in a campaign, the crucial question for a candidate becomes how to apportion it most wisely so as to gain favorable and far-reaching exposure.

In our models, we assumed that the favorableness of a candidate's image in a state depends on his resource expenditures in that state as compared to those of his opponent. To test these models, we have operationalized expenditures of time in terms of the total number of *campaign appearances* that a candidate makes in a state, where campaign appearances are defined to be events at which a candidate either makes some public address to an audience (whether the address takes the form of a major speech or brief remarks, but excluding news conferences) or participates in some public activity like a parade, motorcade, or fair. To be sure, the counting of all campaign appearances in a state as equivalent ignores important differences among them (e.g., size of audience and extent of news coverage), but the distribution of both politically "important" and "unimportant" appearances in each state visited probably makes the aggregated data roughly comparable for the purposes of our analysis.

The campaign appearance data are based on news coverage of the 1960, 1964, 1968, and 1972 presidential campaigns in the *New York Times*, from September 1 to the day before Election Day in November in each year, supplemented by such other sources as congressional reports, *Congressional Quarterly Weekly Reports*, and *Facts on File*.[34] Although these data on campaign appearances are only as accurate as coverage by the media, they are generally highly correlated with data collected by Stanley Kelley, Jr., on the number of hours spent by the presidential and vice-presidential candidates in each state in the 1960 and 1964 elections, which is based in part on the candidates' own personal schedules.[35]

---

33. The most comprehensive source of this information for the 1968 election is Herbert E. Alexander, *Financing the 1968 Election* (Lexington, Mass.: D. C. Heath and Co., 1971).

34. For other data on the itineraries of presidential candidates, see *Source Book of American Presidential Campaign and Election Statistics: 1948–1968*, ed. John H. Runyon, Jennefer Verdini, and Sally S. Runyon (New York: Frederick Ungar Publishing Co., 1971), pp. 139–73. This work does not contain data on the itineraries of vice-presidential candidates, whose campaign appearances were combined with the appearances of presidential candidates in our analysis, for reasons given in the text.

35. These data are summarized in Stanley Kelley, Jr., "The Presidential Campaign," in *The Presidential Election and Transition*, ed. Paul T. David (Washington, D.C.: Brookings Institution, 1961), pp. 57–87; and Stanley Kelley, Jr., "The Presidential Campaign," in *National Election of 1964*, pp. 42–81.

For each party, the appearances made by both its presidential and vice-presidential candidates were combined, and this total was used as an indicator of the candidates' resource allocations to each state in the four campaigns studied. Combining the appearances of each party's two nominees seemed preferable to singling out the time expenditures of only the presidential candidate, since often the two candidates adopt complementary strategies. For example, in the 1960 race, Lyndon Johnson, the Democratic vice-presidential candidate, was assigned the task of holding the South for the Democrats, and he devoted more than twice as much time to this region as his running mate, John Kennedy, or either of the Republican candidates, Richard Nixon and Henry Cabot Lodge.[36] If the campaign appearances of Johnson had not been added to Kennedy's, it would appear that the Democrats did not consider the southern states strategically important to their fashioning a victory, which was manifestly not the case.

Since a detailed analysis of the campaign appearance data has been reported elsewhere,[37] we summarize here only the conclusions of the analysis. Generally speaking, the 3/2's rule was found to fit the campaign appearance data somewhat better than the proportional rule, but Brams and Davis did not promulgate it as an "immutable law":

Like any theoretical consequence of a set of assumptions, its applicability will be limited to those situations that can be reasonable well characterized by these assumptions—particularly the postulated goal of maximizing one's expected electoral vote and the assumption that candidates match each other's campaign expenditures in each state—which are not easy to verify. Further, the instability of the 3/2's rule as an equilibrium point, which makes it vulnerable to only small deviations, may also limit its applicability, especially when candidates resort to adaptive strategies in response to each other's allocations. Finally, another potential source of slippage between the theoretical allocations and the actual campaign behavior of candidates occurs in the reconstruction of campaign itineraries, which is a task fraught with difficulties that certainly contributes to unreliability in the data.[38]

Brams and Davis nonetheless concluded:

Our data do make clear that the candidates generally make disproportionately large expenditures of time in the largest states. While one could always find a better-fitting function than the 3/2's rule for any particular campaign, it would not constitute an explanatory

---

36. Computed from data given in Kelley, "The Presidential Campaign," in *Presidential Election and Transition*, p. 72.

37. Brams and Davis, "The 3/2's Rule in Presidential Campaigning."

38. Brams and Davis, "The 3/2's Rule in Presidential Campaigning," p. 126.

model unless one could derive it from assumptions that are both interpretable and plausible. Fitting a curve to empirical data may help one summarize repeated instances of a phenomenon, but in itself it does not impart a logic to the curve that we consider the hallmark of scientific explanation.[39]

In section 7.10, the allocations prescribed by the 3/2's rule will be given for each state from 1972 through 1980, when the next decennial census will be taken. (Population figures from this census will be used to reapportion electoral votes of the states for the 1984 presidential election.) The calculations will reveal the extent to which the 3/2's rule favors the largest states, even on a per capita basis, despite the two-senator bonus that favors the smallest states.[40]

7.10. CAMPAIGN ALLOCATIONS AND BIASES THROUGH 1980

Using data from the 1970 census and the electoral votes of each state through 1980, we have indicated in Table 7.2 the percentages of time a candidate would spend in each state and the District of Columbia if he allocated his resources proportionally or if he allocated them according to the 3/2's rule. (These figures are also given for large, medium, and small states: Large states have 20 or more electoral votes, medium states between 10 and 19, and small states less than 10.) For example, a candidate's proportional allocation to California on the basis of its forty-five electoral votes represents about an 8 percent commitment of his resources, but the 3/2's rule nearly doubles this commitment to 15 percent. On the other hand, the 3/2's rule would slash a 0.56 percent commitment to Alaska by nearly a factor of five. According to this rule, then, California should receive about twice as much, and Alaska about one-fifth as much, resources *per electoral vote* as would be commensurate with their respective forty-five and three electoral votes.

This is what we call the *electoral bias* (*EB*) of the present system, which is simply the ratio of 3/2's allocations to proportional allocations for all states. These ratios are given in Table 7.2 and show that the nine largest states, with 52 percent of the population, are advantaged by the 3/2's rule (*EB* > 1.00), the remaining forty-one disadvantaged (*EB* < 1.00).

The *individual bias* (*IB*) of the Electoral College is the concept most relevant to assessing the degree to which the 3/2's rule engenders campaign allocations in states that are inconsistent with the egalitarian principle of "one man, one vote." This bias is the ratio of each state's 3/2's percent allocation to its percent share of the total population.

---

39. Brams and Davis, "The 3/2's Rule in Presidential Campaigning," p. 126.

40. In the following analysis we assume that Maine operates under the Electoral College system, though starting with the 1972 presidential election, it adopted a "district plan," whereby its two senatorial electoral votes go to the winner of the state, but its two electoral votes based on population are decided on a district-by-district basis.

TABLE 7.2 ELECTORAL AND INDIVIDUAL BIASES OF 3/2'S CAMPAIGN ALLOCATIONS IN FIFTY STATES AND DISTRICT OF COLUMBIA

State	Electoral votes	Proportional allocation (%)	3/2's allocation (%)	Electoral bias (EB): $\dfrac{3/2's}{Proportional}$	Individual bias (IB): $\dfrac{3/2's}{Population}$
1. Calif.	45	8.36	14.69	1.76	1.50 (1)
2. N.Y.	41	7.62	12.78	1.68	1.43 (2)
3. Pa.	27	5.02	6.78	1.35	1.17 (3)
4. Texas	26	4.83	6.36	1.32	1.15 (5)
5. Ill.	26	4.83	6.34	1.31	1.16 (4)
6. Ohio	25	4.65	5.96	1.28	1.14 (6)
7. Mich.	21	3.90	4.57	1.17	1.05 (7)
Large	211	39.22	57.48	1.47	1.27
8. N.J.	17	3.16	3.33	1.05	0.94 (9)
9. Fla.	17	3.16	3.24	1.02	0.97 (8)
10. Mass.	14	2.60	2.44	0.94	0.87 (10)
11. Ind.	13	2.42	2.17	0.90	0.85 (12)
12. N.C.	13	2.42	2.14	0.89	0.86 (11)
13. Mo.	12	2.23	1.90	0.85	0.82 (15)
14. Va.	12	2.23	1.89	0.85	0.83 (14)
15. Ga.	12	2.23	1.88	0.84	0.83 (13)
16. Wis.	11	2.04	1.69	0.83	0.78 (18)
17. Tenn.	10	1.86	1.45	0.78	0.75 (22)
18. Md.	10	1.86	1.45	0.78	0.75 (21)
19. Minn.	10	1.86	1.43	0.77	0.76 (20)
20. La.	10	1.86	1.40	0.75	0.78 (17)
Medium	161	29.93	26.41	0.88	0.84
21. Ala.	9	1.67	1.22	0.73	0.72 (28)
22. Wash.	9	1.67	1.21	0.73	0.72 (27)
23. Ky.	9	1.67	1.18	0.71	0.75 (23)
24. Conn.	8	1.49	1.02	0.68	0.68 (36)
25. Iowa	8	1.49	0.98	0.66	0.71 (30)
26. S.C.	8	1.49	0.94	0.63	0.74 (25)
27. Okla.	8	1.49	0.94	0.63	0.74 (24)
28. Kans.	7	1.30	0.77	0.59	0.69 (34)
29. Miss.	7	1.30	0.76	0.59	0.70 (33)
30. Colo.	7	1.30	0.76	0.58	0.70 (32)
31. Ore.	6	1.12	0.63	0.57	0.62 (43)
32. Ark.	6	1.12	0.61	0.55	0.64 (41)
33. Ariz.	6	1.12	0.58	0.52	0.67 (39)
34. W. Va.	6	1.12	0.58	0.52	0.67 (38)
35. Neb.	5	0.93	0.45	0.48	0.61 (44)
36. Utah	4	0.74	0.30	0.40	0.58 (49)
37. N.M.	4	0.74	0.29	0.40	0.59 (48)
38. Me.	4	0.74	0.29	0.39	0.60 (47)
39. R.I.	4	0.74	0.29	0.38	0.61 (45)
40. Haw.	4	0.74	0.26	0.35	0.68 (37)
41. N.H.	4	0.74	0.25	0.34	0.69 (35)
42. Idaho	4	0.74	0.25	0.33	0.70 (31)
43. Mont.	4	0.74	0.24	0.33	0.71 (29)
44. S.D.	4	0.74	0.24	0.32	0.73 (26)
45. D.C.	3	0.56	0.19	0.34	0.51 (51)
46. N.D.	3	0.56	0.17	0.31	0.57 (50)
47. Del.	3	0.56	0.16	0.29	0.60 (46)
48. Nev.	3	0.56	0.15	0.28	0.64 (42)
49. Vt.	3	0.56	0.15	0.26	0.67 (40)
50. Wyo.	3	0.56	0.13	0.23	0.77 (19)
51. Alas.	3	0.56	0.12	0.22	0.81 (16)
Small	166	30.86	16.11	0.52	0.68

In substantive terms, *IB* represents the relative proportion of resources a candidate would commit *per person* to each state if he made his allocations according to the 3/2's rule. These per capita allocation ratios are also given in Table 7.2, along with the ranks (in parentheses) of these ratios from the highest to the lowest for the fifty states and the District of Columbia.

The ranking of states on the basis of their *IB* values, unlike their ranking on *EB* values, does not correspond perfectly to their ranking in terms of electoral votes. Thus, for example, whereas an individual voter in the largest state, California, ranks as the most attractive target for a candidate who allocates according to the 3/2's rule—receiving 50 percent more attention (*IB* = 1.50) than he would get if the candidate allocated his time strictly according to the population of each state—a voter in the smallest state, Alaska, with an *IB* of 0.81, is still comparatively well off based on the sixteenth position of Alaska among the fifty states and the District of Columbia. Its *IB* is below the proportional norm of 1.00, but far from the bottom rung of the ladder that citizens who live in Washington, D.C. have the dubious distinction of occupying, with *IB*'s equal to 0.51.

There are two reasons why the *IB* scores of states are only an imperfect reflection of their electoral votes. First, when two states have the same number of electoral votes, like Texas and Illinois with twenty-six each, citizens of the larger state (Texas) will be slightly disadvantaged, because the attention they receive from the candidates according to the 3/2's rule must be divided among more people.

The two-senator bonus accorded to all states more seriously upsets the generally positive relationship between *IB* scores and the electoral votes of a state. This bonus, naturally, is much more significant to a state like Alaska, with one representative, than a state like California, with forty-three representatives: In terms of percentages, the two-senator bonus inflates Alaska's per capita representation by 200 percent, California's by only about 5 percent.

Some critics of the Electoral College have charged that this bonus favors small states, which it obviously does on a proportional basis. On the other hand, proponents of the Electoral College have responded that this favoritism is counteracted by the fact that the large blocs of votes cast by large states in the Electoral College have a greater chance of being decisive, especially in close elections.[41] To what extent do these opposing forces cancel each other out?

The balance between these forces, it turns out, is very one-sided: The large-state bias created by the 3/2's rule swamps the small-state bias result-

---

41. Recall from section 5.6 that the voting power of large states in the Electoral College is only slightly greater than their vote proportions by either the Shapley-Shubik or Banzhaf indices. Although interesting, this result is of limited substantive significance since the real contest for the presidency does not occur *within* the Electoral College, which is largely a ratifying institution. Rather, it is the effects of unit rule *outside* the Electoral College (e.g., on presidential campaigning and the two-party system) which make this institution loom so large; it is precisely some of these effects that we have tried to capture in our models.

ing from the two-senator bonus, giving citizens of the fifteen most populous states the highest *IB* scores. This is not unexpected, since the 3/2's rule compels the candidates to make inordinately large expenditures of resources in the largest states, even out of proportion to their populations.

Yet not all citizens of even these fifteen large states are favored by the 3/2's rule. Only the citizens of the seven very largest states, which comprise less than a majority (45 percent) of the population, receive representation greater than in proportion to their numbers (*IB* > 1.00). Citizens in the remaining forty-three states, plus the District of Columbia (whose sad plight will be recounted subsequently), receive attention below the one-man, one-vote standard from candidates who adhere to the 3/2's allocation rule. Only with Alaska's entry into the sixteenth position of the *IB* ranking does the two-senator bonus begin to help the very small states. Although Wyoming breaks into the *IB* ranking at the nineteenth position, none of the four other very small states with three electoral votes is successful in outdistancing the 3/2's rule by an amount sufficient to rank it above the fortieth position. In the cases of Alaska and Wyoming, the two-senator bonus gives each a per capita electoral-vote representation more than four times greater than that of California, which accounts for their ability to overcome to some degree the large-state bias of the 3/2's rule.

Although the states that the two-senator bonus helps most are among the smallest, small size alone is not the only factor that tends to counteract the 3/2's rule. The ability of a state just to meet the quota for a certain number of electoral votes is also important, for it gives the state higher per capita representation than that of larger states with the same number of electoral votes. Thus, the three smallest states with four electoral votes—Idaho, Montana, and South Dakota—succeed in raising their population rankings of forty-second, forty-third, and forty-fourth by eleven, fourteen, and seventeen notches (one more notch if the District of Columbia is included), respectively, on the *IB* scale. No other states besides these and Alaska and Wyoming are able to better their population rankings by as much as ten positions.

From the perspective of per capita representation, the most unfortunate citizens are those who live in Washington, D.C. The Twenty-third Amendment limits the District of Columbia to electoral representation no greater than the least populous state (Alaska), which has three electoral votes. This limitation is strangely inconsistent with the fact that the District's population is greater than that of four states that each have four electoral votes. With an *IB* of 0.51, its citizens rank below those of all fifty states in per capita representation.

Using Washington, D.C. as a basis for comparison, the ratios of the *IB*'s of all states to the District's *IB* are ranked in Table 7.3. Similarly, taking the lowest-ranking state on the *EB* scale (Alaska), the ratios of all states' *EB* scores to that of Alaska are also ranked. These ratios reveal that the largest state, California, has 2.92 times as great individual representation as Washington, D.C., and 8.13 times as great electoral representation as Alaska. This means that a candidate who campaigns according

TABLE 7.3   RANKINGS OF *EB* AND *IB* RATIOS FOR FIFTY STATES
AND DISTRICT OF COLUMBIA

Rank	EB ratio: EB (state i) EB (Alaska)		IB ratio: IB (state i) IB (D.C.)		Rank	EB ratio: EB (state i) EB (Alaska)		IB ratio: IB (state i) IB (D.C.)	
1.	Calif.	(8.13)	Calif.	(2.92)	27.	Okla.	(2.91)	Wash.	(1.41)
2.	N.Y.	(7.76)	N.Y.	(2.79)	28.	Kans.	(2.73)	Ala.	(1.41)
3.	Pa.	(6.25)	Pa.	(2.28)	29.	Miss.	(2.71)	Mont.	(1.39)
4.	Texas	(6.09)	Ill.	(2.26)	30.	Colo.	(2.70)	Iowa	(1.38)
5.	Ill.	(6.07)	Texas	(2.25)	31.	Ore.	(2.63)	Idaho	(1.37)
6.	Ohio	(5.94)	Ohio	(2.22)	32.	Ark.	(2.52)	Colo.	(1.37)
7.	Mich.	(5.42)	Mich.	(2.04)	33.	Ariz.	(2.42)	Miss.	(1.36)
8.	N.J.	(4.87)	Fla.	(1.89)	34.	W. Va.	(2.40)	Kans.	(1.35)
9.	Fla.	(4.74)	N.J.	(1.84)	35.	Neb.	(2.22)	N.H.	(1.35)
10.	Mass.	(4.34)	Mass.	(1.70)	36.	Utah	(1.87)	Conn.	(1.33)
11.	Ind.	(4.15)	N.C.	(1.67)	37.	N.M.	(1.83)	Haw.	(1.32)
12.	N.C.	(4.10)	Ind.	(1.65)	38.	Me.	(1.81)	W. Va.	(1.32)
13.	Mo.	(3.94)	Ga.	(1.62)	39.	R.I.	(1.77)	Ariz.	(1.31)
14.	Va.	(3.92)	Va.	(1.61)	40.	Haw.	(1.60)	Vt.	(1.30)
15.	Ga.	(3.90)	Mo.	(1.61)	41.	D.C.	(1.58)	Ark.	(1.25)
16.	Wis.	(3.82)	Alas.	(1.58)	42.	N.H.	(1.56)	Nev.	(1.24)
17.	Tenn.	(3.60)	La.	(1.52)	43.	Idaho	(1.54)	Ore.	(1.20)
18.	Md.	(3.60)	Wis.	(1.52)	44.	Mont.	(1.52)	Neb.	(1.19)
19.	Minn.	(3.55)	Wyo.	(1.51)	45.	S.D.	(1.49)	R.I.	(1.19)
20.	La.	(3.47)	Minn.	(1.49)	46.	N.D.	(1.43)	Del.	(1.18)
21.	Ala.	(3.38)	Md.	(1.46)	47.	Del.	(1.35)	Me.	(1.16)
22.	Wash.	(3.36)	Tenn.	(1.46)	48.	Nev.	(1.27)	N.M.	(1.15)
23.	Ky.	(3.26)	Ky.	(1.45)	49.	Vt.	(1.21)	Utah	(1.13)
24.	Conn.	(3.17)	Okla.	(1.45)	50.	Wyo.	(1.05)	N.D.	(1.11)
25.	Iowa	(3.06)	S.C.	(1.44)	51.	Alas.	(1.00)	D.C.	(1.00)
26.	S.C.	(2.93)	S.D.	(1.42)					

to the 3/2's rule would allocate to California more than eight times as many resources per electorate vote as he does to Alaska, and almost three times as many resources per person as he does to Washington, D.C. Although Washington, D.C. has 1.76 times as many electoral votes per person as California, which is why its individual bias of 0.51 is greater than (i.e., deviates less from the standard of 1.00) its electoral bias of 0.34, it would still need about twice as many electoral votes as it has (six), holding all other states constant, to wipe out its individual bias (i.e., make its *IB* ≈ 1.00).

To be sure, Washington, D.C. is the nation's capital, and its residents probably do not suffer from any lack of exposure to presidential candidates. Yet for Alaska, whose electoral votes have only 12 percent of the drawing power of California's (*EB* of 0.22 for Alaska versus 1.76 for California), it may be more than distance that has kept away all presidential and vice-presidential candidates except Richard Nixon in his 1960 campaign. And for the citizens who live in the small and medium states that together comprise 55 percent of the population of the United States, the average *IB* and *EB* scores given in Table 7.2 for these groups of states indicate that they are generally disadvantaged, even if the inequities of the electoral system visited upon them do not match the injustice done to the citizens of Alaska (by the *EB* measure) and the District of Columbia (by the *IB* measure).

So far we have seen that the 3/2's rule greatly favors the large states and generally overwhelms what per capita electoral-vote advantage the small states do have. Of course, the relative neglect of some states by presidential candidates is not only a function of their electoral representation but also the degree to which candidates estimate that their campaigning or other expenditures of resources can make a difference in the probable outcome of the election in these states. If no feasible allocations of time, money, or other resources can possibly change the probable outcome, then a candidate would be foolish to give it anything more than token attention. Richard Nixon, who placed great stress on the symbolic value of visiting all fifty states in his 1960 presidential bid, despite an injury at the beginning of the campaign that kept him sidelined for several days, dumped this strategy in his 1968 and 1972 campaigns. There is no evidence that such "tokenism" brings any payoff in extra votes.

In an actual campaign, a candidate and his advisors would normally pare down the fifty states and the District of Columbia (for which 3/2's allocations are given in Table 7.2) to a smaller list of states in which un-committed voters are likely to have a decisive impact on the election outcome. The states a candidate should ignore by this criterion would include not only those in which his opponent is heavily favored but also those in which he has piled up big early leads prior to the campaign, for it makes as little sense to expend resources in states that seem invulnerable to one's opponent as it does to conduct forays into states that an opponent has virtually locked up. In fact, candidates and their advisors seem to make precisely these calculations, which V. O. Key, Jr., reports also characterize the effects of unit rule in Georgia. There, in party nominations for statewide races, the candidate who wins a plurality in each of Georgia's 159 counties receives all the unit votes of that county (one, two, or three, depending on its population):

> . . . [a practical politician] classifies the counties into three groups: those in which he is sure of a plurality; those in which he has no chance of a plurality; those which are doubtful. He forgets about the first two groups except for routine campaign coverage. He concentrates his resources in the third group; expenditures, appearances by the candidate, negotiations, all the tricks of county politicking.[42]

---

42. V. O. Key, Jr., *Southern Politics in State and Nation* (New York: Alfred A. Knof, 1949), p. 122. This citation is due to Melvin J. Hinich and Peter C. Ordeshook, "The Electoral

In presidential campaigns today, the widespread use of public opinion polls makes it easier than ever before to weed out those states "securely" in the camp of one candidate and pinpoint the "toss-up" states likely to swing either way. Insofar as polls indicate the largest states to be the toss-up states, candidates who act on this information and concentrate almost all their resources in these states will magnify even the large-state bias of the 3/2's rule. This is so because the noncompetitive (small) states would be eliminated from consideration altogether, and allocations that would ordinarily go to them would be concentrated in the competitive (large) states. Indeed, the Democrats reportedly wrote off twenty-four states in 1968,[43] though our campaign appearance data indicate that one or both of the Democratic candidates made appearances in more than half of these states. Apparently, even the best laid plans are not always followed.

7.11. LIMITATIONS AND EXTENSIONS OF THE MODELS

The rationale of the 3/2's rule rests on the idea that states can be pinpointed like military targets and, independent of nearby and faraway other targets, captured with a high probability by a concentration of forces superior to that of one's opponent (given that his allocations are known or can be estimated). As Richard M. Scammon and Ben J. Wattengerg warn, however, this logic may be seriously flawed:

> It is extremely difficult, and probably impossible, to move 32,000 votes in a New Jersey Presidential election without moving thousands and tens of thousands of votes in each of the other forty-nine states. The day of the pinpoint sectional or statewide campaign is gone—if it ever existed—and the fact that votes cannot be garnered in bushels on specific street corners is of crucial significance when one looks at the arithmetic of the future.[44]

Scammon and Wattenberg go on to point out that political rallies usually draw mostly partisans already committed to a candidate. For this reason, relatively few uncommitted voters are likely to observe, much less be

---

College: A Spatial Analysis," *Political Methodology* (forthcoming), in which the effect of the Electoral College on equilibrium strategies, given different distributions of policy preferences of the electorate, is examined. For related research on the Electoral College, see Melvin J. Hinich, Richard Mickelson, and Peter C. Ordeshook, "The Electoral College Vs. A Direct Vote: Policy Bias, Reversals, and Indeterminate Outcomes," *Journal of Mathematical Sociology* (forthcoming); and Claude S. Colantoni, Terrence J. Levesque, and Peter C. Ordeshook, "Campaign Resource Allocations under the Electoral College, *American Political Science Review* (forthcoming).

43. Joseph Napolitan, *The Election Game and How to Win It* (Garden City, N.Y.: Doubleday and Co., 1970), p. 62.

44. Richard M. Scammon and Ben J. Wattenberg, *The Real Majority: An Extraordinary Examination of the American Electorate* (New York: Coward-McCann, 1970), p. 213.

persuaded by, the speech a candidate makes at a rally in their state.[45]
Their choices, Scammon and Wattenberg argue, will depend much more
on national coverage of the campaign, particularly by television.[46]
Yet, these analysts do not dismiss campaigns as insignificant:

> ... [T]he campaign is important in providing the candidate with
> *something to do* that can be televised, photographed, and written
> about for national consumption. . . . Further, if a candidate does *not*
> campaign at all, his opposition will criticize him for "not taking his
> case to the people."
>
> Beyond all that, however, is a certain extremely valuable democratic
> symbolism that underlies all the flesh pressing, baby kissing, hurly-
> burly of a campaign. There is great value in a system that somehow
> demands that a candidate get sweaty and dirty and exhausted, his
> hands bleeding, his hair messed by the masses of people whom he
> wants to represent. The successful candidate in America must *touch*
> the people, figuratively and literally.[47]

How a candidate does this is not well understood, even if we do know that
campaigns often decide the outcomes of elections.[48]
Our interest in this analysis, however, is not in showing how the conduct
of campaigns affects election outcomes.[49] Rather, it is to establish that
candidates act *as if* they believe their conduct matters when it comes to
making campaign allocations. Perhaps "the spectacle of seeing one's

---

45. According to a series of Gallup polls, less than 6 percent of all people saw any one of
the presidential candidates in person in 1968. See Sara Davidson, "Advancing Backward
with Muskie," *Harper's*, June 1972, p. 61.

46. Scammon and Wattenberg, *Real Majority*, pp. 214–16. Moreover, television news shows,
documentaries, and other specials, which are generally beyond the direct control of candidates,
rank far above television advertising as the most important media influences, at least for the
split-ticket voter. See De Vries and Tarrance, *Ticket-Splitter*, p. 78.

47. Scammon and Wattenberg, *Real Majority*, p. 217; italics in original.

48. At the local level, party organization activities do appear to influence election results
favorably, including the vote for president. See William J. Crotty, "Party Effort and Its Impact
on the Vote," *American Political Science Review*, 65 (June 1971), pp. 439–50, and references cited therein; also, Robert J. Huckshorn and Robert C. Spencer, in *The Politics
of Defeat: Campaigning for Congress* (Amherst, Mass.: University of Massachusetts Press,
1971), stress the importance of campaign organization on election results. At the national
level, on the other hand, there is no clear-cut relationship between the campaign spending of
each party and the outcome of presidential elections. See Twentieth Century Fund, *Voters'
Time: Report of the Twentieth Century Fund Commission on Campaign Costs in the Electronic
Era* (New York: Twentieth Century Fund, 1969), pp. 11–13; Congressional Quarterly
Dollar Politics: The Issue of Campaign Spending (Washington D.C.: Congressional Quarterly,
1971), pp. 14, 19–31; and Dunn, *Financing Presidential Campaigns*, p. 9. More opaque,
still, is the relationship between major strategic decisions made by candidates, especially
during the heat of a close race, and their effect on the vote. See Polsby and Wildavsky,
Presidential Elections, pp. 199–206.

49. That the organization and planning of presidential campaigns is critical to a candidate's
eventual success is stressed in Jerry Bruno and Jeff Greenfield, *The Advance Man* (New
York: William Morrow & Co., 1971).

opponent run around the country at a furious pace without following suit is too nerve-racking to contemplate,"[50] but it also appears that the as-if assumption is rooted in the actual beliefs of candidates, who, at least at the state and local level, think their campaigns have an impact on the election outcome.[51] As evidence for the postulated goal that presidential candidates campaign as if they seek to maximize their expected electoral vote, we have already indicated that the 3/2's rule that follows from this goal mirrors quite well the actual campaign allocations of candidates.

The increasing nationalization of presidential campaigns obviously limits the applicability of any resource-allocation model that makes states the units of analysis. It is worth noting, though, that in postulating that candidates maximize their expected electoral vote, we do *not* assume that voting by individuals in different states is statistically independent. If we had postulated for candidates the goal of maximizing their probability of winning a majority of electoral votes, the statistical-independence assumption would be convenient, and probably necessary, to obtain tractable results.

The goal of maximizing one's expected electoral vote may not be equivalent to maximizing one's probability of winning the election. To illustrate, a candidate who concentrates his resources in states with a bare majority of electoral votes may win these states, and the election, with a high probability, but his expected number of electoral votes will be relatively small. On the other hand, if he spreads his resources somewhat thinner over more states, he may increase his expected number of electoral votes, but at the price of lowering the probability that he will win a majority of electoral votes. Thus, the two goals may be logically inconsistent— that is, they may lead to contradictory strategic choices on the part of rational candidates. It does not seem that the implications of these different goals will be seriously contradictory in most cases—at least in two-candidate races—but this question needs to be explored further.

Finally, it should be pointed out that several restrictive assumptions made earlier in the development of the models can be relaxed. The mini-max solution of the popular-vote model, and the (unstable) equilibrium solution of the electoral-vote model, hold *mutatis mutandis* when three or more candidates compete in an election. Furthermore, it is not necessary to assume, as we did initially, that the committed voters in each state are

---

50. Polsby and Wildavsky, *Presidential Elections*, p. 183. If the steeliness of a candidate's nerves can be judged by this standard, Richard Nixon's nerves progressively hardened in his successive bids for the presidency: As an incumbent president in the 1972 campaign, he made only 18 campaign appearances, whereas as a nonincumbent in 1968 and 1960 he made 140 and 228 appearances, respectively. Yet, though he cut down the number of his appearances in successive campaigns, in all three races Nixon made disproportionally greater allocations to the large states than to the medium and small states.

51. John W. Kingdon, *Candidates for Office: Beliefs and Strategies* (New York: Random House, 1968), pp. 109–14. See also Robert A. Schoenberger, "Campaign Strategy and Party Loyalty: The Electoral Relevance of Candidate Decision-Making in the 1964 Congressional Elections," *American Political Science Review*, 63 (June 1969), pp. 515–20.

evenly divided at the outset of a campaign. If a candidate's polls indicate, for example, that he is ahead by particular margins in some states, and behind by particular margins in others, he can use this information to determine his optimal allocations to each state, given that he can estimate the probable allocations of his opponent.

This problem can be conceptualized as one in which a candidate tries to reduce the effect of an opponent's supporters in a state by matching them against his own. If at the start of a campaign one candidate has more supporters in a state than the other candidate, the lesser candidate faces the problem of winning over a sufficient number of uncommitted voters to neutralize both his opponent's extra supporters and those uncommitted voters his opponent is likely to pick up, assuming that commitments once made are not broken.

Given the allocations of the candidates for all states i that define p_i, and the distribution of supporters committed to each candidate in each state, the probability of neutralizing an opponent's support and then going on to win is given by Lanchester's Linear Law, for which Richard H. Brown has provided a useful approximate solution.[52] Optimal allocations can be obtained by maximizing the expected electoral vote based on these probabilities for all states. Holding size constant, it would appear that the most attractive targets by far for campaign allocations in the electoral-vote model are those states in which the committed voters are split roughly fifty–fifty.

Finally, it seems possible that the models we have developed may be applicable to the analysis of resource-allocation decisions in other political arenas, particularly those involving the distribution of payoffs to members of a coalition. For example, the apportionment of foreign aid, the commitment of military forces, the granting of patronage and other favors, and even the assignment of priorities all involve decisions about how to allocate resources among actors (or programs) of different weight or importance. Moreover, these decisions are often influenced by the actions or probable responses of an opponent in an essentially zero-sum environment. Presumably, the extent to which decision making in such competitive situations is conditioned by the kinds of goals postulated in this chapter could be determined by comparing the resource expenditures actually made with the theoretical implications that these goals entail (e.g., the proportional or 3/2's rules).

7.12. SUMMARY AND CONCLUSION

We began our analysis on a critical note by contending that many of the arguments made for and against the Electoral College suffer from a primarily descriptivist and nontheoretical orientation. They have been

52. Richard H. Brown, "Theory of Combat: The Probability of Winning," *Operations Research*, 11 (May–June 1963), pp. 418–25.

responsible for more obfuscation than clarification of the possible consequences that would follow from abandonment of the Electoral College and the substitution of direct popular-vote election of a president in its place.

As a way of making the comparison of alternative electoral systems more rigorous, we assumed at the outset that different electoral systems encourage political actors to seek different goals. By deducing the theoretical consequences that follow from the rational pursuit of these goals, we suggested that the effects of different electoral arrangements—specifically, different vote-aggregation procedures—on political behavior could be illuminated.

This approach to the analysis of political behavior in general, and electoral behavior in particular, is still unorthodox in political science, though the recent development of spatial models of party competition provides one example of the power of this approach. Unlike party-competition models, however, which take the positions of candidates on issues as their focus for analysis, we took these positions as given and instead asked how a presidential candidate should allocate his resources to convey as favorable an impression of his positions and personality as possible. We assumed that the favorableness of his image to uncommitted voters in each state would be a function of the amount of resources he allocated to each state, as compared with the amount allocated by his opponent.

Under a system of direct popular-vote election of a president, we postulated that a candidate would seek to maximize the minimum expected popular vote he could obtain. If this were a candidate's goal, it was shown that his optimal minimax strategy would be to allocate his resources in proportion to the number of uncommitted voters in each state. If one candidate deviated from this strategy, and these deviations were known to his opponent, we indicated how this information could be used by the opponent to capitalize on the weaknesses of the candidate who departed from his minimax strategy.

Under the present Electoral College system, we demonstrated that there is no minimax solution to the resource-allocation problem in pure strategies. We showed, however, that if candidates matched each other's resource expenditures, and if their total resources were the same, then their optimal allocations would be roughly proportional to the 3/2's power of the electoral votes of each state (subsequently refined to include population). This rule encourages disproportionally large expenditures in the largest states, whose plausibility was demonstrated by showing how the potential voting power of a voter, derived from a measure of his decisiveness, is greater for voters in large states than in small states. We also indicated that the 3/2's allocations are inherently unstable, and we described the most devastating strategy that could be used against them by a candidate who exploits information about his opponent's allocations.

As a test of the models, the 3/2's allocation rule was compared to a proportional-allocation rule based on the electoral votes of each state. Using time as a resource, we operationalized allocations in terms of the number of campaign appearances of presidential and vice-presidential candidates in each state in the 1960, 1964, 1968, and 1972 presidential election campaigns.

The 3/2's rule proved to be generally superior to the proportional rule in explaining the time allocations of the candidates. For presidential elections through 1980, we indicated the 3/2's allocations for all states and electoral and individual biases that these allocations engender in each state. We concluded our analysis with a discussion of both theoretical and empirical limitations and extensions of the models.

Probably the most compelling conclusion that emerges from our analysis, at least for noncandidates, does not concern campaign strategy but rather concerns the severely unrepresentative character of presidential campaigns under the Electoral College. By forcing candidates—and after the election, incumbent presidents with an eye on their reelection—to pay much greater attention in terms of their allocations of time, money, and other resources to the largest states, the Electoral College gives disproportionally greater weight to the attitudes and opinions of voters in these states. On a per capita basis, voters in California are 2.92 times as attractive campaign targets as voters in Washington, D.C.; even greater than this ratio of the most extreme individual biases is the most extreme electoral-bias ratio, which makes California 8.13 times as attractive per electoral vote as Alaska.

Although no voter today is actually deprived of a vote by the Electoral College, an aggregation procedure that works to deflate the weight of some votes and inflate the weight of others seems clearly in violation of the egalitarian principle of one man, one vote. The reapportionment decisions of the courts since 1962 have upheld this principle, with reapportionment itself obviating the need for an institution with a large-state "urban bias" like the Electoral College to offset the rapidly fading "rural bias" of the House of Representatives. Furthermore, as a procedure that places a premium on gaining information and responding to an opponent's allocations in each state, the Electoral College tends to encourage manipulative strategic calculations for outmaneuvering the opposition in each state, which may divert a candidate from seeking broad-based nationwide support.

From a normative viewpoint, this is the goal it seems desirable the electoral system should promote. The Electoral College subverts this goal by giving special dispensation to particular states and, additionally, fostering manipulative strategies in them. As an alternative, direct popular-vote election of the president, which would render state boundaries irrelevant, would encourage candidates to maximize their nationwide appeal by tying their support directly to potential votes everywhere on a proportional basis.

To close this book on a philosophical note, we hardly need indicate that the conclusions we have drawn from applications of game theory to politics are not the usual stuff of conventional political theory, whose study is rooted mainly in the traditions of historical inquiry and speculative philosophy. Political theorists have by and large ignored all that smacks of formal, deductive analysis and quantitative empirical research. This seems unfortunate, because questions that relate to such qualities as the representativeness and equality of political institutions, particularly as they are reflected in the strategies used and the power exercised by players and coalitions of players in political games, are fundamental to almost all political thought—and issues of public policy as well.

In this book, we have attempted to show that such questions can be fruitfully attacked not just be making endless observations of the workings of political institutions but by demonstrating that rational actors will behave in certain ways as they try to anticipate and cope with the actions of other actors in different institutional and environmental settings. To understand the behavior of political actors as they pursue particular ends in an evermore complex world necessitates the construction of theoretical models that can help transform vague insights into coherent and logical explanations that clarify the insights and relate them to each other.

In situations of partial or total conflict, game theory can help to make these explanations more rigorous, but it alone is not sufficient; formal analysis must be coupled with empirical research that can substantiate the truth of the logical assertions derived from the analysis. The most relevant and powerful theory depends on the presence of both ingredients, as we have tried to show in this book through our exploration of game-theoretic questions in concrete, empirical settings.

GLOSSARY

This glossary contains definitions of the game-theoretic and related terms used in this book. Concepts that have been developed in detail in the text (e.g., the power indices, measures of bias in the Electoral College), or that would require extended discussion (e.g., successively more desirable strategies), are not included. As in the text, an attempt has been made to define the terms in relatively nontechnical language.[1]

Admissible Strategy. In a voting game in normal form, a strategy that satisfies the following two conditions: There is no other strategy providing

(1) an outcome at least as good, whatever strategies are adopted by the other voters;

(2) a better outcome for some selection of strategies of the other voters.

Characteristic Function. A function v that assigns a value to every subset (or coalition) of players in a game, which is the minimum amount that the coalition can ensure for itself whatever other coalition(s) forms. It must satisfy the condition of superadditivity and the condition,

$$v(\phi) = 0,$$

where ϕ is the empty coalition containing no members.

Coalition. A subset of players; its incentive to form may be to coordinate the selection of joint strategies, to increase the voting power of its members, to enforce the choice of a particular outcome on other players, or to achieve some other end mutually beneficial to its members.

Coalition Rationality. See **Rationality.**

Coalition Structure. In an n-person game in characteristic-function form, a partition τ of the set of players in a game into (disjoint) coalitions. The set of admissible coalitions (as allowed by a rule of admissible coalition changes, ψ) is designated $\psi(\tau)$.

---

1. For an excellent technical glossary that contains references to the mathematical literature, see Gerhard Schwödiauer, "A Glossary of Game Theoretical Terms" (Department of Economics, New York University, Working Paper No. 1, June 1971).

Collective Good. A good that, when supplied to some members of a collectivity, cannot be withheld from other members of that collectivity.

Complete Information. A game is a game with complete information if all players know from the start the rules of the game as described by the extensive or normal forms.

Condorcet Winner. An alternative that defeats all others in a series of pairwise contests.

Constant-Sum (Zero-Sum) Game. A game in normal form in which the value of every coalition and its complement always sums to some constant (or zero), which is the value of the grand coalition; all constant-sum games can be converted to zero-sum games by subtracting the appropriate constant from the payoffs to the players.

Cooperative Game. A game in which preplay communication (and bargaining) can occur and players can make binding and enforceable agreements.

Core. In an n-person game in characteristic-function form, the set of imputations that satisfies coalition rationality; imputations in the core are not dominated by any other imputation with respect to all possible coalitions of players.

Critical Defection. In a weighted majority game, a defection of a player that transforms a minimal winning coalition (with respect to at least one player) into a nonwinning coalition.

Cumulative Voting. A system of voting in which voters in the electorate in multimember constituencies are allowed to cumulate their votes on fewer than the number of candidates to be elected.

Cyclical Majorities. Majorities that can defeat other majorities in a series of pairwise contests; such majorities exist when there is a paradox of voting.

Decision Rule. The rule in a voting body that specifies the number of votes required for collective action to be taken that is binding on all the members; the decision rule defines the size of a minimal winning coalition.

Desirable Strategy. In a voting game in normal form, a strategy that satisfies the following two conditions: There is no other strategy providing

(1) an outcome at least as good, whatever strategies are selected by the other voters without straightforward strategies;

(2) a better outcome for some selection of strategies of the other voters without straightforward strategies.

For the definition of successively more desirable strategies, see p. 75.

Determinate Voting Procedure.   A voting procedure that results in the selection of a single outcome if voting is sophisticated.

Dictator.   In a weighted majority game, a player whose votes by themselves are sufficient to determine the outcome of the voting process; only coalitions in which a dictator is a member are winning in such a game.

Dominant Strategy.   In cumulative voting, a strategy that assures a group in the electorate of winning a particular number of seats, and, in addition, provides for the possibility of its capturing an extra seat above this assured number.

Domination.   In a game in normal form, one strategy of a player dominates another if it leads to outcomes no worse, and, for at least one set of strategy choices of the other players, a better outcome; a strategy that always leads to better outcomes than another strategy *strictly* dominates that other strategy. In a game in characteristic-function form, one imputation $X = (x_1, x_2, \ldots, x_n)$ dominates another imputation $Y = (y_1, y_2, \ldots, y_n)$ with respect to a nonempty coalition S if it is "feasible," i.e.,

$$\sum_{i \text{ in } S} x_i \leq v(S),$$

and gives a better payoff to every member of S, i.e.,

$$x_i > y_i \text{ for all } i \text{ in } S.$$

Dummy.   In a weighted majority game, a player whose votes can never have any effect on the outcome of the voting process.

Equilibrium Choice.   In a voting game in normal form, a collective choice of a voting body that is not vulnerable. (Unless otherwise specified, equilibrium choices are assumed to be of order one, i.e., invulnerable to the substitution of a strategy by just one voter, given the strategy choices of the other voters.)

Equilibrium Strategies.   In a game in normal form, a set of strategies for all players from which it is not to the advantage of any player to deviate if no other player changes his strategy choice. Equilibrium strategies may be stable *locally* or *globally*: In the case of the former, only "small" unilateral deviations are nonoptimal; in the case of the latter, all unilateral deviations are nonoptimal.

Expected Payoff. The payoff that a player receives for every possible outcome of the game, weighted by the probability that that outcome occurs.

Extensive Form. The representation of a game as a sequence of moves specified by a game tree.

External Costs. The costs one bears as a result of the actions of others.

Fair Game. A two-person zero-sum game in normal form whose value is equal to zero; that is, a game in which neither player wins anything from the other when the players choose their optimal strategies.

Finite Game. In extensive form, a game with a finite number of moves and a finite number of alternatives at each move; in normal form, a game with a finite number of pure strategies.

Game. The totality of rules that relate players or coalitions to outcomes.

Game against Nature. See **One-Person Game.**

Game of Chance. A game in which the outcome does not depend at all on the strategy choices of the players but instead on some random or stochastic process determined by a probability distribution.

Game of Strategy. A game in which the outcome depends at least in part on the strategy choices of the players, which may involve some randomized selection procedure (as in the use of mixed strategies).

Game Tree. A topological tree constructed from the rules of a game, where the vertices of the tree represent choice points, or moves, and the branches the alternatives that can be chosen.

Grand Coalition. The coalition containing all players in a game.

Group Rationality. See **Rationality.**

Imputation. A vector $X = (x_1, x_2, \ldots, x_n)$ of payoffs to players in an n-person game that satisfies the conditions of individual rationality and group rationality.

Individual Rationality. See **Rationality; Vote Trade.**

Infinite Game. In normal form, a game in which players have an infinite number of pure strategies from which to choose.

Information Effect. The tendency of incomplete and imperfect information to induce coalitions to form that are larger than minimal winning size as a cushion against uncertainty.

Information Sets. Sets of vertices, or moves, on a game tree that are indistinguishable to the players.

Maximin. Shorthand for the maximum of the minima, which is the payoff that the maximizing (conventionally, row) player can ensure for himself in a two-person constant-sum game; also refers to the strategies of the players associated with a maximin (= minimax, i.e., saddlepoint) in a strictly determined game, to the optimal mixed strategies of players in a nonstrictly determined game.

Maximizing Player. In a two-person constant-sum game, the player (conventionally, row) who maximizes the minima of payoffs that he can ensure for himself.

Metagame. A game derived from a given game in which players successively choose strategies with knowledge (or expectations) about the strategy choices of the other player(s).

Metastrategy. A strategy conditional on knowledge (or expectations) about the strategy choices of the other player(s).

Minimal Winning Coalition. In a simple game, a winning coalition which would be rendered nonwinning were at least one of its members to resign or defect to another coalition; minimal winning coalitions in a game are *distinct* if the membership of no two minimal winning coalitions is identical.

Minimax. Shorthand for minimum of the maxima, which is the payoff that the minimizing (conventionally, column) player can limit the maximizing (conventionally, row) player to in a two-person constant-sum game; also refers to the strategies of the players associated with a minimax (= maximin, i.e., saddlepoint) in a strictly determined game, to the optimal mixed strategies of players in a nonstrictly determined game.

Minimizing Player. In a two-person constant-sum-game, the player (conventionally, column) who minimizes the maxima of payoffs that the other player can obtain.

Mixed-Motive Game. A variable-sum game, i.e., a game in which players have both competitive and complementary interests and in which their gains and/or losses from the play of the game do not always sum to some constant amount.

Mixed Strategy. In a two-person constant-sum game, a strategy that involves the random selection of a strategy from two or more pure strategies according to a particular probability distribution.

Move. In a game in extensive form, a choice point, or vertex, of the game tree, where one from a given set of alternatives is chosen either by a player or by chance.

***N*-Person Game.** A game with more than two players; payoffs from the play of an *n*-person game are usually defined in terms of the strategy choices of the players (normal form) or the values of the game to all possible coalitions (characteristic-function form).

Noncooperative Game. A game in which preplay communication (and bargaining) cannot occur and players cannot make binding and enforceable agreements.

Nonzero-Sum Game. See **Variable-Sum Game.**

Normal Form. Representation of a game by a payoff matrix, where the players are assumed to choose strategies simultaneously and independently.

One-Person Game. A game against nature, wherein "nature" is assumed to be a fictitious player whose choices are not conscious or based on rational calculation, but instead on chance.

Optimal Strategy. In a two-person constant-sum game, a (maximin or minimax) strategy that assures a player he will obtain a payoff (or expected payoff) at least equal to the value of the game.

Outcome. The social state realized from the player of a game; in a game in normal form, an entry in a payoff matrix determined when each player chooses a strategy.

Paradox of New Members. A voting situation where the addition of one or more players to a voting body increases the voting power of at least one of the original players.

Paradox of Quarreling Members. A voting situation where a quarrel between two (or more) members that prevents them from joining together to help form a winning coalition increases their individual and combined voting power.

Paradox of Size. A voting situation where the voting power of a coalition is less than the sum of the voting power of its members.

Paradox of Vote Trading. A vote-trading situation in which all voters benefit from making individually rational vote trades but in which, because of external costs incurred from the trades of others, all voters are worse off after trading.

Paradox of Voting. Given at least two voters and three alternatives, a voting situation in which the preference scales of voters are such that no alternative can defeat all others in a series of pairwise contests if voting is sincere.

Payoff. The utility associated with the outcome of a game for a player or coalition of players.

Payoff Matrix. In a game in normal form, a rectangular array, or matrix, whose entries indicate the outcomes—or payoffs—to each player resulting from their different strategy choices.

Perfect Information. A game in extensive form is called a game with perfect information if every information set on the game tree contains only a single vertex or move. This ensures that each player is fully informed about the previous moves in the game and thus knows exactly where he is on the game tree at all times.

Pivot. In a weighted majority game, the player who is decisive to a coalition's becoming winning as players are successively added to form the grand coalition. The position that the pivot occupies in this sequential buildup is called the *pivotal position*. The probability that a player will be a pivot in a coalition, given that all different sequences in which players join the grand coalition are equiprobable, is called his *pivotalness*.

Play. In a game in extensive form, one selection of alternatives in a game, the last alternative being chosen at a move prior to a termination point, that traces a path through the game tree.

Player. A participant in a game who is capable of rational calculation and can make conscious choices; a player need not be a single individual but may represent any autonomous decision-making unit.

Power Discrepancy. In a weighted majority game, the sum of percent deviations of the voting power indices from the voter weights for the largest and smallest players in the game.

Power Equivalent Games. Weighted majority games with players of different weights and having different decision rules wherein the distribution of voting power among the players is the same.

Preference Scale.   A ranking by a player of possible social states, or outcomes, of a game.

Protocoalition.   In a simple game, a nonwinning coalition S, i.e.,

$$v(S) = 0.$$

In a weighted majority game, the sum of the weights of members in any protocoalition S must be less than the decision rule.

Public Good.   See **Collective Good.**

ψ-Stability.   Given a pair (X, τ), representing an imputation X and a coalition structure τ, the value of any coalition S that could form (according to a rule of admissible coalition changes, ψ), cannot exceed that which all of its members presently receive, i.e.,

$$v(S) \leq \sum_{i \text{ in } S} x_i,$$

and the payoff to player i must be greater than what he can get in a coalition by himself, i.e.,

$$x_i > v(\{i\}),$$

unless $\{i\}$ is in τ.

Pure Strategy.   A single strategy that completely describes a player's actions in all contingencies that may arise.

Rational Player.   In a two-person constant-sum game, a player who chooses an optimal strategy; more generally, a player who chooses the most preferred of his alternatives (e.g., a strategy, coalition partners) as determined by both his own preference scale and the choices of the other players who, under given environmental constraints, are assumed to act similarly.

Rationality.   In an n-person game in characteristic-function form, if $x_i, i = 1, 2, \ldots, n$, is the payoff to player i, v is the characteristic function, and N is the grand coalition, then

$$x_i \geq v(\{i\})$$

is the condition of *individual rationality*,

$$\sum_{i \text{ in } N} x_i = v(N)$$

is the condition of *group rationality*, and

$$\sum_{i \text{ in } S} x_i \geq v(S)$$

is the condition of *coalition rationality*, where S is any coalition.

Rules of a Game. Instructions for playing a game that relate the choices of players to outcomes.

Saddlepoint. An entry in the payoff matrix of a strictly determined game that is simultaneously the minimum of the row in which it occurs and the maximum of the column in which it occurs.

Safe/New-Safe Strategies. In cumulative voting, strategies that assure a group in the electorate of winning a particular number of seats but provide no opportunity for its capturing an extra seat above this assured number.

Salience. In a weighted majority game, the difference in utility that it makes for a player to be in the majority (i.e., win) versus the minority (i.e., lose) on a roll-call vote.

Security Level. In a game in normal form, the minimum payoff that a player can receive for each of his strategy choices.

Separability. A voter's preference scale is separable at a voting division if the top outcome of one subset at the division is not ranked higher than the bottom outcome of the other subset.

Side Payments. Payments involving some common medium of exchange, such as money, that can be transferred among players before or after play in an n-person game.

Simple Game. A game defined by a characteristic function that assigns two values, 0 and 1, to coalitions, which are usually associated with losing and winning, respectively. Since there is zero value associated with losing, each player is motivated to be a member of a winning coalition, giving to such games a winner-take-all character.

Sincere Voting. Voting directly in accordance with one's preferences.

Sincere-Anticipatory Voting. Voting on the substantive outcomes is sincere; voting on the procedures to be used for choosing a substantive outcome is based on an anticipation of the sincere outcomes that would be chosen under each procedure.

Size Principle. In n-person zero-sum games in which side payments are permitted, players are rational, and in which they have perfect information, only minimum winning coalitions occur.

Sophisticated Voting. Voting that forecloses the possibility that a voter's worst outcomes will be chosen—insofar as this is possible—through the adoption of successively more desirable strategies, given that other voters act likewise.

Straightforward Procedure. A voting procedure that affords a voter with a given preference scale a straightforward strategy at all voting divisions, or moves, in the game tree.

Straightforward Strategy. In a voting game in normal form, a strategy that is unconditionally best, or dominant, for a voter.

Strategy. In a game in normal form, a complete set of instructions specifying the choice a player should make at every move in a game tree.

Strictly Competitive Game. A constant-sum game, i.e., a game of pure conflict wherein players have no common interests and what one player wins the other player(s) necessarily loses.

Strictly Determined Game. A two-person constant-sum game in normal form whose payoff matrix contains a saddlepoint.

Superadditivity. One of the two conditions a characteristic function must satisfy that specifies two subsets of players, S and T, that form a coalition and act in concert, provided $S \cap T = \phi$ (S and T are disjoint, i.e., have no members in common), can obtain at least as much and possibly more by combining, rather than acting separately, i.e.,

$$v(S \cup T) \geq v(S) + v(T).$$

Supersophisticated Voting. Voting that reduces the sophisticated strategies of one or more voters to some smaller set when the assumption of complete information is relaxed.

Termination Point. Endpoint of a game tree; associated with each termination point is a sequence of choices that traces a path through the game tree and uniquely characterizes the play of the game that led to that point.

Transitivity. For any three alternatives A, B, and C, preferences are transitive if, given A is preferred to B and B is preferred to C, A is preferred to C.

Two-Person Game. A game with exactly two players; payoffs from the play of a two-person game are usually defined in terms of sequences of moves (extensive form) or strategy choices (normal form) of the players.

Utility. A number indicating the value, or degree of preference, that a player attaches to an outcome in a game.

Value. In two-person constant-sum games in normal form, the best outcome that two players can ensure for themselves (equal to the saddle-point in a strictly determined game); in n-person games, the amount that every coalition can ensure for itself, as given by the characteristic function.

Variable-Sum (Nonzero-Sum) Game. A game in normal form in which the sum of the payoffs to all players is not always equal to some constant (or zero) but varies according to the strategy choices of the players.

Vote Trade. An exchange between voters i and j in which on one roll call voter i (majority position holder) gives his support to voter j (minority position holder) in return for the support of voter j (majority position holder) on another roll call (where voter i is in the minority). A vote trade is *individually rational* for a pair of voters if the minority position for which each voter receives support changes an outcome more salient to him than the outcome of the roll call on which he switches his vote.

Voting Procedure. A way of arranging the outcomes to be voted upon.

Vulnerability. A collective choice of a voting body is vulnerable if another choice

(1) can be obtained from the first by substituting a strategy of at least one voter;

(2) is preferred to it by that voter, or those voters.

Weighted Majority Game. A simple game defined by players of specified weights and a decision rule (e.g., simple majority) that distinguishes between coalitions of winning and losing size.

Zero-Sum Game. See **Constant-Sum Game.**

ANNOTATED
BIBLIOGRAPHY

This is a very selective bibliography of books chosen on the basis of two criteria: (1) that the treatment of game theory be relatively nonspecialized; (2) that the discussion or examples involve applications, or at least be potentially relevant, to politics (broadly conceived). The classification of books by levels means roughly the following:

Elementary: Requires virtually no mathematical background.

Intermediate: Although the mathematics used is not generally advanced, concepts and ideas are developed fairly rigorously and extensive use is often made of mathematical symbolism.

Advanced: Primarily mathematical works that require a high level of mathematical sophistication.

Elementary

Davis, Morton D. *Game Theory: A Nontechnical Introduction.* New York: Basic Books, 1970.
    As its subtitle implies, this book uses almost no mathematics. Yet it offers a rather sophisticated treatment of the subject matter of game theory and also contains many provocative (though mostly hypothetical) examples.

McDonald, John. *Strategy in Poker, Business and War.* New York: W. W. Norton & Co., 1950.
    The first popular treatment of game theory, which also includes a discussion of games of chance, *viz.*, poker; contains some delightful illustrations.

Rapoport, Anatol. *Fights, Games, and Debates.* Ann Arbor, Mich.: University of Michigan Press, 1960.
    About one-third of this book is devoted to a discussion, with examples, of elementary game theory. The contrast between "games," on the one hand, and "fights" and "debates" on the other, provides a useful perspective on different approaches that have been used to study cooperation and conflict in social situations.

Shubik, Martin (ed.). *Game Theory and Related Approaches to Social Behavior*. New York: John Wiley & Sons, 1964.
A rather diffuse collection of articles containing a useful expository essay by the editor.

Singleton, Robert R., and William F. Tyndall. *Games and Programs: Mathematics for Modeling*. San Francisco: W. H. Freeman, 1974.
Includes a description of the relationship between game theory and linear programming.

Intermediate

Farquharson, Robin. *Theory of Voting*. New Haven, Conn.: Yale University Press, 1969.
An elegant treatment of strategic voting that is marred by numerous errors. Many details are omitted from the analysis, making its appearance deceptively simple.

Luce, R. Duncan, and Howard Raiffa. *Games and Decisions: Introduction and Critical Survey*. New York: John Wiley & Sons, 1957.
A superb treatment of applied game theory and related topics, though now somewhat dated.

Rapoport, Anatol (ed.). *Game Theory as a Theory of Conflict Resolution*. Dordrecht-Holland: D. Reidel Publishing Co., 1974.
Although this volume contains some purely mathematical contributions, they—along with the more empirically oriented articles—raise questions of an applied nature relevant to the study of real-life conflicts and their resolution.

Rapoport, Anatol. *N-Person Game Theory: Concepts and Applications*. Ann Arbor, Mich.: University of Michigan Press, 1970.

Rapoport, Anatol. *Two-Person Game Theory: The Essential Ideas*. Ann Arbor, Mich.: University of Michigan Press, 1966.
Both books are characterized by a strenuous effort on the part of the author to explain verbally the intuitive meaning of mathematical concepts.

Rapoport, Anatol, and Albert M. Chammah. *Prisoner's Dilemma: A Study in Conflict and Cooperation*. Ann. Arbor. Mich.: University of Michigan Press, 1965.
A review of mathematical and empirical research up to 1965 on game theory's most famous game.

Riker, William H. *The Theory of Political Coalitions*. New Haven, Conn.: Yale University Press, 1962.
The first major application of *n*-person game theory in political science and now a modern classic that has stimulated numerous subsequent studies.

Schelling, Thomas C. *The Strategy of Conflict.* Cambridge, Mass.: Harvard University Press, 1960.
A critique of classical game theory with suggestions for its reorientation, especially as applied to the analysis of bargaining in international relations.

Advanced

Burger, Ewald. *Introduction to the Theory of Games*, translated by John E. Freund. Englewood Cliffs, N.J.: Prentice-Hall, 1963.
Good coverage of most major topics.

Dresher, Melvin. *Games of Strategy: Theory and Applications.* Englewood Cliffs, N.J.: Prentice-Hall, 1961.
Contains several well-developed examples relevant to the study of defense and military strategy.

Howard, Nigel. *Paradoxes of Rationality: Theory of Metagames and Political Behavior.* Cambridge, Mass.: MIT Press, 1971.
A major extension of classical game theory with applications to the study of international conflict.

Owen, Guillermo. *Game Theory.* Philadelphia: W. B. Saunders Co., 1968.
A widely used mathematics textbook.

Parthasarathy, T., and T. E. S. Raghavan. *Some Topics in Two-Person Games.* New York: American Elsevier Publishing Co., 1971.
The best modern treatment of two-person game theory.

von Neumann, John, and Oskar Morgenstern. *Theory of Games and Economic Behavior.* Princeton, N.J.: Princeton University Press, 1944, 1947, 1953.
The classic study, for which perseverance pays; remarkable in its breadth and depth of coverage, especially as the original work in the field.

INDEX

INDEX